MULTIVARIATE STATISTICAL ANALYSIS

MULTIVARIATE STATISTICAL ANALYSIS

Proceedings of the Research Seminar at Dalhousie University, Halifax, October 5-7, 1979

Edited by

R. P. GUPTA
Dalhousie University

1980

NORTH-HOLLAND PUBLISHING COMPANY
AMSTERDAM · NEW YORK · OXFORD

ISBN: 0 444 86019 3

Publishers:

NORTH-HOLLAND PUBLISHING COMPANY
AMSTERDAM • NEW YORK • OXFORD

Sole distributors for the U.S.A. and Canada:
ELSEVIER NORTH-HOLLAND, INC.
52 VANDERBILT AVENUE,
NEW YORK, N.Y. 10017

PRINTED IN THE NETHERLANDS

PREFACE

This book comprises the proceedings of the research seminar on "Multivariate Statistical Inference" held at Dalhousie University, Halifax, Nova Scotia, Canada, October 5, 6, and 7, 1979. The participants of the conference came from Canada and the United States. Seven one hour invited lectures and thirty-five contributed papers, each of thirty minutes were presented. All papers included in the proceedings have been refereed.

The first conference on "Multivariate Statistical Inference" held at Dalhousie University, was organized by me in March, 1972. This was followed by another conference under the general title "Applied Statistics", which was held in May 1974. The proceedings of both of these conferences were published by North Holland Publishing company. Most papers of the proceedings were appreciated for their content by reviewers in professional journals. Encouraged by the general comments on the importance of these two conferences as successful attempts in focusing the progress achieved in recent research, the second conference on "Multivariate Statistical Inference" was organized.

The emphasis in this conference was on both the theory and applications of statistical distributions with reference to multi-dimensional random variables. Our aim was to provide a workshop-context in which the papers presented could benefit from the informed criticism of conference participants; hence, the planned activities in the conference included main lectures and expositions, seminar lectures and expositions, seminar lectures - study groups. Several papers dealt with concerning parameters of the population concerned.

The main lectures were given by Professors:

Rolf E. Bargmann,	University of Georgia
Allan T. James,	Univ. of Adelaide (on leave at Princeton Univ.)
Robb J. Muirhead,	University of Michigan
Govind S. Mudholkar,	University of Rochester
M.S. Srivastava,	University of Toronto
J.N.K. Rao,	Carleton University
Neil M. Timm	University of Pittsburgh

A popular lecture on "Calator in the classroom" was given by Professor Rolf E. Bargmann. The papers of Professors A.T. James and J.N.K. Rao could not, unfortunately, be included in the proceedings.

ACKNOWLEDGEMENTS

I express my sincere thanks to several persons for their help and encouragement in organisation of this conference. Dr. Henry D. Hicks, President of Dalhousie University opened the conference. Vice-President Guy MacLean and Dean J. Gray expressed warm welcome to the participants at the opening of the conference. It is a pleasure to record my thanks to all of them.

Thanks are also due to my colleagues Drs. C.A. Field, George Gabor, J.B. Garner and Jean Thiébaux, who chaired several sessions and helped me in several ways. Special mention has to be made of Dr. Patrick N. Stewart, Chairman, Mathematics Department who provided with all sorts of facilities, encouragement and sincere cooperation in the organizational matters.

The contributors and referees are to be thanked for the fine spirit of cooperation and the prompt handling of the correspondence.

The conference would not have been possible without the support from the Faculty of Graduate Studies, Dalhousie University and National Science Engineering Research Council of Canada. My heartiest thanks to both.

Last but not least, my thanks go to Miss Smith and Paula Flemming, who not only typed the manuscript with admirable accuracy but also took care of several aspects of the organizational matters.

R.P. Gupta

CONTENTS

Multivariate Statistical Analysis
R.P. Gupta (ed.)

TESTS OF MULTIPLE CORRELATION WITH ADDITIONAL DATA*

Manzoor Ahmad
Department of Mathematics
Université du Québec à Montréal

In this paper we study the problem of testing independence of a variate X with a set of other variates $(\underline{Y}',\underline{Z}')$, when additional data on $(X,\underline{Y}')$ is available. The likelihood ratio test is given and step down procedures have been suggested.

1. INTRODUCTION

In an investigation with severable variables it is not uncommon, specially in Biometrics and econometrics, that a necessity arises to include certain new variables at a stage when some data has already been collected. In such cases it is important that the proper statistical analysis of the problem at hand be done using the data on all the variables as well as those only on the original variables, called incomplete sample or additional data. Recently some work has been done for a few multivariate testing problems with additional data. See for example Bhargava [1], Eaton & Kariya [2], Clement, Giri & Sinha [1977]. In this paper we have studied the problem of testing independence of a variate X with a set of other variates $(\underline{Y}',\underline{Z}')$ on the basis of a complete set of data on $(X,\underline{Y}',\underline{Z}')$ and an additional data only on $(X,\underline{Y}')$. We have derived in Section 1 the likelihood ratio test for this problem. The small sample distribution of the L.R.T. seems to be very complicated and is not in a workable form. In Section 2, we have therefore suggested two step down test procedures for the above problem, which are practical in the sense that their size and power for certain class of alternatives can be computed. It may be mentioned that this problem has also been treated by Eaton and Kariya [4] using additional data only on X. They derived a Conditional Uniformally Most Powerful Invariant (under a suitable group of transformations) test using Wijsman's [9] representation theorem. In a different context it was noted by Lee Geisser [7] that under the setup of Eaton and Kariya, the LRT is independent of the additional data.

*Prepared with the partial support of Natural Sciences and Engineering Research Council Canada Grant A3450.

2. THE LR TEST

We consider a $p = 1 + q + r$-dimensional random vector $\underline{W}$ having a multivariate normal density with a mean vector $\underline{\mu}$ and a positive definite covariance matrix Σ. We partition $\underline{W}$, $\underline{\mu}$ and Σ as

$$\underline{W} = \begin{pmatrix} X \\ \underline{Y} \\ \underline{Z} \end{pmatrix}, \quad \underline{\mu} = \begin{pmatrix} \mu_1 \\ \underline{\mu}_2 \\ \underline{\mu}_3 \end{pmatrix} \quad \text{and} \quad \Sigma = \begin{pmatrix} \sigma_{11} & \Sigma_{12} & \Sigma_{13} \\ \Sigma_{21} & \Sigma_{22} & \Sigma_{23} \\ \Sigma_{31} & \Sigma_{32} & \Sigma_{33} \end{pmatrix}$$

where (μ_1,σ_{11}), $(\underline{\mu}_2,\Sigma_{22})$ and $(\underline{\mu}_3,\Sigma_{33})$ are respectively the parameters of the marginal densities of the components X: 1×1, $\underline{Y}$: $q\times 1$ and $\underline{Z}$: $r\times 1$ of W.

Based on N independent observations $\{\underline{\omega}'_\alpha = (x_\alpha, \underline{y}'_\alpha, \underline{z}'_\alpha)\}_{\alpha=1}^N$ on $\underline{W}' = (X, \underline{Y}', \underline{Z}')$ and M independent observations $\{\underline{a}'_\beta = (u_\beta, \underline{v}'_\beta)\}_{\beta=1}^M$ on $(X, \underline{Y}')$ we treat the testing problem

$$H_o\colon \rho^2_{1.23} = 0 \quad \text{versus} \quad H_1\colon \rho^2_{1.23} > 0 \tag{2.1}$$

where $\rho^2_{1.23}$ is the squared population multiple correlation coefficient between X and $(\underline{Y}', \underline{Z}')$.

The likelihood function based on the above observations is given by

$$L(\underline{\mu},\Sigma) = k|\Sigma_{(11)}|^{-M/2}|\Sigma|^{-N/2} \exp[-\tfrac{1}{2} \sum_{\alpha=1}^{N} (\underline{\omega}_\alpha - \underline{\mu})'\Sigma^{-1}(\underline{\omega}_\alpha - \underline{\mu})$$
$$- \tfrac{1}{2} \sum_{\beta=1}^{M} (\underline{a}_\beta - \underline{\mu}_{(1)})'\Sigma^{-1}_{(11)}(\underline{a}_\beta \cdot \underline{\mu}_{(1)})] \tag{2.2}$$

where

$$k = (2\pi)^{-1/2(Np+Mq+M)}, \quad \underline{\mu}_{(1)} = \begin{pmatrix} \mu_1 \\ \underline{\mu}_2 \end{pmatrix} \quad \text{and} \quad \Sigma_{(11)} = \begin{pmatrix} \sigma_{11} & \Sigma_{12} \\ \Sigma_{21} & \Sigma_{22} \end{pmatrix}.$$

Write

$$s = \sum_{\alpha=1}^{N} (\underline{\omega}_\alpha - \bar{\underline{\omega}})(\underline{\omega}_\alpha - \bar{\underline{\omega}})' = \begin{pmatrix} s_{11} & s_{12} & s_{13} \\ s_{21} & s_{22} & s_{23} \\ s_{31} & s_{32} & s_{33} \end{pmatrix}, \tag{2.3}$$

$$t = \sum_{\beta=1}^{M} (\underline{a}_\beta - \bar{\underline{a}})(\underline{a}_\beta - \bar{\underline{a}})' = \begin{pmatrix} t_{11} & t_{12} \\ t_{21} & t_{22} \end{pmatrix} \quad \text{with}$$

$$N\,\bar{\underline{\omega}} = \sum_{\alpha=1}^{N} \underline{\omega}_\alpha \quad \text{and} \quad M\,\bar{\underline{a}} = \sum_{\beta=1}^{M} \underline{a}_\beta. \tag{2.4}$$

Denoting the component $(\bar{x}, \bar{\underline{y}}')$ of $\bar{\underline{W}}$ by $\bar{\underline{b}}'$ we have

$$L(\underline{\mu},\Sigma) = k|\Sigma_{(11)}|^{-M/2}|\Sigma|^{-N/2}\operatorname{etr}[-\tfrac{1}{2}\,\Sigma_{(11)}^{-1}t]\operatorname{etr}[-\tfrac{1}{2}\begin{pmatrix}\Sigma_{(11)}^{-1} & \underline{0}\\ \underline{0} & \Sigma_{33.1}^{-1}\end{pmatrix} Ds\,D']$$

$$\cdot \operatorname{etr}[-\tfrac{1}{2}\,\Sigma_{(11)}^{-1}\{(M\underline{\bar{a}}+N\underline{\bar{b}})-(M+N)\underline{\mu}_{(1)}\}\{(M\underline{\bar{a}}+N\underline{\bar{b}})-(M+N)\underline{\mu}_{(1)}\}']$$

$$\times \operatorname{etr}[-\tfrac{N}{2}\,\Sigma_{33.1}^{-1}\{\underline{\bar{z}}-\underline{\mu}_3-\Sigma_{3(1)}\Sigma_{(11)}^{-1}(\underline{\bar{b}}-\underline{\mu}_{(1)})\}\{\underline{\bar{z}}-\underline{\mu}_3-\Sigma_{3(1)}\Sigma_{(11)}^{-1}(\underline{\bar{b}}-\underline{\mu}_{(1)})\}'] \qquad (2.5)$$

where

$$\Sigma'_{(1)3} = \Sigma_{3(1)} = (\Sigma_{31},\Sigma_{32})\ , \quad \Sigma_{33.1} = \Sigma_{33} - \Sigma_{3(1)}\Sigma_{(11)}^{-1}\Sigma_{(1)3}\ ,$$

and

$$D = \begin{pmatrix} I_{1+q} & 0\\ -\Sigma_{3(1)}\,\Sigma_{(11)}^{-1} & I_q\end{pmatrix}$$ with I_j as the identity matrix of order $j\times j$.

Maximizing $L(\underline{\mu},\Sigma)$ first with respect to $\underline{\mu}_3$ while keeping other variables fixed and then maximizing with respect to $\underline{\mu}_{(1)}$ we get

$$\max_{\underline{\mu}} L(\underline{\mu},\Sigma) = k|\Sigma_{(11)}^{-1}|^{-M/2}|\Sigma|^{-N/2}\operatorname{etr}[-\tfrac{1}{2}\,\Sigma_{(11)}^{-1}t]$$

$$\times \operatorname{etr}[-\tfrac{1}{2}\,(\begin{matrix}\Sigma_{(11)}^{-1} & 0\\ 0 & \Sigma_{33.(1)}^{-1}\end{matrix})DsD'] = L^*(\Sigma)\ \text{(say)}. \qquad (2.6)$$

Let L_{H_i} denote L under $H_i(i=0,1)$. It is easy to verify from (2.5) and (2.6) that

$$\max_{\underline{\mu}} L_{H_i}(\underline{\mu},\Sigma) = L^*_{H_i}(\Sigma)\ ,$$

where $L^*_{H_i}$ denotes L^* under H_i , $i = 0,1$. In order to facilitate further maximization with respect to Σ under H_i $(i=0,1)$ we write L^* as

$$L^*(\Sigma) = k|\Sigma_{(11)}|^{-\frac{N+M}{2}}|\Sigma_{33.(1)}|^{-\frac{N}{2}}$$

$$\cdot \operatorname{etr}[-\tfrac{1}{2}\,\Sigma_{(11)}^{-1}(t+s_{(11)})]\operatorname{etr}[-\tfrac{1}{2}\,\Sigma_{33.\ (1)}^{-1}s_{33}^{*}]\ , \qquad (2.7)$$

where

$$s_{33}^{*} = s_{33}-s_{3(1)}\Sigma_{(11)}^{-1}\Sigma_{(1)3}-\Sigma_{3(1)}\Sigma_{(11)}^{-1}s_{(1)3}+\Sigma_{3(1)}\Sigma_{(11)}^{-1}s_{(11)}\Sigma_{(11)}^{-1}\Sigma_{(1)3}$$

$$s'_{(1)3} = s_{3(1)} = (s_{31},s_{32}) \text{ and } s_{(11)} = \begin{pmatrix} s_{11} & s_{12}\\ s_{21} & s_{22}\end{pmatrix}\ .$$

We now have

$$L^*_{H_o}(\Sigma) = k|\sigma_{11}|^{-\frac{N+M}{2}} |\Sigma_{22}|^{-\frac{N+M}{2}} |\Sigma_{33.2}|^{-\frac{N}{2}}$$

$$\cdot \exp[-\frac{1}{2}\sigma_{11}^{-1}(t_{11}+s_{11})] \cdot \mathrm{etr}[-\frac{1}{2}\Sigma_{22}^{-1}(s_{22}+t)]$$

$$\cdot \mathrm{etr}[-\frac{1}{2}\Sigma_{33.2}^{-1}s_{33}^{**}] \tag{2.8}$$

since under H_o

$$\Sigma_{33(1)} = \Sigma_{33} - \Sigma_{32}\Sigma_{22}^{-1}\Sigma_{23} = \Sigma_{33.2}$$

$$s_{33}^* = s_{33} - s_{32}\Sigma_{22}^{-1}\Sigma_{23} - \Sigma_{32}\Sigma_{22}^{-1}s_{23} + \Sigma_{32}\Sigma_{22}^{-1}s_{22}\Sigma_{22}^{-1}\Sigma_{23} = s_{33}^{**} \text{ (say)}.$$

The transformation

$$\left.\begin{array}{l}\Sigma_{11}\\ \Sigma_{22}\\ \Sigma_{32}\\ \Sigma_{33}\end{array}\right\} \longrightarrow \left\{\begin{array}{ll}\Sigma_{11} & \\ \Sigma_{22} & \\ \Sigma_{32}\Sigma_{22}^{-1} = \Sigma_{32}^* & \text{(say)}\\ \Sigma_{33.2} = \Sigma_{33}^* & \text{(say)}\end{array}\right.$$

being one to one, we have

$$\underset{(\Sigma_{11},\Sigma_{22},\Sigma_{23},\Sigma_{33})}{\mathrm{Max}} L^*_{H_o}(\Sigma) = \underset{(\Sigma_{11},\Sigma_{22},\Sigma_{23}^*,\Sigma_{33}^*)}{\mathrm{Max}} L^*_{H_o}(\Sigma) = \underset{\Sigma_{23}^*}{\mathrm{Max}}\left[\underset{(\Sigma_{11},\Sigma_{22},\Sigma_{33}^*)}{\mathrm{Max}} L^*_{H_o}(\Sigma)\right]$$

$$= \underset{\Sigma_{23}^*}{\mathrm{Max}}[(\text{const.})(t_{11}+s_{11})^{-\frac{M+N}{2}} |t_{22}+s_{22}|^{-\frac{M+N}{2}} |s_{33}-\Sigma_{32}^*s_{23}-s_{32}\Sigma_{23}^*+\Sigma_{32}^*s_{22}\Sigma_{23}^*|^{-\frac{N}{2}}$$

$$= (\text{const.})(t_{11}+s_{11})^{-\frac{M+N}{2}} |t_{22}+s_{22}|^{-\frac{M+N}{2}} |s_{33}-s_{32}s_{22}^{-1}s_{23}|^{-\frac{N}{2}}, \tag{2.9}$$

since $|(\Sigma_{32}^*-s_{32}s_{22}^{-1})s_{22}(\Sigma_{32}^*-s_{32}s_{22}^{-1})|$ is minimum if and only if $\Sigma_{32}^* = s_{32}s_{22}^{-1}$. Using similar arguments we also get

$$\underset{\Sigma}{\mathrm{Max}}\ L^*_{H_1}(\Sigma) = (\text{const.})|t+s_{(11)}|^{-\frac{M+N}{2}} |s_{33}-s_{3(1)}s_{(11)}^{-1}s_{(1)3}|^{-\frac{N}{2}} \tag{2.10}$$

From (2.9) and (2.10) it follows that the likelihood ratio test will reject H_o for large values of

$$\lambda = \frac{(t_{11}+s_{11})^{\frac{M+N}{2}} |t_{22}+s_{22}|^{\frac{M+N}{2}} |s_{33.2}|^{\frac{N}{2}}}{|t+s|^{\frac{M+N}{2}} |s_{33.(1)}|^{\frac{N}{2}}} \tag{2.11}$$

where $s_{33.2} = s_{33}-s_{32}s_{22}^{-1}s_{23}$ and $s_{33.(1)} = s_{33}-s_{3(1)}s_{(11)}^{-1}s_{(1)3}$. It can be easily shown that

$$\lambda = \frac{(1-r_{1.2N}^2)^{\frac{N}{2}}}{(1-r_{1.2N+M}^2)^{\frac{N+M}{2}} (1-r_{1.23N}^2)^{\frac{N}{2}}} \tag{2.12}$$

where $r_{1.2N}^2$ and $r_{1.2N+M}^2$ are squared sample multiple correlation coefficient between X and $\underline{Y}$ based on $\{(x_\alpha,\underline{y}_\alpha')\}_{\alpha=1}^N$ and $\{(x_\alpha,\underline{y}_\alpha')\}_{\alpha=1}^V \cup \{((y_\beta,\underline{v}_\beta)\}_{\beta=1}^M$ respectively, and $r_{1.23N}^2$ is the squared multiple correlation coefficient between X and $(\underline{Y}',\underline{Z}')$ based on $\{(x_2,\underline{y}_\alpha',\underline{z}_\alpha')\}_{\alpha=1}^N$.

3. STEP DOWN TEST PROCEDURES

The idea behind the step down test procedures lies in the fact that the problem (1.1) is equivalent to first testing

$$H_{01}: \Sigma_{12} = \underline{0} \text{ against } H_{11}: \Sigma_{12} \neq \underline{0} . \tag{3.1}$$

and afterwards testing

$$H_{02}: \Sigma_{13} = \underline{0} \text{ against } H_{12}: \Sigma_{13} \neq \underline{0} \tag{3.2}$$

in case H_{01} is accepted, or vice versa.

For problem (3.1) we will consider tests based on two independent statistics $R_{1.2M}^2$ and $R_{1.2N}^2$ which are the squared sample multiple correlation coefficients between X and $\underline{Y}$ based on $\{(U_\beta,\underline{V}_\beta')\}_{\beta=1}^M$ and $\{(X_\alpha,\underline{Y}_\alpha')\}_{\alpha=1}^N$ respectively. A usual technique of combining the two would be to base the test on the tail probabilities

$$P\{R_{1.2i}^2 > r_{1.2i}^2 | H_{01}\} \quad , \quad i = N,M$$

rather than on the actual values $r_{1.2N}^2$, $r_{1.2M}^2$ of $R_{1.2N}^2$ and $R_{1.2M}^2$. However the resulting test has a very complicated distribution even under H_{01} , and hence is not at all practical. A different technique recently suggested by Monti and Sen [1976], yields a simpler test procedure which is locally optimal. In our context, we will derive a locally most powerful test among those based on $R_{1.2M}^2$ and $R_{1.2N}^2$. For simplicity we shall denote $R_{1.2M}^2$ and $R_{1.2N}^2$ by G and H , and denote the following non-central beta density

$$(1-\Theta)^{\frac{\ell}{2}} \sum_{i=1}^{\infty} \frac{\Theta^i}{i!} \frac{\Gamma\left(\frac{\ell}{2}+i\right)}{\Gamma\left(\frac{\ell}{2}\right)} \frac{g^{\frac{\delta}{2}+i-1}(1-g)^{\frac{\ell-\delta}{2}-1}}{B\left(\frac{\delta}{2}+i, \frac{\ell-\delta}{2}\right)}$$

$$(0 \le \Theta < 1, 0 \le g \le 1) \text{ by } \beta\left(g; \frac{\ell}{2}, \frac{\ell-\delta}{2}, \Theta\right). \tag{3.3}$$

When $\Theta = 0$, (3.3) becomes a central beta density which we shall denote by $\beta(g; \frac{\ell}{2}, \frac{\ell-\delta}{2})$. It is well known (Giri [1977]) that

$$f_G(g;\rho_{1.2}^2) = \beta\left(g; \frac{m}{2}, \frac{m-q}{2}, \rho_{1.2}^2\right), \text{ and} \tag{3.4}$$

$$f_H(h;\rho_{1.2}^2) = \beta\left(g; \frac{n}{2}, \frac{n-q}{2}, \rho_{1.2}^2\right), \text{ where} \tag{3.5}$$

$\rho_{1.2}^2$ is the population squared multiple correlation coefficient between X and $\underline{Y}$. The locally most powerful test based on G and H rejects H_{01} for large values of

$$U = \left(\frac{1}{f_G}\frac{\partial}{\partial\tau} f_G + \frac{1}{f_H}\frac{\partial}{\partial\tau} f_H\right)_{\tau=0} \tag{3.6}$$

with $\tau = \rho_{1.2}^2$ (see for example Ferguson [5]. Using (2.3), (2.4) and (2.5), a direct computation of (3.6) yields

$$U = -\frac{m+n}{2} + \frac{m^2}{q} g + \frac{n^2}{q} h .$$

Hence the locally most powerful test among those based on $R_{1.2M}^2$ and $R_{1.2N}^2$ is given by

$$\text{Reject } H_{01} \text{ if } \frac{n^2}{m^2} R_{1.2N}^2 + R_{1.2M}^2 \text{ is large .} \tag{3.7}$$

Once H_{01}' is accepted we proceed to a test of (3.2) by considering the statistic

$$W = \frac{R_{1.23N}^2 - R_{1.2N}^2}{1 - R_{1.2N}^2}$$

which can be interpreted as the sample multiple partial correlation coefficient between X and $\underline{Z}$, eliminating the effect of linear regression of X on $\underline{Y}$, (Das Gupta [3]). Under H_{02}, W follows a central beta distribution $\beta(\omega; \frac{q}{2}, \frac{n-q}{2})$ and is independent of $R_{1.2N}^2$.

Having obtained the tests for component hypotheses (3.1) and (3.2), now we select critical regions A_1 and A_2 of suitably chosen levels α_1 and α_2, and propose $A_1 \cup A_2$ as a α-level critical region for our problem, the levels being related as $(1-\alpha_1)(1-\alpha_2) = (1-\alpha)$, so that a choice of α_1 will determine α_2 or vice-versa. Set

$$A_1 \equiv \{(g,h) \mid \frac{n^2}{m^2} g + h \geq k_1\}$$

and

$$A_2 \equiv \{\omega \mid \omega > k_2\} ,$$

where k_1 and k_2 are determined by

$$\iint_{A_1} \beta(h; \tfrac{q}{2}, \tfrac{n-q}{2}) \beta(g; \tfrac{q}{2}, \tfrac{m-q}{2}) dh dg = \alpha_1$$

and

$$\int_{\omega \geq k_2} \beta(\omega; \tfrac{r}{2}, \tfrac{n-q-r}{2}) dw = \alpha_2 .$$

Note that the constant k_1 can be found for a given value of α_1 by using the equations (3.2) and (3.10) of Monti and Sen [8]. The power function of the critical region $A_1 \cup A_2$ is given by

$$P(A_1 \cup A_2 | H_1) = 1 - \int_{\{\frac{n^2}{m^2} g + h \leq k_1\} \cap \{\omega \leq k_2\}} f_{W|G}(\omega | G = g; \rho^2_{1.2}, \rho^2_{1.23}) f_G(g; \rho^2_{1.2}) f_H(h; \rho^2_{1.2}) d\omega\, dg dh \tag{3.8}$$

where $f_{W|G}$ denotes the conditional density of W given G. Under the alternatives $\rho^2_{1.2} = 0$, $\rho^2_{1.23} > 0$, (3.8) can be written using Das Gupta [3] (his equation 3.12) and standard arguments regarding change of order of summation and integration, as

$$= 1 - \int_0^{\frac{m^2}{n^2} k_1} [\beta(g; \tfrac{q}{2}, \tfrac{n-q}{2}) \int_0^{k_1 - \frac{n^2}{m^2} g} \beta(h; \tfrac{q}{2}, \tfrac{m-q}{2}) dh$$

$$\cdot (1-\Theta(g))^{\frac{n}{2}} \sum_{i=0}^{\infty} \frac{\Theta(g)^i}{i!} \frac{\Gamma(\frac{n}{2} + i)}{\Gamma(\frac{n}{2})} \int_0^{k_2} \beta(\omega; \tfrac{r}{2} + i, \tfrac{n-q-r}{2}) d\omega] dg \tag{3.9}$$

where $\Theta(g) = \dfrac{\rho^2_{1.23}(1-g)}{1-g\rho^2_{1.23}}$ with $\rho^2_{1.32} = \dfrac{\Sigma_{13}\Sigma^{-1}_{33.2}\Sigma_{31}}{\sigma_{11}}$.

Here we first test (3.2) using $R^2_{1.3N}$, rejecting H_{o2} for large values of $R^2_{1.3N}$ as usual. For testing H_{o1} when H_{o2} is accepted, we have once more two independent test statistics $G = (R^2_{1.2M})$ and

$$H^* = \frac{R^2_{1.23N} - R^2_{1.3N}}{1 - R^2_{1.3N}}$$

which is the sample multiple partial correlation coefficient between X and $\underline{Y}$ eliminating the linear regression of X on $\underline{Z}$. Under the condition $\Sigma_{23} = 0$, we are in the same setup as Monti and Sen [8] and as before, using the distribution of H^* (Das Gupta [3], equation (3.12)] and that of G, a locally most powerful test can be shown to be the following

$$\text{Reject } H_{o2} \text{ for large values of } \frac{(n-r)^2}{m^2} H^* + G . \qquad (3.11)$$

Even if $\Sigma_{23} \neq 0$, we will propose (2.11) as our test of H_{o2}. Denote by A_3 a critical region of the type $\{U^* = R^2_{1.3N} > k_3\}$ of level α_1, and by A_4 a critical region of the type $\{\frac{(n-r)^2}{m^2} H^* + G \geq k_4\}$ of level α_2. We propose $A_3 \cup A_4$ as a α-level critical region for our problem by relating α with α_1 and α_2 as

$$(1-\alpha_1)(1-\alpha_2) = (1-\alpha) .$$

Here the constants k_3 and k_4 are such that

$$\int_{u>k_3} \beta(u, \tfrac{r}{2}, \tfrac{n-r}{2})du = \alpha_1$$

and

$$\iint_{\frac{(n-r)^2}{m^2} h+g>k_4} \beta(h; \tfrac{q}{2}, \tfrac{n-q-r}{2})\beta(g; \tfrac{q}{2}, \tfrac{m-q}{2})dhdg = \alpha_2 .$$

As before, for a given α_2 we can find k_4 using Monti and Sen [1976]. As to the power of this test, we consider alternatives of the form $(\rho^2_{1.23} > 0, \rho^2_{1.23} = 0)$ and readily obtain [using equation 3.12 of Das Gupta [1977]

$$P(A_3 \cup A_4 \mid \rho^2_{1.23} > 0, \rho^2_{1.3} = 0)$$

$$= 1 - \int_0^{k_3} \beta(u; \frac{r}{2}, \frac{n-r}{2})\{\int_0^{\min[k_4,1]} \beta(g; \frac{q}{2}, \frac{m-q}{2}, \lambda^*)(\int_0^{\min[\frac{m^2}{(n-q)^2}(k_4-g),1]} \beta(h; \frac{r}{2}, \frac{n-q-r}{2}, \xi(u))dh)dg\}du \quad (3.12)$$

where

$$\xi(u) = \frac{\lambda^*(1-u)}{1-\lambda^* u} \quad \text{and} \quad \lambda^* = \frac{\Sigma_{12}(\Sigma_{22}-\Sigma_{23}\Sigma_{33}^{-1}\Sigma_{32})^{-1}\Sigma_{21}}{\sigma_{11}} .$$

ACKNOWLEDGEMENT

The author is thankful to Dr. B.K. Sinha for his useful comments and suggestions.

REFERENCES

[1] Bhargava, R.P. (1975). Some one sample hypothesis testing problems when there is a monotone sample from a multivariate normal population. Ann. Inst. Statist. Math. 27, 327-340.

[2] Clement, B., Giri, N., Sinha, B.K. (1977). Tests for means Rapport technique de l'Ecole Polytechnique de Montréal.

[3] Das Gupta, S. (1977). Tests on multiple correlation coefficient and multiple partial correlation coefficient. J. Multivariate Anal. 7, 82-88,

[4] Eaton, M.L. and Kariya, T. (1974). Testing for independence with additional I information. University of Minnesota Technical Report, No. 238.

[5] Ferguson, T.S. (1967). Mathematical Statistics, Academic Press, New York.

[6] Giri, N. (1977). Multivariate Statistical Inference. Academic Press, New York.

[7] Lee, J.C.,and Geisser, S. (1972). Growth Curve Prediction. Sankhya A34, 393-412.

[8] Monti, K.L., et Sen, P.K. (1976). The locally optimal Combination of Independent test Statistics. J. Am. Statist. Assoc. 71, 903-911.

[9] Wijsman, R.A. (1967). Cross-sections of orbits and their applications. Fifth Berk. Symp. Math. Statist. Prob. 1, 389-400.

Multivariate Statistical Analysis
R.P. Gupta (ed.)

ON D-,E-,D_A- and D_M- OPTIMALITY PROPERTIES OF TEST PROCEDURES OF HYPOTHESES CONCERNING THE COVARIANCE MATRIX OF A NORMAL DISTRIBUTION

P.K. Banerjee
Department of Mathematics
University of New Brunswick
Fredericton, New Brunswick, Canada

N. Giri
Department of Mathematics
University of Montreal
Montreal, P.Q. Canada

Das Gupta [1977] considered the problem of testing the hypotheses concerning the covariance matrix. In this paper we study various optimum properties of these tests from the view point of Isaacson's type D and type E properties.

1. INTRODUCTION

Let $\underline{X}^\alpha, \alpha = 1,\ldots,n$ be a random sample of size n from a p-variate normal population with unknown mean μ and unknown positive definite covariance matrix Σ. Write $n\bar{X} = \sum_{\alpha=1}^{n} X^\alpha$, $S = \sum_{\alpha=1}^{n} (X^\alpha-\bar{X})(X^\alpha-\bar{X})'$, $N = n-1$, $b_{[i]}$ for the i-vector consisting of first i-components of a vector b and $C_{[i]}$ for $i \times i$-upper left-hand submatrix of a matrix C. We shall also write for any $p \times p$ matrix $C = (C_{ij})$

$$C = \begin{pmatrix} C_{11} & C_{(12)} & C_{(13)} \\ C_{(21)} & C_{(22)} & C_{(23)} \\ C_{(31)} & C_{(32)} & C_{(33)} \end{pmatrix} \tag{1}$$

where $C_{(22)}, C_{(33)}$ are of dimensions $p_1 \times p_1$, $p_2 \times p_2$ respectively satisfying $p_1 + p_2 = p-1$. Let

$$\rho_1^2 = \Sigma_{(12)}\, \Sigma_{(22)}\, \Sigma_{(21)}/\Sigma_{11} ,$$

$$\rho_1^2 + \rho_2^2 = \rho^2 = (\Sigma_{(12)}\ \Sigma_{(13)}) \begin{pmatrix} \Sigma_{(22)} & \Sigma_{(23)} \\ \Sigma_{(32)} & \Sigma_{(33)} \end{pmatrix}^{-1} (\Sigma_{(21)} \Sigma_{(31)})/\Sigma_{11} ,$$

$$\bar{R}_1 = S_{(12)} S_{(22)}^{-1} S_{(21)}/S_{11} ,$$

$$\bar{R}_1 + \bar{R}_2 = R^2 = (S_{(12)}S_{(13)}) \begin{pmatrix} S_{(22)} & S_{(23)} \\ S_{(32)} & S_{(33)} \end{pmatrix}^{-1} (S_{(21)}\ S_{(31)})/S_{11} . \quad (2)$$

Obviously $\rho_i^2 \geq 0$, $\bar{R}_i \geq 0$, $i=1,2,\rho$ is the population multiple correlation coefficient between first and the remaining $p-1$ components of the normal p-vector and R is the corresponding sample multiple correlation coefficient.

We shall treat here the following 3 problems.

(A) To test the null hypothesis $H_{10}: \rho_1^2 = 0$, $\rho_2^2 = 0$ against the alternatives $H_{11}: \rho_1^2 > 0$, $\rho_2^2 = 0$ where μ,Σ are unknown.

(B) To test the null hypothesis $H_{20}: \rho_1^2 = 0$, $\rho_2^2 = 0$ against the alternatives $H_{21}: \rho_1^2 = 0$, $\rho_2^2 > 0$ when μ,Σ are unknown.

(C) To test the null hypothesis $H_{30}: \rho_2^2 = 0$ against the alternatives $H_{31}: \rho_2^2 > 0$ when μ,Σ are unknown.

These problems have earlier been considered by Das Gupta (1977) and Giri (1979). They derived the LRT and studied its optimum properties including locally minimaxity.

We shall study here within the class of invariant tests the status of these various tests from the point of view of Isaacson's Type D and type E properties (Isaacson (1951)) and Giri and Kiefer's type D_A and type D_M properties (1964A). Specifically, we shall show that for problem (A) the likelihood ratio test is not type D among all G_T-invariant tests and hence is not of type D_A , type D_M or type E among all tests. For problem (B) the likelihood ratio test and the locally minimax test are not of type D among all G_T-invariant tests. When $p_2 = 1$ the locally minimax test is unique type D among G_T-invariant tests. For problem (C) the likelihood ratio test is not type E among all G_T-invariant tests unless $p_1 = 1$, in which case the likelihood ratio test is type E unique among all G_T-invariant tests.

2. OPTIMUM TESTS

Let $F = (F_{ij})$ be a unique lower triangular matrix with positive diagonal elements of the form

$$F = \begin{pmatrix} F_{11} & 0 & 0 \\ 0 & F_{(22)} & 0 \\ 0 & F_{(32)} & F_{(33)} \end{pmatrix} \quad (3)$$

such that $\Sigma = FF'$ and let

$$\theta = \begin{pmatrix} F_{(22)} & 0 \\ F_{(32)} & F_{(33)} \end{pmatrix}^{-1} (\Sigma_{(21)}\Sigma_{(31)})/F_{11} \quad , \tag{4}$$

and $\theta' = (\theta'_{(1)}, \theta'_{(2)})$ with $\theta_{(i)}$, $p_i \times 1$, $i = 1,2$. Clearly $\theta'\theta = \rho_1^2 + \rho_2^2$ with $\theta'_{(1)}\theta_{(1)} = \rho_1^2$ and $\rho_1^2 = 0$ if and only if $\theta_{(1)} = 0$, $\rho_2^2 = 0$ if and only if $\theta_{(2)} = 0$.

The problems under consideration obviously remain invariant under the group of translation and also under the multiplicative group G_T of $p \times p$ non-singular matrices of the form (3). This transforms the sufficient statistic $(\bar{X},S)$ and the parameters (μ,Σ) in the following manner:

$$g(\mu,\Sigma,\bar{X},S) \rightarrow (g\mu + b, g\Sigma g', g\bar{X} + b, g\delta g') \quad .$$

A maximal invariant under the above group in the smaple space is a $(p_1 + p_2)$-dimensional statistic $R = (R_1,\ldots,R_{p_1+p_2})$ defined by (Giri and Kiefer, 1964b)

$$\sum_1^i R_j = (S_{(12)}S_{(13)})_{[i]} \begin{pmatrix} S_{(22)} & S_{(23)} \\ S_{(32)} & S_{(33)} \end{pmatrix}^{-1}_{[i]} (S_{(21)}S_{(31)})_{[i]}/S_{11} \quad .$$

$$i = 1,\ldots,p_1+p_2 \quad . \tag{5}$$

Thus $R_i \geq 0$ and $\sum_1^{p_1} R_j = \bar{R}_1$, $\sum_1^{p_1+p_2} R_j = \bar{R}_1 + \bar{R}_2$. A corresponding maximal invariant $\Sigma = (\delta_1,\ldots,\delta_{p_1+p_2})$ in the parametric space is defined by

$$\sum_{j=1}^{i} \delta_j = (\Sigma_{(12)}\Sigma_{(13)})_{[i]} \begin{pmatrix} \Sigma_{(22)} & \Sigma_{(23)} \\ \Sigma_{(32)} & \Sigma_{(33)} \end{pmatrix}^{-1}_{[i]} (\Sigma_{(21)}\Sigma_{(31)})_{[i]}/\Sigma_{11} \quad . \tag{6}$$

Thus $\delta_i \geq 0$ and $\sum_1^{p_1} \delta_j = \rho_1^2$, $\sum_1^{p_1+p_2} \delta_j = \rho_1^2 + \rho_2^2$. The joint density of the maximal invariant is given by (Giri and Kiefer, 1964a).

$$f_{\underset{\sim}{\Delta}}(r) = \frac{(1-\rho^2)^{N/2}(1-\bar{r}_1-\bar{r}_2)^{\frac{1}{2}(N-p-1)}}{1+\sum_{i}^{p_1+p_2} r_i(\frac{1-\rho^2}{\gamma_i}-1)^{\frac{1}{2}}\,\Gamma(\frac{1}{2}(N-p+1)\,\pi^{\frac{1}{2}(p-1)}}$$

$$\times \frac{1}{\prod_{1}^{p_1+p_2}\{r_i^{\frac{1}{2}}\,\Gamma(\frac{1}{2}(N-i+1))\}} \sum_{\beta_1=0}^{\infty} \cdots \sum_{\beta_{p_1+p_2}=0}^{\infty}$$

$$\prod_{1}^{p_1+p_2} \frac{\{\Gamma(\frac{1}{2}(N-i+1)+\beta_i)\}}{(2\beta_i)!} \left(\frac{4\, r_i\alpha_i}{1+\sum_{1}^{p_1+p_2} r_i(\frac{1-\rho^2}{\gamma_i}-1)}\right)^{\beta_i}, \tag{7}$$

where

$$\gamma_i = 1-\sum_{1}^{i}\delta_j\ ,\ i=1,\ldots,p_1+p_2,\ \gamma_0=1,\ \alpha_i^2=\frac{\delta_i(1-\rho^2)}{\gamma_{i-1}\gamma_i}\ .$$

From Basu and Giri (1973) the Lebesgue density function of $\bar{R}$ on $S=\{\bar{r},\ \bar{r}_i \geq 0\ ,\ \bar{r}_i < 1\}$ is given by

$$f_{\underset{\sim}{\Delta}}(r) = K(1-\rho^2)^{\frac{N}{2}}\,(1-\bar{r}_1-\bar{r}_2)^{\frac{1}{2}(N-p-1)}\prod_{1}^{2}\bar{r}_1^{\frac{1}{2}p_i-1}$$

$$\times\,[1+\sum_{1}^{2}\bar{r}_1\,(\frac{1-\rho^2}{\gamma_i}-1)]^{-\frac{N}{2}}\sum_{\beta_1=0}^{\infty}\sum_{\beta_2=0}^{\infty}\Gamma(\beta_1+\beta_2+\frac{1}{2}N)$$

$$\prod_{1}^{2}\frac{\Gamma(\frac{1}{2}(N+p_i-\sigma_i)+\beta_i)\ \Gamma(\frac{1}{2}+\beta_i)}{\Gamma(\frac{1}{2}p_i+\beta_i)}\left(\frac{4\,\bar{r}_i\,\bar{\alpha}_i^2}{1+\sum_{1}^{2}\bar{r}_i\,(\frac{1-\rho^2}{\gamma_i}-1)}\right)^{\beta_i}, \tag{8}$$

where $\bar{\gamma}_i = 1-\sum_{1}^{i}\bar{\delta}_j,\ \sigma_i=\sum_{1}^{i}p_j,\ \bar{\alpha}_i^2=\dfrac{\bar{\delta}_i(1-\rho^2)}{\bar{\gamma}_i\ \bar{\gamma}_{i-1}}$

and K is the normalizing constant independent of parameter values.

In the problems (A) - (C) the parameter set $\Omega = \{\theta,\Sigma)$ with θ as defined in (4) where F is the unique element in G_T with positive diagonal elements such that $FF' = \Sigma$. Then $\textcircled{H} = \{\theta\}$ is Euclidean p-space. The group G_T operates trivially on $\textcircled{H}$ and transitively on H = {positive definite symmetric Σ} i.e. $g(\theta,\Sigma) \to (\theta, g\Sigma g')$, $g \in G_T$. Obviously problem (A) has the hypothesis $\theta_{(1)} = 0$ and the alternatives $\theta_{(1)} \neq 0$ in the presence of the nuisance parameter Σ (given that $\theta_{(2)} = 0$). The problems (B) and (C) have the same hypothesis $\theta_{(2)} = 0$ and have the same alternatives $\theta_{(2)} \neq 0$ in the presence of nuisance parameter Σ. The only difference between these two problems is that in the first one $\theta_{(1)} = 0$ is given while in the second one $\theta_{(1)}$ is unknown and is therefore a part of the nuisance parameter set. We shall now treat these three problems separately.

Problem A. Here $\Delta = (\delta_1,\ldots,\delta_{p_1},0,\ldots,0)$, $\rho^2 = \rho_1^2 = \theta_1^2 + \ldots + \theta_{p_1}^2$.

From (7)

$$\frac{f_\Delta(r|H_{11})}{f_\Delta(r|H_{10})} = 1 + \frac{N\rho^2}{2}\Big(-1 + \sum_1^{p_1} r_j\big(\sum_{i>j} \eta_i + (N-j+1)\eta_j\big) + B(r,\eta,\rho^2)$$

$$= 1 - \frac{N\rho^2}{2} + \frac{N}{2}\sum_1^{p_1} \delta_i\big(\sum_{j<i} r_j + (N-i+1)r_i\big) + B(r,\eta,\rho^2) \qquad (9)$$

where $B(r,\eta,\rho^2) = o(\rho^2)$ uniformly in r as $\rho^2 \to 0$. Differentiating under the integral sign, the power function $\beta_\emptyset(\Delta)$ of any G_T invariant level α test $\emptyset(r)$ is given by

$$\beta_\emptyset(\Delta) = \alpha\Big(1 - \frac{N}{2}\rho^2\Big) + \frac{N}{2}\sum_1^{p_1} \delta_i \int \emptyset(r)\Big[\sum_1^{i-1} r_j + (N-i+1)r_i\Big] f_\Delta(r|H_{10})\, dr + o(\rho^2) \qquad (10)$$

The matrix $B_\emptyset(0)$ of second partial derivatives of the power function $\beta_\emptyset(\Delta)$ of every G_T-invariant test $\emptyset(r)$ with respect to θ_i, $i = 1,\ldots,p_1$ at $\theta_{(1)}=0$ (i.e. at $\Delta = 0$) is a diagonal matrix with i-th diagonal element $b_{\emptyset,i}(0)$, $i = 1,\ldots,p_1$ given by

$$b_{\emptyset,i}(0) = -\alpha + N\int \emptyset(r)\Big[\sum_1^{i-1} r_j + (N-i+1)r_i\Big] f_\Delta(r|H_{10})dr \qquad (11)$$

From Giri and Kiefer (1964a) we conclude that a G_T-invariant level α test

$\emptyset^*(r)$ is type D (and hence type D_A and D_M among all tests) among G_T-invariant tests $\emptyset(r)$ if and only if $\emptyset^*$ is of the form (apart from a set of Lebesgue measure 0)

$$\emptyset^*(r) = \begin{cases} 1 , & \text{if } \sum_1^{p_1} q_i\{ \sum_1^{i-1} r_i + (N-j+1)r_i\} > C \\ 0 , & \text{otherwise} \end{cases} \tag{12}$$

in which the q_i^{-1} are proportional to $b_{\emptyset^*,i}(0)$. If $\emptyset$ is a test whose power function depends on Δ only through ρ_1^2 then with the above parametrization for θ , $B_\emptyset(0)$ is a multiple of the identity matrix and then q_i^{-1} are all equal. The likelihood ratio test of this problem has rejection region $\bar{r}_1 > C_\alpha$, C_α depending on the size α of the test. But when the q_i are equal, the rejection region corresponding to (12) is of the form $\sum_1^{p_1}(N+p_1-2j+1)r_j \geq C_\alpha$ which is not the rejection region of the likelihood ratio test unless $p_1 = 1$. Hence we conclude the following:

Theorem 1. For $0 < \alpha < 1 < p_1 < N$ the likelihood ratio test for problem (A) is not of type D among all G_T-invariant tests and hence is not of type D_A , D_M or E among all tests.

Problem B. Here $\Delta = (0,\ldots,0,\delta_{p_1+1},\ldots,\delta_{p_1+p_2})$, $\rho_2^2 = \rho^2 = \theta'_{(2)}\,\theta_{(2)} = \theta^2_{p_1+1} + \ldots + \theta^2_{p_1+p_2}$. From (7)

$$\frac{f_\Delta(r|H_{21})}{f_\Delta(r|H_{20})} = 1 + \frac{N\rho^2}{2}\left(-1 + \bar{r}_1 + \sum_{j=p_1+1}^{p_1+p_2}\left[\sum_{i>j}\eta_i + (N-j+1)\eta_j\right]^2\right.$$
$$+ B(r,\eta,\rho^2)$$
$$= 1 - \frac{N\rho^2}{2} + \frac{1}{2}\sum_{i=p_1+1}^{p_1+p_2}\delta_i\left(\sum_{j=1}^{i-1} r_j + (N-i+1)r_i\right)$$
$$+ B(r,\eta,\rho^2) \tag{13}$$

where $B(r,\eta,\rho^2) = o(\rho^2)$ uniformly in r as $\rho^2 \to 0$. As in (A) the power function of any G_T-invariant test $\emptyset(r)$ can be written in the form

$$\beta_{\emptyset}(\Delta) = \alpha(1-\frac{N}{2}\rho^2) + \frac{1}{2}N \sum_{i=p_1+1}^{p_1+p_2} \delta_i \int \emptyset(r) \left[\sum_1^{i-1} r_j + (N-j+1)r_i \right] + o(\rho^2) \tag{14}$$

In this case, of course, the matrix $B_{\emptyset}(0)$ of the second partial derivatives of the power function $\beta_{\emptyset}(\Delta)$ of every G_T-invariant test $\emptyset(r)$ with respect to θ_i , $i = p_1+1,\ldots,p_1+p_2$ at $\theta_{(2)} = 0$ is a diagonal matrix with the i-th diagonal element

$$b_{\emptyset,i}(0) = -\alpha + N\int\emptyset(r)\left[\sum_1^{p_1+i-1} r_j + (N-p_1-i+1)r_{p_1+i}\right] f_{\Delta}(r|H_{20})dr \tag{15}$$

Now from Giri and Kiefer (1964a) we conclude that a G_T-invariant level α test $\emptyset^*$ is type D among all G_T invariant tests if and only if $\emptyset^*$ has the form (apart from a set of Lebesgue measure 0)

$$\emptyset^*(r) = \begin{cases} 1, & \text{if } \sum_1^{p_2} q_i \left[\sum_1^{p_1+i-1} r_j + (N-p_1-i+1)r_{p_1+i}\right] > C \\ 0, & \text{otherwise} \end{cases} \tag{16}$$

where the q_i^{-1} are proportional to $b^*_{\emptyset,i}(0)$. If $\emptyset$ is a test whose power function depends on Δ only through $\rho_2^2 = \rho^2$, $B_{\emptyset}(0)$ is a multiple of the identity matrix. Since then, the q_i are all equal, such a test $\emptyset$ in order to be type D must be of the form (apart from a set of Lebesgue measure o)

$$\emptyset(r) = \begin{cases} 1, & \text{if } p_2\bar{r}_1 + \sum_1^{p_2} (N-p_1-i+1)r_{p_1+i} > C \\ 0, & \text{otherwise} \end{cases} \tag{17}$$

On the otherhand any test $\emptyset^*$ (say) which depends only on $\bar{R}_1,\bar{R}_2$ has power function depending only on ρ_2^2 (see (7)). Such a test $\emptyset^*$ were it type D, would have to be of the form (17). But the test in (17) does not depend on $R_{(2)} = (R_{p_1+1},\ldots,R_{p_1+p_2})$ only through $\bar{R}_2$ unless $p_2 = 1$ when (17) reduces to the locally minimax test. Thus we have the following theorem.

Theorem 2. For $0 < \alpha < 1 < p_1 < N$. The likelihood ratio test and the locally minimax test (all being based on $\bar{r}_1,\bar{r}_2$) are not of type D among all G_T - invariant tests. When $p_2 = 1$ the locally minimax test is unique type D (locally most powerful) among all G_T-invariant tests and hence admissible in this class.

Problem C. To characterize G_T-invariant type E tests in the presence of nuisance parameter, $\delta_{(1)} = (\delta_1,\ldots,\delta_{p_1})$ we first observe that any such test $\emptyset$ must be similar. Since $R_{(1)} = (R_1,\ldots,R_{p_1})$ is a complete sufficient statistic for $\delta_{(1)}$ under H_{30}, $\emptyset$ has Neymann structure, that is, $\emptyset$ is conditionally of size α given $R_{(1)} = r_{(1)}$. From (7) the conditional distribution of $R_{(2)}$ given $R_{(1)} = r_{(1)}$ depends on $R_{(1)}$ only through $\bar{R}_1$. Thus we conclude that a G_T-invariant type E test is a function of $\bar{R}_1$ and $R_{(2)}$ only. The power function $\beta_\emptyset(\rho_1^2,\delta_{(2)})$ of such a test $\emptyset(\bar{R}_1,R_{(2)})$ which depends only on ρ_1^2 and $\rho_{(2)}$ is given by (14) with $\emptyset(r)dr$ replaced by $\emptyset(\bar{r}_1,r_{(2)})d\bar{r}_1\, dr_2$ and $f_\Delta(r|H_{20})$ replaced by $f_\Delta(\bar{r}_1,r_{(2)}|\rho_1^2,o)$ (see Basu and Giri (1973)).

$$f_\Delta(\bar{r}_1,r_{(2)}|\rho_1^2,o) = K(1-\rho_1^2)^{N/2}(1-\bar{r}_1-\bar{r}_2)^{\frac{1}{2}(N-p-1)}(\bar{r}_1)^{\frac{p_1}{2}-1}$$

$$\times \prod_{i=p_1+1}^{p_1+p_2} r_i^{-1/2} \sum_{\beta_1=o}^{\infty} \frac{\Gamma^2(\frac{1}{2}\beta_1+\frac{N}{2})}{(2\beta)!}(\bar{r}_1\,\rho_1^2)^{\beta_i} \tag{18}$$

where K is the normalizing constant.

Therefore for a specified ρ_1^2 we can characterize the form of a G_T-invariant Neyman structure level α type E test in the same line as problem (B) and we obtain the same characterization as in (16) with q_i replaced by $q_1(\rho_1^2)$ which is proportional to $b_{\emptyset^*,i}(\rho_1^2)$ where $b_{\emptyset^*,i}(\rho_1^2)$ is given by (15) by replacing $\emptyset(r)f_\Delta(r|H_{20})dr$ by $\emptyset(\bar{r}_1,r_{(2)})f_\Delta(\bar{r}_1,r_{(2)}|\rho_1^2,o)\,d\bar{r}_1\,dr_{(2)}$ and C by $C(\rho_1^2,\bar{r}_1)$, $C(\rho_1^2,\bar{r}_1)$ being determined by the conditional size α condition. Thus as in problem (B) if we consider a test $\emptyset$ depending only on $\bar{R}_1,\bar{R}_2$, its power will depend, apart from ρ_1^2, on $\delta_{(2)}$ only though ρ_2^2 and hence $B_\emptyset(\rho_1^2)$ is a multiple of the identity matrix. Since the $q_1(\rho_1^2)$ are all equal such a test is type E if and only if it is of the form (17) with C replaced by $C(\bar{r}_1)$ which is determined by the condition that the conditional size is α. It is easy to note taht $C(\bar{r}_1)$ does not depend ρ_1^2 as $\bar{R}_1$ is sufficient for ρ_1^2 as $\bar{R}_1$ is sufficient for ρ_1^2 under H_{30}. From (17), $\emptyset(\bar{r}_1,r_{(2)})$ does not depend on $r_{(2)}$ only through $\bar{r}_2$ unless $p_2 = 1$. In this case it reduces to the likelihood ratio test which rejects H_{30} whenever $r_{p_1+1}(1-\bar{r}_1)^{-1} \geq C_\alpha$ (Note that Z is independent of $\bar{R}_1$ under H_{30}). Hence we have the following theorem.

Theorem 3. For $0 < \alpha < 1$ no test $\emptyset(\bar{R}_1,\bar{R}_2)$ and in particular the likelihood ratio test is type E among all G_T-invariant tests unless $p_2 = 1$ in which case the likelihood ratio test is type E unique among all G_T-invariant tests and hence admissible in this class.

REFERENCES

[1] Basu, S.K. and Giri, N. (1973. On the invariance test of a hypothesis concerning the covariance matrix of a multivariate normal distribuiton, Technical report, University of Montreal.

[2] Das Gupta, S. (1977). Tests on multiple correlation coefficient and multiple partial correlation coefficient, Journal of multivariate analysis, 7, pp. 82-88.

[3] Giri, N. (1979). Locally minimax tests for multiple correlations; to appear in the Canadian Journal of Statistics, Vol. 7, No. 1.

[4] Giri, N., and Kiefer, J. (1964a). Local and asymptotic minimax properties of multivariate tests, Annal. Math. Statistics, pp. 21-35.

[5] Giri, N. and Kiefer, J. (1964b). Minimax character of R^2-test in the simplest case, Annal. Math. Statist. 35, pp. 1475-1490.

[6] Kiefer, J. (1957). Invariance, minimax and sequential estimation and continuous time processes, Annal. Math. Statist. 28, pp. 573-601.

[7] Lehmann, E.L. (1959). Testing statistical hypothesis, Wiley, N.Y.

Multivariate Statistical Analysis
R.P. Gupta (ed.)

SCALING OF MULTI-DIMENSIONAL CONTINGENCY TABLES BY UNION-INTERSECTION

Rolf E. Bargmann
Department of Statistics
University of Georgia
Athens, Georgia, 30602
U.S.A.

When the levels of two response variables are categorized (nominal), such levels or states need to be reordered or, better still, scaled, i.e., the states 1,2,3...r of each variable need to be replaced by scale values $a_1, a_2, \ldots a_r$. H.O. Lancaster showed that, if the states are replaced by dummy variables ($x_i=1$ if an observation is in state i of the first variable, 0 otherwise, etc.), and the canonical correlation is obtained, then the canonical weights are the best scale values in the sense that the resulting two-dimensional histogram yields the closest approach to a bivariate normal distribution. To extend this technique to three or more categorized variables, it is necessary to maximize the union-intersection statistic for the test of internal independence, $(\lambda_\ell - \lambda_s)/(\lambda_\ell + \lambda_s)$, where λ_ℓ and λ_s are the largest and smallest characteristic roots of the correlation matrix. This technique is described and demonstrated with illustrations in which a correlated trivariate normal distribution is subdivided into slices of unequal width, with the slices shuffled in a random permutation. The original situation and, especially, the underlying correlation matrix are well reproduced by this extended Lancaster technique.

STATEMENT OF THE PROBLEM

If the levels of some response variable represent an ordinal scale (e.g., a three-point or five-point scale for preference ratings) replacement of the arbitrary numbers, say 1 - 5 , by scaled numbers, $a_1, a_2, \ldots a_5$ is often desirable, especially if such variables are to be correlated with other variables of a similar kind. Techniques of normalization were introduced in the 19th century (Fechner (1)). A standard normal distribution is sliced in the same proportion as the observed values, and some midpoint (e.g., the expected value under each slice) is chosen as the scale value $a_1, a_2, \ldots a_r$ corresponding to the original ordinal scale values 1,2,...r . Without a technique of this kind, the famous box problem in factor analysis (Thurstone (9)) could not have been resolved for non-linear combinations of factors.

Numerous attempts have been made to approximate the histogram corresponding to a two-dimensional contingency table by a bivariate normal surface. As is readily seen, such a solution does not even require that the levels or states of each factor be ordered. These variables can have categorized (nominal) levels. The scaling problem was solved by Lancaster (5) who proved that the best approach to bivariate normality is attained by considering two sets of dummy variables $x_1, x_2, \ldots x_r$; $y_1, y_2, \ldots y_s$ such that, if an observation falls into cell (i,j) of a contingency table, $x_i = 1$ and $y_j = 1$, all other x and y being zero. The canonical correlation of these two sets of variables must be obtained and the weights (coefficients) of the canonical variables are the desired scale values. It should be noted that this straightforward approach permits reordering of rows and columns in a contingency table so that the association (as measured by correlation) between the two categorized variables is a maximum. The algorithm can be found on pp. 585-587 of Kendall and Stuart (3); a more efficient one is applied in Kundert and Bargmann (4).

In order to extend the Lancaster principle to three or more dimensions it is necessary to employ a union-intersection statistic for the test of independence in three or more sets of variables; (note that the canonical correlation is the union-intersection test statistic of independence of two sets). For example, in three-dimensional contingency tables we would have dummy variables $x_1, x_2, \ldots x_r$, $y_1, y_2, \ldots y_s$, $z_1, z_2, \ldots z_t$, and the corresponding canonical variables $u = \underline{a}_1'\underline{x}$, $v = \underline{a}_2'\underline{y}$, and $w = \underline{a}_3'\underline{z}$ would be those which maximize the union-intersection test statistic for the hypothesis:

$$\begin{bmatrix} 1 & \rho_{uv} & \rho_{uw} \\ \rho_{uv} & 1 & \rho_{vw} \\ \rho_{uw} & \rho_{vw} & 1 \end{bmatrix} = I$$

As shown by Schuenemeyer and Bargmann (7) , this statistic is the maximum eccentricity of a correlation ellipsoid, $(\lambda_\ell - \lambda_s)/(\lambda_\ell + \lambda_s)$, where λ_ℓ and λ_s are, respectively, the largest and smallest characteristic root of the <u>correlation</u> matrix (not covariance or Wishart matrices). This is quite true for any number of sets. Given a vector of p random variables, $\underline{y}$ (and the corresponding p by p correlation matrix), the two linear combinations which produce the largest possible correlation (other than one) are $(\underline{e}_\ell' + \underline{e}_s')\underline{y}$ and $(\underline{e}_\ell' - \underline{e}_s')\underline{y}$, where $\underline{e}_\ell$ and $\underline{e}_s$ are the eigenvectors associated with the largest and smallest root of the correlation matrix. This paper deals with the description and application of an algorithm which obtains such scale values for three or more categorized variables. A computer program (FORTRAN) has been developed which performs this analysis for up to five categorized variables, with a maximum of five levels per factor. All illustrations were analyzed by this program.

THE ALGORITHM

The present description is for the three-dimensional case, with obvious extension to higher dimensions (they are described in Schuenemeyer (6)). The entry n_{ijk} in cell (i,j,k) of the contingency table is translated into n_{ijk} equal vectors (length $r + s + t$) in which $x_i = 1$, $y_j = 1$, $z_k = 1$, and all other entries are zero. From this we can obtain matrices of corrected sums of squares and products ("Wishart" matrices):

$$E_{11} = D_{n_{i..}} - \frac{1}{n} \underline{n}_{i..} \underline{n}'_{i..} \tag{1}$$

where $\underline{n}'_{i..} = (n_{1..}, n_{2..}, \ldots, n_{r..})$ is the margin for Factor 1 ("Rows") of the contingency table, and $D_{n_{i..}}$ has the same elements in the principal diagonal of a diagonal matrix; similarly

$$E_{22} = D_{n_{.j.}} - \frac{1}{n} \underline{n}_{.j.} \underline{n}'_{.j.} \tag{2}$$

and

$$E_{33} = D_{n_{..k}} - \frac{1}{n} \underline{n}_{..k} \underline{n}'_{..k} \tag{3}$$

in terms of the margins for Factor 2 ("Columns") and Factor 3 ("Layers"). Obviously, these matrices are singular, of rank $(r-1)$, $(s-1)$, and $(t-1)$, respectively; n denotes the total number of observations.

$$E_{12} = N_{12} - \frac{1}{n} \underline{n}_{i..} \underline{n}'_{.j.} \tag{4}$$

where the matrix N_{12} contains elements $n_{ij.}$, the marginal total for the Factor 1 vs. Factor 2 contingency table (summed over Factor 3). Similarly

$$E_{13} = N_{13} - \frac{1}{n} \underline{n}_{i..} \underline{n}'_{..k} \tag{5}$$

where N_{13} contains elements $n_{i.k}$ (Factor 1 vs. 3, summed over 2) and

$$E_{23} = N_{23} - \frac{1}{n} \underline{n}_{.j.} \underline{n}'_{..k} \tag{6}$$

where N_{23} contains elements $n_{.jk}$ (Factor 2 vs. 3, summed over 1). These are placed into a matrix

$$E = \begin{array}{c} \\ (\underline{x}) \\ (\underline{y}) \\ (\underline{z}) \\ \\ \end{array} \begin{array}{c} \begin{array}{ccc} (\underline{x}') & (\underline{y}') & (\underline{z}') \end{array} \\ \begin{bmatrix} E_{11} & E_{12} & E_{13} \\ E'_{12} & E_{22} & E_{23} \\ E'_{13} & E'_{23} & E_{33} \end{bmatrix} \\ \begin{array}{ccc} (r) & (s) & (t) \end{array} \end{array} \begin{array}{c} \\ (r) \\ (s) \\ (t) \\ \\ \end{array} \tag{7}$$

where the left and top designators indicate the variable sets, and the right and bottom designators indicate the order of the matrices. The canonical weights on these dummy variables, i.e.,

$\underline{a}_1' = (a_{11}, a_{21}, \ldots, a_{r1})$; $\underline{a}_2' = (a_{12}, a_{22}, \ldots, a_{s2})$; $\underline{a}_3' = (a_{13}, a_{23}, \ldots a_{t3})$

are the desired scale values. They are those vectors that produce maximum eccentricity in the correlation matrix of $u = \underline{a}_1'\underline{x}$, $v = \underline{a}_2'\underline{y}$, and $w = \underline{a}_3'\underline{z}$. Of course they can and should be standardized, by applying a shift and multiplier to each set so that (u,v,w) have zero means and unit variances. This does not affect the correlations.

In order to reduce this E matrix to a (non-singular) correlation matrix, we perform Choleski decompositions (or any other one for the positive semidefinite matrices):

$$E_{11} = T_1 T_1' \tag{8}$$

where T_1 has r rows and $(r-1)$ columns ; $E_{22} = T_2 T_2'$ (T_2 is s by $(s-1)$), and $E_{33} = T_3 T_3'$ (T_3 is t by $(t-1)$). A conditional inverse from the left is obtained for T_1, i.e., some matrix $T_1^{(-1)}$ of order $(r-1)$ by r, such that $T_1^{(-1)} T_1 = I$ (of order $(r-1)$). The same procedure is applied to T_2 and T_3. We thus obtain a correlation matrix

$$R = \begin{array}{c} \\ (\underline{x}^*) \\ (\underline{y}^*) \\ (\underline{z}^*) \\ \\ \end{array} \begin{array}{c} \begin{array}{ccc} (\underline{x}^{*\prime}) & (\underline{y}^{*\prime}) & (\underline{z}^{*\prime}) \end{array} \\ \begin{bmatrix} I & R_{12} & R_{13} \\ R_{12}' & I & R_{23} \\ R_{13}' & R_{23}' & I \end{bmatrix} \\ \begin{array}{ccc} (r-1) & (s-1) & (t-1) \end{array} \end{array} \begin{array}{c} \\ (r-1) \\ (s-1) \\ (t-1) \\ \\ \end{array} \tag{9}$$

where $R_{12} = T_1^{(-1)} E_{12} (T_2^{(-1)})'$

$R_{13} = T_1^{(-1)} E_{13} (T_3^{(-1)})'$

and

$$R_{23} = T_2^{(-1)} E_{23} (T_3^{(-1)})' \tag{10}$$

This is a simplification of a procedure introduced by Steel [8].

Thus, let $u = \underline{b}_1'\underline{x}^*$, $v = \underline{b}_2'\underline{y}^*$, $w = \underline{b}_3'\underline{z}^*$, (11)

where $\underline{x}^*$, $\underline{y}^*$, $\underline{z}^*$ are the transformed dummy variables, and $\underline{b}_i'\underline{b}_i = 1$.
Then est.corr$(u,v) = \underline{b}_1' R_{12} \underline{b}_2$, est.corr$(u,w) = \underline{b}_1' R_{13} \underline{b}_3$ and

$$\text{est.corr}(v,w) = \underline{b}_2' R_{23} \underline{b}_3 \tag{12}$$

The desired $\underline{a}_i$ are related to the $\underline{b}_i$ by the obvious transformation:

$$\underline{a}_i = (T_i^{(-1)})'\underline{b}_i \quad (i = 1,2,3) \tag{13}$$

The problem is solved when the $\underline{b}_i$ have been found in such a way that the maximum eccentricity of the correlation matrix

$$C = \begin{bmatrix} 1 & c_{12} & c_{13} \\ c_{12} & 1 & c_{23} \\ c_{13} & c_{23} & 1 \end{bmatrix} \tag{14}$$

(where $c_{ij} = \underline{b}_i' R_{ij} \underline{b}_j$) is a maximum, subject to the constraints ($\underline{b}_i'\underline{b}_i = 1, i=1,2,3$). Equivalently (since we use a descent method for the maximization), we may let the $\underline{b}_i$ have arbitrary length, and find them in such a way that the maximum eccentricity of the correlation matrix with elements $c_{ij} = \underline{b}_i' R_{ij} \underline{b}_j / (\underline{b}_i'\underline{b}_i \; \underline{b}_j'\underline{b}_j)^{1/2}$ is maximized (without constraints on the $\underline{b}_i$).

We use the method of Fletcher and Powell (2), to find the $\underline{b}_i$ which maximize $\Psi = (\lambda_\ell - \lambda_s)/(\lambda_\ell + \lambda_s)$, where λ_ℓ and λ_s are the largest and smallest characteristic root of the matrix C. Thus we need expressions for the gradients. Quite generally (k sets, not just three), we obtain

$$\partial\Psi/\partial b_{mi} = \{(\partial\lambda_\ell/\partial b_{mi})(1-\Psi) - (\partial\lambda_s/\partial b_{mi})(1+\Psi)\}/(\lambda_\ell+\lambda_s) \quad , \tag{15}$$

where b_{mi} is the m'th component of $\underline{b}_i$, and ℓ and s refer to the largest and smallest roots.

$$\partial\lambda_\ell/\partial_{mi} = (2e_{i\ell}/h_i)\{\sum_{\substack{j=1\\ j\neq i}}^{k} (e_{j\ell}/h_j) \sum_{\beta=1}^{p_j-1} b_{\beta j} r_{m\beta}^{(i,j)} - (e_{i\ell}/h_i)(\lambda_\ell - 1)b_{mi}\} \tag{16}$$

(and $\partial\lambda_s/\partial b_{mi}$ the same expression with ℓ replaced by s), where

b_{mi} = m'th component of $\underline{b}_i$

$e_{i\ell}$ = i'th component of the largest eigenvector of the correlation matrix C

h_i = $(\underline{b}_i'\underline{b}_i)^{1/2}$ (length of $\underline{b}_i$, unconstrained)

k = dimensionality of the contingency table

p_j = number of levels in Factor j (Note that the sum stops at one less)

$r_{m\beta}^{(i,j)}$ = the (m,β) element of R_{ij} (see(10)).

After attainment of the maximum of Ψ (all components of the gradient vector $< 10^{-6}$ in absolute value, in our program), the $\underline{b}_i$ are normalized to unit length. Then

$$\underline{a}_i = (T_i^{(-1)})'\underline{b}_i \tag{17}$$

are the desired scale values for the levels of Factor i. To standardize we obtain, e.g., for Factor 1:

$$a^*_{m1} = (a_{m1} - c_1) \quad (m=1,2,\ldots r)$$
$$a^{**}_{m1} = d_1 a^*_{m1} \tag{18}$$

where

$$c_1 = \sum_{i=1}^{r} n_{i..} a_{i1} / n$$

and

$$d_1 = \{ \sum_{i=1}^{r} n_{i..} (a^*_{i1})^2 / n\}^{-1/2}$$

The analogous formulas for Factors 2 and 3 are obtained by replacement of $n_{i..}$ by $(n_{.j.}, n_{..k})$ and a_{i1} by (a_{j2}, a_{k3}) . Generalizations are obvious.

Initial values for the $\underline{b}_i$, for the iterative procedure, can be obtained in a variety of ways. In our program we chose (because of its simplicity) a weighted average of the k-1 canonical vectors for Set i (when compared with each of the other sets). Somewhat more time-consuming is initialization by "multiple-canonical" correlation, i.e., as the first guess of $\underline{b}_i$ we use the canonical weight of Factor i vs. all the other sets combined. Such initial weights are very close to the maximum maximum-eccentricity solution, and require only very few iterations by Fletcher Powell.

ILLUSTRATION

Contingency Tables were obtained as follows:

Set 1: 3,000 random normal numbers were obtained (N(0,1)) and stored as a 1,000 by 3 matrix Z .

Set 2: 15,000 random normal numbers were obtained (N(0,1)) and stored as a 5,000 by 3 matrix Z .

The i'th row of each matrix was transformed as follows:

$$u_{i1} = z_{i1}$$
$$u_{i2} = -0.6z_{i1} + 0.8z_{i2}$$
$$u_{i3} = 0.8z_{i2} + 0.6z_{i3}$$

Thus, the rows of U, $\underline{u}_i' = (u_{i1}, u_{i2}, u_{13})$ (i=1 to 1000 in Set 1, i = 1 to 5000 in Set 2) represent sample vectors from a trivariate normal distribution with mean vector (0,0,0) and dispersion matrix

$$\Sigma = \begin{bmatrix} 1 & -.6 & 0 \\ -.6 & 1 & .64 \\ 0 & .64 & 1 \end{bmatrix}$$

The actual sample mean vectors and dispersion matrices were:

Set 1 : $(.010\ ,\ -.026\ ,\ -.055)\ \begin{bmatrix} 1.002 & -.658 & -.003 \\ -.658 & 1.086 & .696 \\ -.003 & .696 & 1.063 \end{bmatrix}$

Set 2 : $(.012\ ,\ -.012\ ,\ -.006)\ \begin{bmatrix} 1.014 & -.601 & .008 \\ -.601 & 1.006 & .650 \\ .008 & .650 & 1.024 \end{bmatrix}$

The three variables were subdivided ("sliced") as follows:

A : ($<$ -1.5 , -1.5 to -0.5 , -0.5 to +0.8 , $>$ +0.8)

B : ($<$ -0.5 , -0.5 to 0 , $>$ 0)

C : ($<$ -1.2 , -1.2 to -0.2 , -0.2 to +0.9 , $>$ +0.9)

Thus, the expected values under each slice are:

A : -1.885 , -0.936 , 0.130 , 1.367

B : -1.141 , -0.161 , 0.798

C : -1.688 , -0.644 , 0.316 , 1.446

Set 1 :

The level designations in the contingency table are as follows

Factor A: Level 1 = slice 2 ; level 2 = slice 3 ; level 3 = slice 4 ; level 4 = slice 1 .

Factor B: Level 1 = slice 1 ; level 2 = slice 3 ; level 3 = slice 2 .

Factor C: Level 1 = slice 2 ; level 2 = slice 3 ; level 3 = slice 1 ; level 4 = slice 4 .

Contingency Table

C=1

A \ B	1	2	3
1	6	39	20
2	70	31	57
3	62	0	7
4	0	18	0

C=2

A \ B	1	2	3
1	0	81	5
2	30	100	51
3	49	11	23
4	0	37	0

C=3

A \ B	1	2	3
1	16	2	10
2	48	2	9
3	38	0	0
4	3	7	1

C=4

A \ B	1	2	3
1	0	54	0
2	0	58	8
3	4	27	7
4	0	9	0

Results:

	Slice	Level	Initial	Max. max eccentr.	Expected	
A	1	4	-1.584	-1.813	-1.885	
	2	1	-1.236	-1.064	-0.936	
	3	2	0.195	0.111	0.130	
	4	3	1.386	1.457	1.367	
B	1	1	1.268	1.269	1.141	(sign reversed)
	2	3	0.279	0.274	0.161	
	3	2	-0.984	-0.983	-0.798	
C	1	3	0.194	1.726	1.688	(sign reversed)
	2	1	0.478	0.644	0.644	
	3	2	0.508	-0.472	-0.316	
	4	4	-2.222	-1.509	-1.446	

Correlations from Contingency Table

Initial

1	.510	.038
	1	.329
		1

Max. Eccentr.

1	.510	.057
	1	.547
		1

Expected

1	.640	0
	1	.600
		1

Set 1 : Results

Set 2: Contingency Table

C=1

A \ B	1	2	3
1	106	46	223
2	32	296	3
3	11	3	99
4	255	363	159

C=2

A \ B	1	2	3
1	1	0	210
2	46	37	130
3	0	0	57
4	38	3	412

C=3

A \ B	1	2	3
1	29	3	443
2	112	233	52
3	1	0	125
4	229	147	525

C=4

A \ B	1	2	3
1	40	57	16
2	2	137	0
3	12	6	20
4	40	231	0

Results

	Slice	Level	Initial	Max. max eccentr.	Expected	
A	1	3	-1.575	-1.686	-1.885	
	2	1	-1.156	-1.112	-0.936	
	3	4	0.089	0.082	0.130	
	4	2	1.546	1.547	1.367	
B	1	2	1.320	1.317	1.141	(reversed sign)
	2	1	0.289	0.298	0.161	
	3	3	-0.942	-0.943	-0.798	
C	1	4	-2.255	-1.781	-1.688	
	2	1	0.311	-0.737	-0.644	
	3	3	0.826	0.444	0.316	
	4	2	-0.832	1.444	1.446	

Correlations from Contingency Table

1	.485	-.039	1	.485	-.007	1	.640	0
	1	-.175		1	-.524		1	-.600
		1			1			1

Set 2: Level Designations: A: Level 1 = slice 2 ; level 2 = slice 4;
level 3 = slice 1 ; level 4 - slice 3.

B : Level 1 = slice 2 ; level 2 = slice 1 ; level 3 = slice 3 .

C: Level 1 = slice 2 ; level 2 = slice 4 ; level 3 = slice 3 ;
level 4 = slice 1 .

REFERENCES

[1] Fechner, G.T. (1860) Elemente der Psychophysik. Leipzig: Breitkopf und Haertel.

[2] Fletcher, R. and Powell, N.J.D. (1963) "A rapidly convergent descent method for minimization", The Computer Journal, 6, 163-168.

[3] Kendall, M.G. and Stuart, A. (1961) The Advanced Theory of Statistics. New York: Hafner Publishing Co.

[4] Kundert, K.R. and Bargmann, R.E. (1972) "Tools of analysis for pattern recognition", THEMIS report No. 22, Univ. of Georgia, 18-24.

[5] Lancaster, H.O. (1957) "Some properties of the bivariate normal distribution considered in the form of a contingency table", Biometrika, 44, 289-292.

[6] Schuenemeyer, J.H. (1975) "Maximum eccentricity as a union-intersection test statistic in multivariate analysis". Ph.D. dissertation, Univ. of Georgia, Athens, Ga.

[7] Schuenemeyer, J.H. and Bargmann, R.E. (1978) "Maximum eccentricity as a union-intersection test statistic in multivariate analysis", Journ. Multivar. Anal., 8, 268-273.

[8] Steel, R.G.D. (1957) "Minimum generalized variance for a set of linear functions", Ann. Math. Stat., 41 , 456-460.

[9] Thurstone, L.L. (1947) Multiple Factor Analysis, Chicago: Univ. of Chicago Press.

Multivariate Statistical Analysis
R.P. Gupta (ed.)

THE NEGATIVE BINOMIAL POINT PROCESS AND ITS INFERENCE

R.T. Burnett and M.T. Wasan
Department of Mathematics and Statistics
Queen's University
Kingston, Ontario

In this paper, The Negative Binomial Random Point Process, is defined and its properties are investigated. This point process is characterized as a Marked Poisson process and both estimation and hypothesis testing problems are discussed with respect to such a characterization. A Multivariate Negative Binomial point process is defined and characterized as a Doubly Stochastic Poisson point process. The asymptotic properties for both processes are considered.

1. RANDOM POINT PROCESSES

In recent years a new and interesting concept in the theory and applications of stochastic processes has arisen. This is the concept of a random point process. This paper illustrates some of the properties of random point processes through the Negative Binomial law.

Section one deals with the most general definition of a random point process and gives the necessary mathematical tools which will be used throughout the paper.

Section two deals with the univariate Negative Binomial random point process which is characterized as a Marked Poisson point process. A limit theorem is given and estimation and hypothesis testing problems are considered.

Section three defines and examines the Multivariate Negative Binomial random point process.

1.1. Foundations. Let S be a fixed locally compact second countable Hausdorff topological space, i.e. the real line. Let $\mathcal{L}$ be the σ-algebra generated by the space S, and let $\mathcal{B}$ be the ring consisting of all bounded (i.e. relatively compact) sets in $\mathcal{L}$. Denote $\mathcal{F}$ to be the class of all $\mathcal{L}$-measurable functions $f: S \to R_+ = [0,\infty]$.

Definition 1.1.1. $\mathcal{T}$ is said to be a DC-semiring if the semiring $\mathcal{T} \subset \mathcal{B}$ has the property that given any $B \in \mathcal{B}$ and any $\varepsilon > 0$, there exist some finite cover of B by $\mathcal{T}$-sets of diameter less than ε.

Recall that a semiring is a class $\mathcal{T}$ of sets which is closed under finite intersections and such that any proper difference between $\mathcal{T}$-sets may be written as a finite disjoint union of sets in $\mathcal{T}$.

Remark 1. On the real line, DC-semirings are families of bounded intervals.

A measure μ on $(S,\mathcal{B})$ is Radon if $\mu(B) < \infty$ for all $B \in \mathcal{B}$. Let $\mathcal{N}$ denote the class of all such measures with $\mu(B) \in Z_+ = \{0,1,...\}$, for all $B \in \mathcal{B}$. Let $\sigma(\mathcal{N})$ be the σ-algebra in $\mathcal{N}$ generated by the mappings $\mu \to \mu(B)$, $B \in \mathcal{B}$.

Definition 1.1.2. A random point process N is any measurable mapping of some fixed probability space (Ω,A,P) into $(\mathcal{N},\sigma(\mathcal{N}))$.

Remark 2. Consider the function $N(B,w)$, where $B \in \mathcal{T}$ and $w \in \Omega$. If B is fixed, $N(B,\cdot)$ is a Z_+-valued random variable. If w is fixed, then $N(\cdot,w)$ is a Z_+-valued measure.

Any random point process can be written in terms of its atoms. If S is the real line, then these atoms can be interpreted as the occurrence of events of the process. The following lemma gives this decomposition, see Kallenberg [3].

Lemma 1.1.1. Every measure $N \in \mathcal{N}$ may be written in the form

$$N = \sum_{j=1}^{k} b_j \mathcal{S}_{t_j}$$

where $k \in Z_+ \cup \{\infty\}$, $b_1,b_2,... \in Z_+$ and $t_1,t_2,... \in S$. $\mathcal{S}_{t_j}(B) = I_B(t_j)$, $B \in \mathcal{B}$.

Thus we can represent a point process by its atoms.

Definition 1.1.3. The distribution of a random point process N is by definition the probability measure PN^{-1} on $(\mathcal{N},\sigma(\mathcal{N}))$ given by

$$(PN^{-1})M = P(N^{-1}M) = P\{w\colon N(w)\varepsilon M\} \quad \text{for } M \in \sigma(\mathcal{N}) \text{ and } w \in \Omega$$

Definition 1.1.4. The L-transform, L_N, of a point process N is defined as

$$L_N(f) = E\exp\{-\int_S f(s)N(ds)\} \quad \text{for } f \in \mathcal{F}.$$

Remark 3. If $f(s) = uI_A(s)$, where $s,u \in R_+$ and $A \in \mathcal{T}$ then

$$L_N(f) = L_{N(A)}(u)$$

where $L_{N(A)}(u)$ is the Laplace transform of the random variable $N(A)$.

We say that two point processes N and E are equal in distribution, written $N \overset{d}{=} E$, if and only if $P\{w\colon N(w)\varepsilon M\} = P\{w\colon E(w)\varepsilon M\}$, for all $M \in \sigma(\mathcal{N})$. For the following theorem see Kallenberg [3].

Theorem 1.1.1. $N \stackrel{d}{=} E$ if and only if $L_N(f) = L_E(f)$, $f \in F$.

Thus we can determine the distribution of a point process by its L-transform.

Definition 1.1.5. A point process N is said to have independent increments if for any m-tuple $B_1,\ldots,B_m$ of disjoint sets in T and non-negative integers $k_1,\ldots,k_m$ we have

$$P\{N(B_1) = k_1,\ldots,N(B_m) = k_m\} = \prod_{i=1}^{m} P\{N(B_i) = k_i\}$$

Definition 1.1.6. A point process N is said to have stationary increments if for any m-tuple $B_1,\ldots,B_m$ of sets in T and any $s \in S$ and non-negative integers $k_1,\ldots,k_m$ we have

$$P\{N(B_1) = k_1,\ldots,N(B_m) = k_m\} = P\{N(B_1+s) = k_1,\ldots,N(B_m+s) = k_m\}$$

where

$$B_i+s = \{x+s : x\in B_i\} \quad , \quad i = 1,\ldots,m .$$

Definition 1.1.7. Let λ be a positive, non-atomic measure on S . Then the point process N^* is said to be a Poisson point process with parameter measure λ if it has stationary independent increments and distribution

$$P\{N^*(B)=k\} = \frac{\lambda(B)^k}{k!} \exp\{-\lambda(B)\} \quad \text{for} \quad B \in T \quad \text{and} \quad k = 0,1,2,\ldots \quad .$$

For the following lemma see Kallenberg [3].

Lemma 1.1.2. Let N^* be a Poisson point process with parameter measure λ on S . Then the L-transform of N^* is

$$L_{N^*}(f) = \exp\{-\int_S \{1-e^{-f(s)}\}\lambda(ds)\} .$$

Remark 4. If $S = [0,\infty)$ and $f(s) = uI_A(s)$, for $u,s \in R_+$ and $A \in T$, then

$$L_{N^*}(f) = L_{N^*(A)}(u) = \exp\{-\lambda\mu(A)(1-e^{-u})\}$$

where $\lambda > 0$ and μ is the Lebesque measure.

Definition 1.1.8. Let N^* be a Poisson point process. Let $\{M_n\}$, for $n = 1,2,\ldots$ be a sequence of independent, identically distributed Z_+-valued random variables, called marks, independent of N^* . Then N is said to be a

Marked Poisson process if it can be represented by the random sum

$$N = \sum_{n=1}^{N^*} M_n \ .$$

For the proofs of the following theorems see Kallenberg [3].

Theorem 1.1.2. Let N be a Marked Poisson process. Then N has stationary independent increments and L-transform,

$$L_N(f) = L_{N^*}(-\log L_M(f)) \ , \ f \in F \ ,$$

where N^* is a Poisson point process and L_M is the common L-transform of the marks $\{M_n\}$.

Remark 5. If $S = [0,\infty)$ and $f(s) = uI_A(s)$, where $u,s \in R_+$ and $A \in T$, then

$$L_N(f) = L_{N^*(A)}(-\log L_M(u)) = \exp\{-\lambda\mu(A)(1-L_M(u))\}$$

Given any point process N on S define

$$\mathcal{B}_N = \{B \in \mathcal{B}: N(\partial B) = 0 \ \text{ a.s.}\}$$

where ∂B denotes the boundary of B .

Theorem 1.1.3. Let $N, N_1, N_2, \ldots$ be point processes on S . Let $T \subset \mathcal{B}_N$. Then N_n converges in distribution to N , written $N_n \xrightarrow{d} N$, if and only if

$$(N_n(I_1),\ldots,N_n(I_k)) \xrightarrow{d} (N(I_1),\ldots,N(I_k))$$

$$\text{for } k \in Z_+ \text{ and } I_1,\ldots,I_k \in T \ .$$

Theorem 1.1.5. Let $\{\{N_{nj}\}\}$ be a double sequence of row-wise independent point processes, for $j=1,\ldots,n$. Let N be a Poisson point process with parameter measure λ . If $A \subset \mathcal{B}_\lambda$ is fixed, then $\sum_{j=1}^{n} N_{nj} \xrightarrow{d} N$ if and only if

1) $\lim_{n\to\infty} \max_{1\le j\le n} P\{N_{nj}(A)>0\} = 0 \quad , \quad A \in T$

2) $\lim_{n\to\infty} \sum_{j=1}^{n} P\{N_{nj}(A) = 1\} = \lambda(A) \quad , \quad A \in T$

3) $\lim_{n\to\infty} \sum_{j=1}^{n} P\{N_{nj}(A)>1\} = 0 \quad , \quad A \in T \ .$

2. THE NEGATIVE BINOMIAL RANDOM POINT PROCESS

2.1. Introduction. In this section we will assume that our topological space S is the non-negative reals, the DC-semiring T is the class of all bounded intervals and the L-measurable functions f are of the form

$$f(s) = uI_A(s)$$

where $s,u \in R_+$ and $A \in T$.

Thus we are restricting the process to the bounded interval A.

Definition 2.1.1. A random point process N on $[0,\infty)$ is said to be a Negative Binomial random point process with parameter $\lambda > 0$ if N is a point process with stationary independent increments and

1) $P\{N(0) = 0\} = 1$

2) $P\{N(A) = k\} = \dfrac{\Gamma(\mu(A)+k)}{\Gamma(\mu(A))(1+\lambda)^{\mu(A)}} \dfrac{(\lambda/1+\lambda)^k}{k!}$ for every $\lambda > 0$,

$A \in T$ and $k = 0,1,2,\ldots$.

This process has, for $A \in T$, mean $\lambda\mu(A)$, variance $\lambda(1+\lambda)\mu(A)$ and L-transform $(1+\lambda-\lambda e^u)^{-\mu(A)}$.

2.2. Negative Binomial as a Marked Poisson Poisson Process. Since the -ve Binom. point process is not uniformly analytically orderly (see Daley [1]) it allows more than one atom of the process to fall on a point in $[0,\infty)$. One way to describe the occurrence of atoms of this process is by a 2-tuple. The first element of the 2-tuple is the point on the real line where the atom fell and the second is the number of atoms that fell on that particular point. The occurrence point of an atom is described by a Poisson point process and the number of atoms at that point by some non-negative integer valued random variable. This type of characterization is called a Marked Poisson process where the above-mentioned random variables are the marks.

Example. Consider the following hypothetical example of a Marked Poisson process.

We would like to model the number of people involved in car accidents on Highway 401 say, over a one month period. The time at which each accident occurs is recorded along with the number of people involved in that particular accident. The occurrence times are to be modelled as a Poisson point process and the number of people involved in each accident as some positive integer valued random variable, called a mark, which is independent of the Poisson point process. Thus the total number of people involved in car accidents is modelled as a Marked

Poisson process.

Theorem 2.2.1. Let N^* be a Poisson point process with parameter $\ell n(1+\lambda)$, for $\lambda > 0$. Let $\{M_n\}$, $n = 1,2,\ldots$ be a sequence of independent, identically distributed random variables, independent of N^*, with distribution function

$$P\{M_n=k\} = \frac{(\lambda/1+\lambda)^k}{k\ \ell n(1+\lambda)}$$

for every $\lambda > 0$, $k = 1,2,\ldots$ and $n = 1,2,\ldots$.

Then the corresponding Marked Poisson process, N, is a Negative Binomial point process with parameter $\lambda > 0$.

Proof. From Theorem 1.1.2, the L-transform of N becomes

$$L_{N(A)}(u) = L_{N^*(A)}(-\log L_{M_n}(u)) \quad , \text{ for } n = 1,2,\ldots$$

where

$$L_{M_n}(u) = \frac{-\ell n(1+\lambda-\lambda e^u)}{\ell n(1+\lambda)} + 1 .$$

Lemma 1.1.2 and Remark 4 gives

$$L_{N^*(A)}(u) = \exp\{-\ell n(1+\lambda)\mu(A)(1-e^{-u})\}$$

Setting $u = -\log L_{M_n}(u)$ we have

$$\begin{aligned} L_{N(A)}(u) &= \exp\{-\ell n(1+\lambda)\mu(A)(1 + \frac{\ell n(1+\lambda-\lambda e^u)}{\ell n(1+\lambda)} - 1) \\ &= \exp\{-\mu(A)\ell n(1+\lambda-\lambda e^u)\} \\ &= (1+\lambda-\lambda e^u)^{-\mu(A)} \end{aligned}$$

which is the L-transform of the Negative Binomial.

Remark 6. Since the Negative Binomial can be characterized as a Marked Poisson process it verifies the fact that it has stationary independent increments.

This representation will be used for estimation purposes.

2.3. A Limit Theorem. We show that a sum of Negative Binomial point processes converges in distribution to the Poisson point process.

Theorem 2.3.1. Let $\{\{N_{nj}\}\}$ be a double sequence of row-wise independent Negative Binomial point processes with parameter λ/n, for $j = 1,\ldots,n$. Then $\sum_{j=1}^{n} N_{nj} \xrightarrow{d} N$, where N is a Poisson point process with parameter λ.

Proof: The proof consists of checking the conditions of Theorem 1.1.5.

The first condition yields, for $A \in T$,

$$\lim_{n\to\infty} \max_{1\le j\le n} P\{N_{nj}(A)>0\} = \lim_{n\to\infty}\{1-1/(1+\lambda/n)^{\mu(A)}\} = 0 .$$

The second condition implies, for $A \in T$,

$$\lim_{n\to\infty} \sum_{j=1}^{n} P\{N_{nj}(A) = 1\} = \lim_{n\to\infty} n\{(\lambda/n)\mu(A)/(1+\lambda/n)^{\mu(A)}\} = \lambda\mu(A)$$

and condition 3 gives, for $A \in T$,

$$\lim_{n\to\infty} \sum_{j=1}^{n} P\{N_{nj}(A)>1\} = \lim_{n\to\infty} n\{1-1/(1+\lambda/n)^{\mu(A)}-(\lambda/n)\mu(A)/(1+\lambda/n)^{\mu(A)+1}\}$$

$$= \lim_{n\to\infty} n \frac{(1+\lambda/n)^{\mu(A)}-1-(\lambda/n)\mu(A)(1+\lambda/n)^{-1}}{(1+\lambda/n)^{\mu(A)}}$$

$$= \lim_{n\to\infty} n \left\{\frac{1+(\lambda/n)\mu(A)+\mathcal{B}(n)-1-(\lambda/n)\mu(A)(1+\lambda/n)^{-1}}{(1+\lambda/n)^{\mu(A)}}\right\}$$

where $\lim_{n\to\infty} n\,\mathcal{B}(n) = 0$.

The above line becomes

$$\lim_{n\to\infty} \frac{\lambda\mu(A)+n\mathcal{B}(n)-\lambda\mu(A)(1+\lambda/n)^{-1}}{(1+\lambda/n)^{\mu(A)}} = 0 .$$

The conditions of Theorem 1.1.5 are satisfied and thus the theorem is proved.

2.4. Inference. We find the maximum likelihood estimate of the parameter λ of a Negative Binomial random point process. This estimate is shown to be an optimal minimum variance unbiased estimate of λ .

The estimate will be based on the observation of the atoms of the process over a fixed bounded interval of the non-negative reals. The atoms can be thought of as the occurrence of a previously determined event in an experiment and the non-negative reals can be interpreted as time.

Throughout the rest of this section we will only consider the bounded interval $[0,t)$. Since the processes involved are stationary, there is no loss of generality by considering this interval.

In order to use maximum likelihood estimation a likelihood function must be defined. Here, the likelihood function will be the probability of one realization of the point process over the interval $[0,t)$, and, in general, will be called the sample function density.

Since the Negative Binomial can be characterized as a Marked Poisson process we will derive the sample function density for such a process and use it for our estimation purposes.

Definition 2.4.1. Let N be a Marked Poisson process on $[0,t)$, where the marks, M_n, are non-negative integer valued random variables. Let N^* be a Poisson point process which counts the atoms of N regardless of their marks. Then $F[\{N(\sigma): 0\le\sigma\le t\}]$ will denote the sample function density of N and is defined to be:

$$F[\{N(\sigma): 0\le\sigma\le t\}] = \begin{cases} P\{N^*(t)=0\} & \text{if } N^*(t)=0 \\ P\{N^*(t)=n|T,M\}\cdot P\{M|T\}F^*(T) & \text{if } N^*(t)=n\ge 1 \end{cases}$$

where T is the vector of the n occurrence times of the atoms and M is the vector of the corresponding marks. $F^*(T)$ is the sample function density of the Poisson point process N^*.

We will denote the sample function density of N, $F[\{N(\sigma): 0\le\sigma\le t\}]$, by $F_{N(t)}$.

Remark 7. Due to the independence properties of the Marked Poisson process the sample function density becomes

$$F_{N(t)} = P\{N^*(t)=n|T\}F^*(T)\prod_{i=1}^{n} P\{M_i\} .$$

Theorem 2.4.1. (Snyder [4], page 147) Let N be a Marked Poisson process on $[0,t)$ with marks $\{M_n\}$ and Poisson counting process N^*. Then its sample function density is:

$$F_{N(t)} = \begin{cases} \exp\{-\lambda t\} & \text{if } N^*(t)=0 \\ \lambda^n \prod_{i=1}^{n} P\{M_i=m_i\}\exp\{-\lambda t\} & \text{if } N^*(t)=n\ge 1 \end{cases}$$

where λ is the parameter of N^*.

From Theorem 2.4.1 and setting the parameter of N^* to be $\ell n(1+\lambda)$, $\lambda>0$ and the mark distribution to be

$$P\{M_i=m_i\} = \frac{(\lambda/1+\lambda)^{m_i}}{m_i\ \ell n(1+\lambda)} , \quad \text{for } i=1,\ldots,n$$

the sample function density for the Negative Binomial becomes

$$F_{N(t)} = \begin{cases} \exp\{-\ell n(1+\lambda)t\} & \text{if } N^*(t) = 0 \\ \prod_{i=1}^{n} \frac{(\lambda/1+\lambda)^{m_i}}{m_i} \exp\{-\ell n(1+\lambda)t\} & \text{if } N^*(t) = n \geq 1 \end{cases} .$$

Set the likelihood function $L(\lambda)$ to be $F_{N(t)}$. Thus the maximum likelihood estimate of λ, $\hat{\lambda}_{mL}$, turns out to be

$$\hat{\lambda}_{mL} = \frac{N(t)}{t} , \quad t > 0 .$$

Remark 8. In order to obtain this estimate the occurrence times of the process were needed, but the estimate itself does not involve these times. This is reasonable since $N(t)$ is sufficient for λ.

$\hat{\lambda}_{mL}$ has expectation λ, thus unbiased, and variance $\frac{\lambda(1+\lambda)}{t}$ which is inversely proportional to t. Thus, the longer one observes the process the less the variance of the estimate.

Definition 2.4.2. A function g of the point process N on $[0,t)$, $g(N(t))$ say, is said to be optimal if

$$\text{Var } g(N(t)) = [E\left\{\frac{d\ \ell n\ F_{N(t)}}{d\lambda}\right\}^2]^{-1} .$$

Since

$$\frac{d\ \ell n\ F_{N(t)}}{d\lambda} = \frac{N(t)-\lambda t}{\lambda(1+\lambda)}$$

and

$$E\left\{\frac{d\ \ell n\ F_{N(t)}}{d\lambda}\right\}^2 = \frac{t}{\lambda(1+\lambda)}$$

we conclude that $\hat{\lambda}_{mL}$ is optimal.

Remark 9. Since the Negative Binomial distribution is complete, any unbiased estimate of λ will be unique and thus have minimum variance. Therefore, $\frac{N(t)}{t}$, is an optimal minimum variance unbiased estimate for λ.

Hypothesis Testing. After observing the Negative Binomial point process N on $[0,t)$ we would like to test the simple hypothesis that $\lambda = \lambda_1$ against $\lambda = \lambda_2$, where $\lambda_1 < \lambda_2$.

The uniformly most powerful α level test, ϕ, where $0 < \alpha < 1$, of this hypothesis is:

$$\phi(F_{N(t)})) = 1 \quad \text{if} \quad \frac{F_{N(t)}(\lambda_2)}{F_{N(t)}(\lambda_1)} \geq K$$

$$= 0 \quad \text{otherwise}$$

where $F_{N(t)}(\cdot)$ is the sample function density of N evaluated at a particular value of λ. K is a constant.

Remark 10. The above test was proven to be most powerful by the Neyman-Pearson Lemma and is adapted here to the point process case.

Therefore, we reject the hypothesis that $\lambda = \lambda_1$ if $N(t) \geq K^*$ where K^* is determined by

$$P_{\lambda_1}\{N(t) \geq K^*\} = \alpha .$$

Of course $N(t)$ has a Negative Binomial distribution.

3. THE MULTIVARIATE NEGATIVE BINOMIAL POINT PROCESS

Introduction. In this section we define the Multivariate Negative Binomial Point Process and characterize it as a Doubly Stochastic Poisson Process. We also show that a sequence of sums of such processes converges to the multivariate Poisson point process with independent components.

3.1. Preliminaries. We define the Multivariate Negative Binomial point process, find its mean vector and covariance matrix and give its marginal and conditional distributions.

Definition 3.1.1. Let $N^{(M)}(A) = \{N^1(A),\ldots,N^M(A)\}$ be a M-variate random point process for $A \in T$. Then $\{N^{(M)}(A): A \in T\}$ is said to be a Multivariate Negative Binomial point process with parameters $\lambda_i > 0$, for $i = 1,\ldots,M$ if

1) $P\{N^{(M)}(0) = 0\} = 1$;

2) $\{N^{(M)}(A): A \in T\}$ has independent stationary increments; and

3) $P\{N^1(A) = k_1,\ldots,N^M(A) = k_M\} = \dfrac{\Gamma(\bar{k}+\mu(A))}{\Gamma(\mu(A))(1+\bar{\lambda})^{\mu(A)+\bar{k}}} \prod_{i=1}^{M} \dfrac{\lambda_i^{k_i}}{k_i!}$

for non-negative integers $k_1,\ldots,k_M$, and $\lambda_i > 0$ for

$$i = 1,\ldots,M , \quad \bar{k} = \sum_{i=1}^{M} k_i \quad \text{and} \quad \bar{\lambda} = \sum_{i=1}^{M} \lambda_i .$$

Definition 3.1.2. (Snyder, [4], 1975). Let $\{X(t): t \geq 0\}$ be a real valued stochastic process. Then for any $A \in T$ such that $\mu(A) = t \geq 0$, $\{N(A): A \in T\}$ is said to be a Doubly Stochastic Poisson point process, with parameter process $\{\lambda X(t): t \geq 0\}$ if for almost every given sample path of the process $\{X(t): t \geq 0\}$,

$N(A)$ is a Poisson point process with parameter $\lambda X(t)$. That is, $\{N(A): A\varepsilon\mathcal{B}\}$ is conditionally a Poisson point process with parameter $\lambda X(t)$ given $\{X(t): t \geq 0\}$.

Therefore

$$P\{N(A) = k|X(t)\} = \frac{[\lambda X(t)]^k}{k!} \exp\{-\lambda X(t)\}$$

for $k = 0,1,2,\ldots$ and $\mu(A) = t$ for $A \varepsilon T$.

Let $\{N^{(M)}(A): A\varepsilon T\}$ be a Multivariate Negative Binomial point process with parameters $\lambda_i > 0$, for $i = 1,\ldots,n$. Then its mean vector is,

$$\mu = \begin{bmatrix} E\{N^1(A)\} \\ \vdots \\ E\{N^M(A)\} \end{bmatrix} = \mu(A) \begin{bmatrix} \lambda_1 \\ \vdots \\ \lambda_M \end{bmatrix}$$

and covariance matrix,

$$\Sigma = \begin{bmatrix} Var\{N^1(A)\} & E\{N^1(A)\cdot N^2(A)\} \ldots & E\{N^1(A)\cdot N^M(A)\} \\ E\{N^2(A)\cdot N^1(A)\} & Var\{N^2(A)\} \ldots & E\{N^2(A)\cdot N^M(A)\} \\ \vdots & & \vdots \\ E\{N^M(A)\cdot N^1(A)\} \ldots & & Var\{N^M(A)\} \end{bmatrix}$$

$$= \mu(A) \begin{bmatrix} \lambda_1(1+\lambda_1) & \lambda_1\lambda_2 & \ldots & \lambda_1\lambda_M \\ \lambda_2\lambda_1 & \lambda_1(1+\lambda_2) & \ldots & \lambda_2\lambda_M \\ \vdots & & & \vdots \\ \lambda_M\lambda_1 & & \ldots & \lambda_M(1+\lambda_M) \end{bmatrix}$$

The Multivariate Negative Binomial point process has marginal distributio-

$$P\{N^j(A) = k_j\} = \frac{\Gamma(\mu(A)+k_j)}{\Gamma(\mu(A))(1+\lambda_j)^{\mu(A)+k_j}} \frac{\lambda_j^{k_j}}{k_j!} \quad \text{for} \quad j = 1,\ldots,M$$

which is again Negative Binomial.

The conditional distribution is, for $i \neq j$, $P\{N^i(A) = k_i|N^j(A) = k_j\}$

$$P\{N^i(A) = k_i|N^j(A) = k_j\} = \frac{\Gamma(\mu^*(A)+k_i)}{\Gamma(\mu^*(A))} \left\{\frac{1+\lambda_j}{1+\lambda_i+\lambda_j}\right\}^{\mu^*(A)} \left\{\frac{\lambda_i}{1+\lambda_i+\lambda_j}\right\}^{k_i} \frac{1}{k_i!}$$

where $\mu^*(A) = \mu(A) + k_j$ and $i = 1,\ldots,M$, $j = 1,\ldots,M$.

3.2. The Multivariate Negative Binomial as a Doubly Stochastic Poisson Process.

The Multivariate Negative Binomial point process can be characterized in the following manner.

Let $N^1(A),\ldots,N^M(A)$ be M independent doubly stochastic Poisson processes with corresponding parameter functions $\lambda_i X(t)$, for $i = 1,\ldots,M$ and $\lambda_i > 0$.

Let $\{X(t): t\geq 0\}$ have the following density,

$$\frac{dP}{dX}\{X(t) < x] = \frac{x^{t-1}\exp\{-x\}}{\Gamma(t)} \quad \text{for} \quad x>0 \quad , \quad t>0 .$$

Then the joint distribution of $N^1(A),\ldots,N^M(A)$, where $\mu(A) = t$, given $X(t)$ is

$$P\{N^1(A) = k_1,\ldots,N^M(A) = k_M | X(t)\} = \prod_{i=1}^{M} \frac{[\lambda_i X(t)]^{k_i}}{k_i!} \exp\{-\lambda_i X(t)\}$$

for non-negative integers $k_1,\ldots,k_M$. Therefore,

$$P\{N^1(A) = k_1,\ldots,N^M(A) = k_M\}$$

$$= \int_0^\infty P\{N^1(A) = k_1,\ldots,N^M(A) = k_M | X(t)\} dP\{X(t) = x\}$$

$$= \int_0^\infty \left\{\prod_{i=1}^{M} \frac{(\lambda_i x)^{k_i}}{k_i!} e^{-\lambda_i x}\right\} \frac{x^{t-1}e^{-x}}{\Gamma(t)} dx$$

$$= \frac{\prod_{i=1}^{M} \frac{\lambda_i^{k_i}}{k_i!}}{\Gamma(t)} \int_0^\infty \exp\{-x(1+\lambda_1+\ldots+\lambda_M)\} x^{k_1+\ldots+k_m+t-1} dx$$

$$= \frac{\Gamma(\bar{k}+\mu(A))}{\Gamma(\mu(A))(1+\bar{\lambda})^{\mu(A)+\bar{k}}} \prod_{i=1}^{M} \frac{\lambda_i^{k_i}}{k_i!}$$

where $\mu(A) = t$, and $\bar{k} = \sum_{i=1}^{M} k_i$, $\bar{\lambda} = \sum_{i=1}^{M} \lambda_i$.

Thus the Multivariate Negative Binomial can be characterized as a Multivariate Doubly Stochastic Poisson process with independent components.

3.3. A Multivariate Limit Theorem. We show that the Multivariate Negative Binomial point process converges to the Multivariate Poisson process with independent components.

Theorem 3.3.1. (Grigelionis [2], 1972). Let $N_{nj}^{(M)}(A) = \{N_{nj}^{1}(A),\ldots,N_{nj}^{M}(A)\}$ for $A \varepsilon T$ and $j = 1,\ldots,n$ be independent M-variate point processes and $P\{N_{nj}^{i}(0) = 0\} = 1$ for $i = 1,\ldots,M$. Let

$$\underline{N}_{n}^{(M)}(A) = \{\sum_{j=1}^{n} N_{nj}^{1}(A),\ldots, \sum_{j=1}^{n} N_{nj}^{M}(A)\}$$

and denote by $N_{0}^{(M)}(A) = \{N_{0}^{1}(A),\ldots,N_{0}^{M}(A)\}$ for $A \varepsilon T$, a M-variate Poisson process with independent components and mean function

$$\underline{\lambda}(A) = \{E[N_{0}^{1}(A)],\ldots,E[N_{0}^{M}(A)]\} = \{\lambda_{1}\mu(A),\ldots,\lambda_{M}\mu(A)\} .$$

Then under the following conditions, for every fixed $A \varepsilon T$,

1) $\lim\limits_{n\to\infty} \max\limits_{1\le j\le n} P\{\sum_{i=1}^{M} N_{nj}^{i}(A) > 0\} = 0$

2) $\lim\limits_{n\to\infty} \sum_{j=1}^{n} P\{N_{nj}^{i}(A) > 0\} = \lambda_{i}\mu(A)$, for $i = 1,\ldots,M$

and

3) $\lim\limits_{n\to\infty} \sum_{j=1}^{n} P\{\sum_{i=1}^{M} N_{nj}^{i}(A) > 1\} = 0$

$\underline{N}_{n}^{(M)}(A)$ converges to $N_{0}^{(M)}(A)$.

Let $N_{nj}^{(M)}(A)$ be a Multivariate Negative Binomial point process with parameters $\frac{\lambda_1}{n},\ldots,\frac{\lambda_M}{n}$ for $j = 1,\ldots,n$. Then

$$\underline{N}_{n}^{(M)}(A) = \{\sum_{j=1}^{n} N_{nj}^{1}(A),\ldots, \sum_{j=1}^{n} N_{nj}^{M}(A)\}$$

converges to a Multivariate Poisson point process with independent components and mean function

$$\underline{\lambda}(A) = \{\lambda_{1}\mu(A),\ldots,\lambda_{M}\mu(A)\} , \text{ for } A \varepsilon T .$$

This can be verified by checking the conditions of Theorem 3.3.1. Therefore we have,

1) for any fixed $A \varepsilon T$,

$$\lim_{n\to\infty} \max_{1\le j\le n} P\{\sum_{i=1}^{M} N_{nj}^{i}(A)>0\} = \lim_{n\to\infty} P\{\sum_{i=1}^{M} N_{nj}^{i}(A)>0\} \le \lim_{n\to\infty} \sum_{i=1}^{M} P\{N_{nj}^{i}(A)>0\}$$

$$= \lim_{n\to\infty} \sum_{i=1}^{M} \{1-1/(1+\lambda_i/n)^{\mu(A)}\}$$

$$= \sum_{i=1}^{M} \lim_{n\to\infty} \frac{(1+\lambda_i/n)^{\mu(A)}-1}{(1+\lambda_i/n)^{\mu(A)}}$$

$$= 0 .$$

Thus condition (1) is satisfied.

2) for any fixed $A \in T$,

$$\lim_{n\to\infty} \sum_{j=1}^{n} P\{N_{nj}^{i}(A) > 0\} = \lim_{n\to\infty} n\{((1+\lambda_i/n)^{\mu(A)}-1)/(1+\lambda_i/n)^{\mu(A)}\}$$

$$= \lim_{n\to\infty} n\left\{\frac{1 + \frac{\lambda_i}{n}\mu(A) + B(n)-1}{(1 + \frac{\lambda_i}{n})^{\mu(A)}}\right\}$$

where $\lim_{n\to\infty} nB(n) = 0$.

The above limit becomes,

$$\lim_{n\to\infty} \left\{\frac{\lambda_i\mu(A) + nB(n)}{(1+\lambda_i/n)^{\mu(A)}}\right\} = \lambda_i\mu(A) \text{ , for } i = 1,\dots,M$$

as desired.

3) for any fixed $A \in T$

$$\lim_{n\to\infty} \sum_{j=1}^{n} P\{\sum_{i=1}^{M} N_{nj}^{i}(A) > 1\}$$

$$< \lim_{n\to\infty} n \sum_{i=1}^{M} \{1-1/(1+\lambda_i/n)^{\mu(A)} - \frac{\frac{\lambda_i\mu(A)}{n}}{(1 + \frac{\lambda_i}{n})^{\mu(A)+1}}\}$$

$$= \sum_{i=1}^{M} \lim_{n\to\infty} n \{\frac{(1 + \frac{\lambda_i}{n})^{\mu(A)} - 1 - \frac{\lambda_i\mu(A)}{n}(1 + \frac{\lambda_i}{n})^{-1}}{(1 + \frac{\lambda_i}{n})^{\mu(A)}}\}$$

$$= \sum_{i=1}^{M} \lim_{n\to\infty} n \{\frac{1 + \frac{\lambda_i\mu(A)}{n} + B(n) - 1 - \frac{\lambda_i\mu(A)}{n}(1 + \frac{\lambda_i}{n})^{-1}}{(1 + \frac{\lambda_i}{n})^{\mu(A)}}\}$$

where $\lim_{n\to\infty} n\mathcal{B}(n) = 0$.

The above limit becomes,

$$\sum_{i=1}^{M} \lim_{n\to\infty} \left\{\frac{\lambda_i\mu(A) + n\mathcal{B}(n) - \lambda_i\mu(A)(1 + \frac{\lambda_i}{n})^{-1}}{(1 + \frac{\lambda_i}{n})\mu(A)}\right\} = 0 .$$

Thus the conditions of Theorem 3.3.1 are satisfied and therefore $\underline{N}^{(M)}(t)$ converges to a Multivariate Poisson point process with independent components.

REFERENCES

[1] Daley, D.J. (1974). Various concepts of orderliness for point processes. Stochastic Geometry. Wiley, New York.

[2] Grigelionis, B. (1972). On weak convergence of the sums of multivariate stochastic point processes. Stochastic point processes: Statistical analysis, theory and appliations. Wiley, New York.

[3] Kallenberg, O. (1975). Random Measures. Akademie-Verlag. Berlin.

[4] Snyder, D.L. (1975). Random Point Processes. John Wiley and Sons, New York.

Multivariate Statistical Analysis
R.P. Gupta (ed.)

TESTS OF LOCATION BASED ON PRINCIPAL COMPONENTS

E.M. Carter

Department of Mathematics and Statistics
University of Guelph
Guelph, Ontario

In this paper the asymptotic distribution of the U statistic in MANOVA is derived when the number of characteristics is first reduced by the method of principal component analysis based on the sample covariance matrix.

1. INTRODUCTION

In experiments where many characteristics are measured on each subject multivariate analysis of variance tests are often insignificant even though individual t-tests are significant. By summarizing the data into q main components we reduce the number of characteristics to reasonable number. MANOVA tests can then be performed on the new set of variables. Such a procedure ignores the information from the components with smaller variances. However if these components have little practical significance then this drawback can be ignored. The distribution of MANOVA test statistics based on the first q components depends on the variances of the last $p-q$ components and, hence, the test procedures must be adjusted accordingly. In this paper we give the asymptotic distribution of the U statistic upto $O(n^{-2})$ for the case of q main components.

2. ASYMPTOTIC DISTRIBUTION OF $U_{q,p,m,n}$

Let $\underline{x}_i \sim N_p(\underline{\mu}_i,\Sigma)$, $i = 1,\ldots,m$, and $S \sim W_p(n^{-1}\Sigma,n)$. Then S may be diagonalized as

(1) $$S = \Gamma D_\ell \Gamma'$$

where $D_\ell = \text{diag}(\ell_1,\ldots,\ell_p)$, $\ell_1 > \ldots > \ell_p > 0$, and $\Gamma\Gamma' = I_p$. We define transformed variables

(2) $$\underline{z}_i = D_\ell^{-\frac{1}{2}}\Gamma'\underline{x}_i \quad , \quad i = 1,\ldots,m \ .$$

The U statistic for testing the hypothesis

(3) $$H : \underline{\mu}_1 = \ldots = \underline{\mu}_m = 0 \quad \text{vs} \quad A : \underline{\mu}_i \neq 0 \quad , \quad \text{some } i \ , \ i = 1,\ldots,m \ ,$$

is given by

(4) $$U_{p,m,n} = |nS|/|nS + \sum_{i=1}^{m} \underline{x}_i \underline{x}_i'| \quad .$$

We then define the U statistic based on the first q principal components $y_i = (z_{1i},\ldots,z_{qi})$, $i = 1,\ldots,m$, to be

(5) $$U_{q,p,m,n} = |I + n^{-1} \sum_{i=1}^{m} \underline{y}_i \underline{y}_i'|^{-1} \quad .$$

We now derive the asymptotic distribution of $\underline{z}_1,\ldots,\underline{z}_m$ and, hence, of $U_{q,p,m,n}$ under the null hypothesis (3). The joint density of $\underline{x}_1,\ldots,\underline{x}_m$ and S is given by

(6) $$C_1|S|^{\frac{1}{2}(n-p-1)}|\Sigma|^{-\frac{1}{2}(n+m)} \exp\{-\tfrac{1}{2}[\sum_{i=1}^{m} \underline{x}_i' \Sigma^{-1} x_i] - \frac{n}{2} \operatorname{tr} S \Sigma^{-1}\} \quad ,$$

where $C_1 = (2\pi)^{-p(p-1)/4-m/2} \prod_{i=1}^{p} \Gamma^{-1}(\frac{1}{2}(n - i + 1))$. Making the transformations on (1) and (2) we obtain the joint density of $\underline{x}_1,\ldots,\underline{x}_m$, Γ , and D_ℓ as

(7) $$C_2|\Sigma|^{-\frac{1}{2}(n+m)} \prod_{1\le i<j\le p} (\ell_i - \ell_j) \prod_{i=1}^{p} \ell_i^{\frac{1}{2}(n+m-p-1)}$$

$$\exp\{-\frac{n}{2} \operatorname{tr} \Sigma^{-1} [\Gamma D_\ell^{\frac{1}{2}} (I + \sum_{i=1}^{m} \underline{z}_i \underline{z}_i' n^{-1}) D_\ell^{\frac{1}{2}}\Gamma']\} \quad ,$$

where $C_2 = (2\pi)^{-p(p-1)/2-m/2} \pi^{\frac{1}{2}p^2} \prod_{i=1}^{p} \Gamma^{-1}(\frac{1}{2}(n - i + 1))\Gamma^{-1}(\frac{1}{2}(p - i + 1))$.

Integrating Γ with respect to the Haar invariant measure $\delta\Gamma$ we obtain the marginal density of $\underline{z}_1,\ldots,\underline{z}_m$ and D_ℓ as

(8) $$C_2 \prod_{1\le i<j\le p} (\ell_i - \ell_j) \prod_{i=1}^{p} \ell_i^{\frac{1}{2}(n+m-p-1)}|\Sigma|^{-\frac{1}{2}(n+m)}$$

$$_0F_0(-\frac{n}{2}\Sigma^{-1} \ , \ D_\ell(I + \sum_{i=1}^{m} \underline{z}_i \underline{z}_i' n^{-1})) \quad ,$$

where $_0F_0(A,B)$ is defined in James (1964) in terms of zonal polynomials of symmetric matrix argument. The asymptotic expansion of the $_0F_0$ hypergeometric functions for two matrices is given in Anderson (1965). As only the roots of Σ are involved in (8) we let $\Sigma = \operatorname{diag}(\lambda_1,\ldots,\lambda_p)$, $\lambda_1 > \ldots > \lambda_p > 0$. The expression for the $_0F_0$ hypergeometric function then becomes

$$ {}_0F_0\left(-\tfrac{n}{2}D_\lambda^{-1}\ ,\ D_\ell(I + n^{-1}\sum_{i=1}^{m} \underline{z}_i\, \underline{z}_i')\right) = $$

(9) $$ C_3 \prod_{1\le i<j\le p} [(\lambda_i - \lambda_j)^{-\frac12} (\theta_i - \theta_j)^{-\frac12}] \prod_{i=1}^{p} \lambda_i^{-(p-1)} $$

$$ \exp[-\tfrac{n}{2}\sum_{i=1}^{p} \theta_i \lambda_i^{-1}] \{1 + n^{-1} \sum_{1\le i<j\le p} \lambda_i \lambda_j (\lambda_i - \lambda_j)^{-1}(\theta_i - \theta_j)^{-1} $$

$$ + O(n^{-2})\} $$

where $\theta_1 > \dots > \theta_p$ are the eigenvalues of $D_\ell(I + n^{-1}\sum_{i=1}^{m} \underline{z}_i\, \underline{z}_i')$,

and $C_3 = (2\pi n^{-1})^{\frac12 p(p-1)}\ \pi^{-\frac12 p^2} \prod_{i=1}^{p} \Gamma(\tfrac12(p - i + 1))$.

In order to express θ_i in terms of ℓ_i and $\underline{z}_1,\dots,\underline{z}_m$ we need the following lemma.

<u>LEMMA</u>. Let $\underline{z}_i = (z_{1i},\dots,z_{pi})'$, $i = 1,\dots,m$, be p vectors and let $D_\ell = \text{diag}(\ell_1,\dots,\ell_p)$, $\ell_1 > \dots > \ell_p$ then for large n the eigenvalues $\theta_1 > \dots > \theta_p$ of $D_\ell(I + n^{-1}\sum_{i=1}^{m} \underline{z}_i\, \underline{z}_i')$ are given by

$$ \theta_i = \ell_i + \ell_i n^{-1} \sum_{j=1}^{m} z_{ij}^2 + n^{-2} \sum_{j\ne i} \ell_i \ell_j (\ell_i - \ell_j)^{-1} (\sum_{k=1}^{m} z_{ik}\, z_{jk})^2 $$
$$ + O(n^{-3}) . $$

<u>Proof</u>. The eigenvalue θ_i satisfies the determinantal equation

$$ |D_\ell(I + n^{-1}\sum_{i=1}^{m} \underline{z}_i\, \underline{z}_i') - \theta_i I_p| = 0 . $$

Letting

$$ \theta_1 = \ell_1 + n^{-1}\, \ell_1 \sum_{k=1}^{m} z_{1k}^2 + n^{-2}\gamma $$

we obtain the expression upto $O(n^{-1})$

$$\left\| \begin{matrix} -\gamma & \ell_1 \sum_{k=1}^{m} z_{1k}z_{2k} & & \cdots & \ell_1 \sum_{k=1}^{m} z_{1k}z_{pk} \\ \ell_2 \sum_{k=1}^{m} z_{1k}z_{2k} & \ell_2 - \ell_1 & 0 & \cdots & 0 \\ & 0 & & & \\ \vdots & \vdots & & \ddots & \\ & & & & 0 \\ \ell_p \sum_{k=1}^{m} z_{1k}z_{pk} & 0 & & \cdots & \ell_p - \ell_1 \end{matrix} \right\| = 0 .$$

Solving the equation for γ, we obtain the desired result for θ_1. By symmetry the proof holds for $\theta_2,\ldots,\theta_p$.

Using the results of the lemma, (8) and (9) we obtain the joint density of $\ell_1,\ldots,\ell_p$, $\underline{z}_i = (z_{1i},\ldots,z_{pi})$, $i = 1,\ldots,m$, as

$$(10) \quad (2\pi)^{-\frac{1}{2}m} \prod_{i=1}^{p} \Gamma^{-1}(\tfrac{1}{2}(n - i + 1)) \prod_{1\le i<j\le p} (\ell_i - \ell_j)^{\frac{1}{2}}(\lambda_i - \lambda_j)^{-\frac{1}{2}}$$

$$\prod_{i=1}^{p} \{(\ell_i/\lambda_i)^{\frac{1}{2}(n+m-p-1)} \lambda_i^{-1} \exp [- \tfrac{n}{2} \ell_i\lambda_i^{-1} - \tfrac{1}{2} \sum_{k=1}^{m} z_{ik}^2 \ell_i\lambda_i^{-1}\}$$

$$\{1 + n^{-1} \sum_{1\le i<j\le p} [\lambda_i\lambda_j(\ell_i - \ell_j)^{-1}(\lambda_i - \lambda_j)^{-1}$$

$$+ \tfrac{1}{2} \ell_i\ell_j(\lambda_i - \lambda_j)(\ell_i - \ell_j)^{-1} \lambda_i^{-1}\lambda_j^{-1}(\sum_{k=1}^{m} z_{ik}z_{jk})^2$$

$$- \tfrac{1}{2} (\ell_i - \ell_j)^{-1} \sum_{k=1}^{m} (\ell_i z_{ik}^2 - \ell_j z_{jk}^2)] + O(n^{-2})\} .$$

By making the transformations

$$(11) \quad \ell_i = \lambda_i(1 + u_i n^{-\frac{1}{2}}) , \quad i = 1,\ldots,p$$

and integrating over u_i we obtain the following theorem

<u>THEOREM 1</u>. Let $\underline{x}_i \sim N_p(\underline{0},\Sigma)$, $i = 1,\ldots,m$, and let $S \sim N_p(n^{-1}\Sigma,n)$. Let $\ell_1 > \ldots > \ell_p$ be the eigenvalues of S and let the columns of Γ be the eigenvectors. If $\lambda_1 > \ldots > \lambda_p$ are the eigenvalues of Σ then the distribution of $(z_{1i},\ldots,z_{pi})' = \underline{z}_i = D_\ell^{-\frac{1}{2}}\Gamma'\underline{x}_i$, $i = 1,\ldots,m$ is given by

$$(2\pi)^{-\frac{1}{2}m}[\exp - \frac{1}{2}\sum_{i=1}^{p} \underline{z}_i' \underline{z}_i] \{1 + n^{-1}[\tfrac{1}{2}(p - q - 1) \sum_{i=1}^{p} \underline{z}_i' \underline{z}_i$$

$$+ \frac{1}{4}\sum_{i=1}^{p} (\sum_{i=1}^{m} z_{1k}^2)^2 + \frac{1}{2}\sum_{1\le i<j\le p} (\sum_{i=1}^{m} z_{ik}z_{jk})^2$$

$$- \sum_{1\le i<j\le p} (\lambda_i - \lambda_j)^{-1} (\sum_{i=1}^{m} \lambda_i z_{ik}^2 - \lambda_j z_{jk}^2) + \frac{1}{4} mp(m - p + 1)$$

$$+ O(n^{-2})\} .$$

<u>THEOREM 2</u>. Let $y_i = (z_{1i},\ldots,z_{2i})'$. Let $U_{q,p,m,n} = |I + n^{-1}\sum_{i=1}^{m} \underline{}_i \underline{}_i'|^{-1}$. Then asymptotically for large n

$$-(n - \tfrac{1}{2}(q - m + 1) + 2q^{-1}\sum_{i=1}^{q}\sum_{j=q+1}^{p} \ell_j(\ell_i - \ell_j)^{-1}) \ln U_{q,p,m,n} \simeq \chi_{qm}^2 .$$

<u>Proof</u>. The proof is obtained by looking at

$$\phi(t) = E\{\exp t(n - \gamma) \ln |I + n^{-1}\sum_{i=1}^{m} \underline{y}_i \underline{y}_i'|\}$$

$$= E\{\exp t(n - \gamma) \operatorname{tr} \ln (I + n^{-1}\sum_{i=1}^{m} \underline{y}_i \underline{y}_i')\} .$$

By expanding the above expression in terms of n^{-1} and integrating with respect to the density of $\underline{z}_1,\ldots,\underline{z}_m$ given in Theorem 1 we obtain

$$\phi(t) = (1 - 2t)^{-\frac{1}{2}mq}\{1 + n^{-1}[\tfrac{1}{4}(q - m + 1)mq - m \sum_{i=1}^{q}\sum_{j=q+1}^{p} \lambda_j(\lambda_i - \lambda_j)^{-1}]$$

$$- n^{-1}(1 - 2t)^{-1}[\tfrac{1}{4}(q - m + 1)mq - m \sum_{i=1}^{q}\sum_{j=q+1}^{p} \lambda_j(\lambda_i - \lambda_j)^{-1}$$

$$+ tmq\gamma] + O(n^{-2})\}$$

$$= (1 = 2t)^{-\frac{1}{2}mq} \{1 + n^{-1}2t(1 - 2t)^{-1} [\tfrac{1}{4}(q - m + 1)mq$$

$$- m \sum_{i=1}^{q}\sum_{j=q+1}^{p} \lambda_j(\lambda_i - \lambda_j)^{-1} - \tfrac{1}{2}mq\gamma] + O(n^{-2})\} .$$

Setting $\gamma = \frac{1}{2}(q - m + 1) - 2q^{-1}\sum_{i=1}^{q}\sum_{j=q+1}^{p} \lambda_i(\lambda_i - \lambda_j)^{-1}$ we obtain $\phi(t) = (1 - 2t)^{-\frac{1}{2}mq} + O(n^{-2})$. As $\lambda_1,\ldots,\lambda_p$ are unknown we use the consistent estimators $\ell_1,\ldots,\ell_p$ instead, and obtain the desired result.

COROLLARY. For the case $m = 1$ we obtain the joint density of the T^2 statistic, $T^2 = \underline{y}_1' \underline{y}_1$, as

$$(n - \tfrac{1}{2}q + 2q^{-1} \sum_{i=1}^{q} \sum_{j=q+1}^{p} \ell_j(\ell_i - \ell_j)^{-1}) \ln (1 + n^{-1}T^2) \simeq \chi_q^2 .$$

3. EXTENSIONS

This paper only considers tests based on principal component analysis of a covariance matrix. Similar results are under investigation for tests based on principal components analysis of a correlation matrix and for tests based on factor scores. Also under investigation are tests involving principal component analysis followed by multiple regression analysis.

It should be noted that alhtough the asymptotic distribuiton of $\underline{z}_1,\ldots,\underline{z}_p$ requires $\lambda_1 > \ldots > \lambda_p$, the final expression for the asymptotic distribution of $U_{q,p,m,n}$ only requires $\lambda_q > \lambda_{q+1}$. Step down procedures have been suggested by Rao (1964) using principal component analysis based on the total sum of squares matrix.

REFERENCES

[1] Anderson, G.A. (1965). An asymptotic expansion for the distribution of latent roots of the estimated covariance matrix. Ann. Math. Statist. 36, 1153-1173.

[2] James, A.T. (1964). Distribution of matrix variants and latent roots derived from normal samples. Ann. Math. Statist. 35, 475-501.

[3] Rao, C.R. (1964). The use and interpretation of principal component analysis in applied research. Sankhyā A 26, 329-358.

Multivariate Statistical Analysis
R.P. Gupta (ed.)

INVARIANT POLYNOMIALS WITH MATRIX ARGUMENTS AND THEIR APPLICATIONS

Yasuko Chikuse
Department of Mathematics & Statistics
University of Pittsburgh
and
Kagawa University, Japan

A class of homogeneous polynomials $C_\phi^{\kappa_1,\kappa_2,\ldots,\kappa_r}(X_1,X_2,\ldots,X_r)$ with a general number r of matrix arguments $(r \geq 1)$, invariant under the orthogonal group, is proposed through the theory of polynomial representations of the linear group. The class of polynomials is a generalization of the invariant polynomials with lower numbers of matrix arguments, worked out previously up to the case of three matrix arguments. Fundamental properties and relations of the polynomials $C_\phi^{\kappa_1,\kappa_2,\ldots,\kappa_r}$ are shown, and some applications in multivariate distribution theory are indicated.

1. INTRODUCTION

The theory of group representations was applied in defining the zonal polynomials of the real positive definite symmetric matrix argument. The zonal polynomials have been discussed and utilized in the area of multivariate distribution theory in an extensive literature (see e.g., Constantine [4] and James [10] as survey references, and many others in the recent literature).

Davis [6], [7] proposed a class of invariant polynomials with two matrix arguments using the group representation theory, extending the zonal polynomials. Some difficulties in distributional problems which could not be solved in terms of zonal polynomials have been solved by these polynomials; these are the distributions of the latent roots of the noncentral Wishart matrix, the noncentral quadratic form and the doubly noncentral multivariate F matrix (Davis [6], [7]), and the distributions of the sums of the two or three (noncentral) Wishart matrices and the roots of an 'extended' central MANOVA matrix W in the multivariate Behrens-Fisher discriminant analysis (Chikuse [2]). Chikuse [3] furthermore generalized Davis' polynomials to a class of invariant polynomials with three matrix arguments, which resolves some of the difficulties still remained in multivariate distributional problems.

In Section 2, we shall define a class of invariant polynomials with a general number r of matrix arguments $(r \geq 1)$ as a generalization of the invariant polynomials with lower numbers of matrix arguments developed previously. The arguments employed are direct extensions of those for the zonal polynomials and Davis' polynomials. Some fundamental properties and relations satisfied by these polynomials are shown in Section 3, generalizing the results of Chikuse [3, Sections 3 to 7].

Section 4 applies the polynomials to some problems in multivariate normal distribution theory. The distributions of the sums of a general number of the central and noncentral Wishart matrices are obtained. We shall derive, as applications for $r = 3$, the distributions of the latent roots of the doubly noncentral multivariate F matrix with unequal covariance matrices and the latent roots of the 'extended' noncentral MANOVA matrix W in the multivariate Behrens-Fisher discriminant analysis.

2. INVARIANT POLYNOMIALS WITH r MATRIX ARGUMENTS

The arguments in Davis [6, Sections 2 and 4] are directly extended for defining the invariant polynomials with general r matrix arguments. The discussion for $r = 3$ is given in Chikuse [3, Section 2] along this line, and in this section we shall give a brief summary for general r to make sure the procedure.

Let $P_k[X]$ and $P_{k_1,k_2,\dots,k_r}[X_1,X_2,\dots,X_r]$ be the classes of homogeneous polynomials of degree k in the elements of X and of degrees $k_1,k_2,\dots,k_r$ in the elements of $X_1,X_2,\dots,X_r$ respectively. By the theory of polynomial representations of the linear group $G\ell(m,R)$ (Boerner [1, Chapter 5] and James (unpublished lecture notes)), we have the decompositions into irreducible invariant subspaces

$$P_k[X] = \bigoplus_{\kappa} V_\kappa[X] ,$$

$$P_{k_1,k_2,\dots,k_r}[X_1,X_2,\dots,X_r] = P_{k_1}[X_1] \otimes P_{k_2}[X_2] \otimes \dots \otimes P_{k_r}[X_r]$$

$$= \bigoplus_{\kappa_1} \bigoplus_{\kappa_2} \dots \bigoplus_{\kappa_r} \bigoplus_{\substack{\phi=2\phi \\ \phi\in\kappa_1\cdot\kappa_2\cdot\dots\cdot\kappa_r}} V_\phi^{\kappa_1,\kappa_2,\dots,\kappa_r}[X_1,X_2,\dots,X_r] . \tag{2.1}$$

Here, κ and $\kappa_1,\kappa_2,\dots,\kappa_r$ and ϕ denote partitions of k and $k_1,k_2,\dots,k_r$ and $f = \sum_{i=1}^{r} k_i$ respectively into not more than m parts, and ϕ , $\kappa_1\cdot\kappa_2\cdot\dots\cdot\kappa_r$ signifies that the irreducible representation of $G\ell(m,R)$ indexed by 2ϕ occurs in the decomposition of the Kronecker product $2\kappa_1 \times 2\kappa_2 \times \dots \times 2\kappa_r$ of the irreducible representations indexed by $2\kappa_1,2\kappa_2,\dots,2\kappa_r$. The decompositions of the Kronecker products are determined by the rule due to Robinson [12, Sections 3.3]. $V_\kappa[X]$ and $V_\phi^{\kappa_1,\kappa_2,\dots,\kappa_r}[X_1,X_2,\dots,X_r]$ contain exactly one one-dimensional subspaces generated by the zonal polynomial $C_\kappa(X)$ and the polynomials in $X_1,X_2,\dots,X_r$, $\Gamma_\phi^{\kappa_1,\kappa_2,\dots,\kappa_r}(X_1,X_2,\dots,X_r)$, respectively, invariant under the transformations $X \to H'XH$ and $X_i \to H'X_iH$, $i = 1,\ldots,r$, for $H \in O(m)$, the orthogonal group. A representation 2ϕ may occur in (2.1) with multiplicity greater than one for a given $\kappa_1,\kappa_2,\ldots,\kappa_r$ and, whence, the $V_{2\phi}^{\kappa_1,\kappa_2,\ldots,\kappa_r}$ and the corresponding $\Gamma^{\kappa_1,\kappa_2,\ldots,\kappa_r}$ are not uniquely defined. Utilizing the arguments of James [9, Section 4] and Saw [13] yields the invariant polynomials $C_\phi^{\kappa_1,\kappa_2,\ldots,\kappa_r}(X_1,X_2,\ldots,X_r)$. These polynomials are constructed as linear combinations of $\Gamma_{\phi'}^{\kappa_1,\kappa_2,\ldots,\kappa_r}$ for $\phi' \equiv \phi$ and '$\Delta_{k_1,k_2,\ldots,k_r}$-orthogonal', and are generated by the set of all distinct products of traces

$$(\mathrm{tr}\, X_1^{a_1}X_2^{a_2}\ldots X_r^{a_r}X_1^{b_1}X_2^{b_2}\ldots)^{p_1}(\mathrm{tr}\, X_1^{g_1}X_2^{g_2}\ldots X_r^{g_r}X_1^{h_1}X_2^{h_2}\ldots)^{p_2}\ldots$$

of total degrees $k_1,k_2,\ldots,k_r$ in the elements of $X_1,X_2,\ldots,X_r$ respectively. The polynomials $C_\phi^{\kappa_1,\kappa_2,\ldots,\kappa_r}(X_1,X_2,\ldots,X_r)$ satisfy the basic relation

$$\int_{O(m)} C_{\kappa_1}(A_1H'X_1H)C_{\kappa_2}(A_2H'X_2H)\ldots C_{\kappa_r}(A_rH'X_rH)dH$$

$$= \Sigma_{\phi\in\kappa_1\cdot\kappa_2\cdot\ldots\cdot\kappa_r} C_\phi^{\kappa_1,\kappa_2,\ldots,\kappa_r}(A_1,A_2,\ldots,A_r)C_\phi^{\kappa_1,\kappa_2,\ldots,\kappa_r}(X_1,X_2,\ldots,X_r)/C_\phi(I). \tag{2.2}$$

and the property of the invariance under the simultaneous transformations

$$X_i \to H'X_iH \ , \ i = 1,\ldots,r \quad \text{for} \quad H \in O(m) \ . \tag{2.3}$$

3. PROPERTIES AND RELATIONS OF THE $C_\phi^{\kappa_1,\kappa_2,\ldots,\kappa_r}$

The results in Chikuse [3, Sections 3 to 7] for $r = 3$ can be easily extended to our case for general r . Hence the results with general terms are only stated here without proofs except for a few cases. In the following, there is no loss of generality in discussing for a subset $\{1,\ldots,q\}$ consisting of the first q elements out of the whole set $\{1,\ldots,r\}$ consisting of the first r nonnegative integers $(1 < q < r)$.

<u>Elementary properties</u>:

$$C_\phi^{\kappa_1,\ldots,\kappa_r}(X_1,\ldots,X_q,I,\ldots,I)$$

$$= \Sigma_{\sigma\in\kappa_1\cdot\ldots\cdot\kappa_q} \alpha_\sigma^{\kappa(q,(r-q));\phi}\, C^{\kappa_1,\ldots,\kappa_q}(X_1,\ldots,X_q) \tag{3.1}$$

for a suitable choice of the α , where $\kappa(q,(r-q))$ denotes the abbreviation for '$\kappa_k,\ldots,\kappa_q$, $(\kappa_{q+1}),\ldots,(\kappa_r)$' . When $X_1 = \ldots = X_q = X$, we have

$$C_\phi^{\kappa_1,\ldots,\kappa_r}(X,\ldots,X,I,\ldots,I) = \Sigma_{\sigma\in\kappa_1\cdot\ldots\cdot\kappa_q} \alpha_\sigma^{\kappa(q,(r-q));\phi}\, \theta_\sigma^{\kappa_1,\ldots,\kappa_q} C_\sigma(X) , \tag{3.2}$$

where

$$\theta_\sigma^{\kappa_1,\ldots,\kappa_q} = C_\sigma^{\kappa_1,\ldots,\kappa_q}(I,\ldots,I)/C_\sigma(I) ; \tag{3.3}$$

in particular

$$C_\phi^{\kappa_1,\ldots,\kappa_r}(X,\ldots,X) = \theta_\phi^{\kappa_1,\ldots,\kappa_r}\, C_\phi(X) , \tag{3.4}$$

$$C_\phi^{\kappa_1,\ldots,\kappa_r}(X,I,\ldots,I) = [C_\phi^{\kappa_1,\ldots,\kappa_r}(I,\ldots,I)/C_{\kappa_1}(I)]C_{\kappa_1}(X) . \tag{3.5}$$

$$C_\phi^{\kappa_1,\ldots,\kappa_q,0,\ldots,0}(X_1,\ldots,X_q,X_{q+1},\ldots,X_r) \overset{\text{def}}{=} C_\phi^{\kappa_1,\ldots,\kappa_q}(X_1,\ldots,X_q) . \tag{3.6}$$

$(\phi \in \kappa_1\cdot\ldots\cdot\kappa_q)$

$$C_\phi^{\kappa_1,\ldots,\kappa_r}(X,\ldots,X,X_{q+1},\ldots,X_r)$$
$$= \Sigma_{\sigma\in\kappa_1\cdot\ldots\cdot\kappa_q} \beta_\sigma^{\kappa(q,(r-q));\phi}\, C_\phi^{\sigma,\kappa_{q+1},\ldots,\kappa_r}(X,X_{q+1},\ldots,X_r) , \tag{3.7}$$

where the β are defined such that

$$\alpha_\sigma^{\kappa(q,(r-q));\phi}\, C_\sigma^{\kappa_1,\ldots,\kappa_q}(I,\ldots,I) = \beta_\sigma^{\kappa(q,(r-q));\phi}\, C_\phi^{\sigma,\kappa_{q+1},\ldots,\kappa_r}(I,\ldots,I) . \tag{3.8}$$

$$C_\phi^{\kappa_1,\ldots,\kappa_r}(\alpha_1X_1,\ldots,\alpha_rX_r) = \alpha_1^{k_1}\ldots\alpha_r^{k_1}\, C_\phi^{\kappa_1,\ldots,\kappa_r}(X_1,\ldots,X_r) . \tag{3.9}$$

$$\prod_{i=1}^r C_{k_i}(X_i) = \Sigma_{\phi\in\kappa_1\cdot\ldots\cdot\kappa_r} \theta_\phi^{\kappa_1,\ldots,\kappa_r}\, C_\phi^{\kappa_1,\ldots,\kappa_r}(X_1,\ldots,X_r) . \tag{3.10}$$

$$C_{\sigma}^{\kappa_1,\ldots,\kappa_q}(X_1,\ldots,X_q)C_{\tau}^{\kappa_{q+1},\ldots,\kappa_r}(X_{q+1},\ldots,X_r)$$

$$= \Sigma_{\phi\epsilon\kappa_1\cdot\ldots\cdot\kappa_q\cdot\kappa_{q+1}\ldots\kappa_r} [\gamma_{\sigma}^{\kappa(q,(r-q));\phi}\,\alpha_{\tau}^{(\sigma),\kappa_{q+1},\ldots,\kappa_r;\phi}$$

$$C_{\sigma}(I)C_{\tau}(I)/C_{\phi}(I)]C_{\phi}^{\kappa_1,\ldots,\kappa_r}(X_1,\ldots,X_q,X_{q+1},\ldots,X_r) \ , \qquad (3.11)$$

where the γ and α are given as in (3.13) below and (3.1).

Integrals over $O(m)$:

$$\int_{O(m)} C_{\phi}^{\kappa_1,\ldots,\kappa_r}(A'H'X\ HA,\ldots,A'H'X_qHA,A_{q+1},\ldots,A_r)dH$$

$$= \Sigma_{\sigma\epsilon\kappa_1\cdot\ldots\cdot\kappa_q}\ \gamma_{\sigma}^{\kappa(q,(r-q));\phi}\ C_{\phi}^{\sigma,\kappa_{q+1},\ldots,\kappa_r}(A'A,A_{q+1},\ldots,A_r)$$

$$C_{\sigma}^{\kappa_1,\ldots,\kappa_q}(X_1,\ldots,X_q) \ , \qquad (3.12)$$

where the γ are defined such that

$$\gamma_{\sigma}^{\kappa(q,(r-q));\phi}\ C_{\phi}^{\sigma,\kappa_{q+1},\ldots,\kappa_r}(I,\ldots,I) = \alpha_{\sigma}^{\kappa(q,(r-q));\phi} \ ; \qquad (3.13)$$

in particular,

$$\int_{O(m)} C^{\kappa_1,\ldots,\kappa_r}(A'H'X_1HA,\ldots,A'H'X_rHA)dH$$

$$= C_{\phi}^{\kappa_1,\ldots,\kappa_r}(X_1,\ldots,X_r)C_{\phi}(A'A)/C_{\phi}(I) \ , \qquad (3.14)$$

$$\int_{O(m)} C^{\kappa_1,\ldots,\kappa_r}(A'H'X_1HA,A_2,\ldots,A_r)dH$$

$$= C_{\phi}^{\kappa_1,\ldots,\kappa_r}(A'A,A_2,\ldots,A_r)C_{\kappa_1}(X_1)/C_{\kappa_1}(I) \ . \qquad (3.15)$$

Multinomial expansions:

$$C_{\phi}(\sum_{i=1}^{r} X_i) = \Sigma_{\substack{\kappa_1\ \ldots\ \kappa_r \\ (\phi\epsilon\kappa_1\cdot\ldots\cdot\kappa_r)}} (f!/\prod_{i=1}^{r} k_i!)\theta_{\phi}^{\kappa_1,\ldots,\kappa_r}C_{\phi}^{\kappa_1,\ldots,\kappa_r}(X_1,\ldots,X_r) \ ; \qquad (3.16)$$

in particular,

$$C_f\Big(\sum_{i=1}^{r} X_i\Big) = \sum_{\substack{k_1,\dots,k_r \\ (\sum_{i=1}^{r} k_i=f)}} \Big(f!/\prod_{i=1}^{r} k_i!\Big) C^{k_1,\dots,k_r}(X_1,\dots,X_r) , \tag{3.17}$$

$$C_{1^f}\Big(\sum_{i=1}^{r} X_i\Big) = \sum_{\substack{k_1,\dots,k_r \\ (\sum_{i=1}^{r} k_i=f)}} \Big(f!/\prod_{i=1}^{r} k_i!\Big) C_{1^f}^{1^{k_1},\dots,1^{k_r}}(X_1,\dots,X_r) . \tag{3.18}$$

$$C_\phi^{\sigma,\kappa_{q+1},\dots,\kappa_r}\Big(\sum_{i=1}^{q} X_i, X_{q+1},\dots,X_r\Big)$$

$$= \sum_{\substack{\kappa_1,\dots,\kappa_r \\ (\sigma\in\kappa_1\cdot\dots\cdot\kappa_r)}} \Big(s!/\prod_{i=1}^{q} k_i!\Big)\beta_\sigma^{\kappa(q,(r-q));\phi}\, C_\phi^{\kappa_1,\dots,\kappa_r}(X_1,\dots,X_r) . \tag{3.19}$$

$$C_\phi^{\kappa_1,\dots,\kappa_r}(I+A_1,\dots,I+A_q,A_{q+1},\dots,A_r)/C_\phi(I)$$

$$= \sum_{r_1=0}^{k_1} \cdots \sum_{r_q=0}^{k_q} \sum_{\substack{\rho_1,\dots,\rho_q \\ (\zeta\in\rho_1\cdot\dots\cdot\rho_q)}} b_{\rho_1,\dots,\rho_q,\kappa_{q+1},\dots,\kappa_r;\zeta}^{\kappa_1,\dots,\kappa_q,\kappa_{q+1},\dots,\kappa_r;\phi}$$

$$C_\zeta^{\rho_1,\dots,\rho_q,\kappa_{q+1},\dots,\kappa_r}(A_1,\dots,A_r)/C_\zeta(I) , \tag{3.20}$$

for a suitable choice of b .

Laplace transforms:

$$\int_{R>0} \operatorname{etr}(-WR)|R|^{a-p}\, C_\phi^{\kappa_1,\dots,\kappa_r}(A_1R,\dots,A_rR)dR$$

$$= \Gamma_m(a,\phi)|W|^{-a}\, C_\phi^{\kappa_1,\dots,\kappa_r}(A_1W^{-1},\dots,A_rW^{-1}) , \tag{3.21}$$

and in particular

$$E_V C_\phi^{\kappa_1,\dots,\kappa_r}(V'A_1V,\dots,V'A_rV) = 2^f(\tfrac{1}{2}n)_\phi C_\phi^{\kappa_1,\dots,\kappa_r}(W'A_1W,\dots,W'A_rW) , \tag{3.22}$$

where $VV' \sim W_m(n,\Sigma)$, $WW' = \Sigma$, $p = \tfrac{1}{2}(m+1)$.

$$\int_{R>0} \operatorname{etr}(-WR)|R|^{a-p} C_\phi^{\kappa_1,\ldots,\kappa_r}(A_1RA_1',A_2,\ldots,A_r)dR$$
$$= \Gamma_m(a,\kappa_1)|W|^{-a} C_\phi^{\kappa_1,\ldots,\kappa_r}(A_1W^{-1}A_1',A_2,\ldots,A_r) . \tag{3.23}$$

$$\int_{R>0} \operatorname{etr}(-WR)|R|^{a-p} C_\phi^{\kappa_1,\ldots,\kappa_r}(A_1R^{-1},\ldots,A_rR^{-1})dR$$
$$= \Gamma_m(a,-\phi)|W|^{-a} C_\phi^{\kappa_1,\ldots,\kappa_r}(A_1W,\ldots,A_rW) , \tag{3.24}$$

where $\Gamma_m(a,-\phi) = (-1)^f\Gamma_m(a)/(-a+p)_\phi$.

$$\int_{R>0} \operatorname{etr}(-WR)|R|^{a-p} C_\phi^{\kappa_1,\ldots,\kappa_r}(A_1R^{-1}A_1',A_2,\ldots,A_r)dR$$
$$= \Gamma_m(a,-\kappa_1)|W|^{-a} C_\phi^{\kappa_1,\ldots,\kappa_r}(A_1WA_1',A_2,\ldots,A_r) . \tag{3.25}$$

Beta-type integrals:

$$\underset{0<\sum_{i=1}^{r-1} T_i<I}{\int_0^I \cdots \int_0^I} \prod_{i=1}^{r-1} |T_i|^{a_i-p} |I - \sum_{i=1}^{r-1} T_i|^{a_r-p} C_\phi^{\kappa_1,\ldots,\kappa_r}(T_1,\ldots,T_{r-1},I - \sum_{i=1}^{r-1} T_i) \prod_{j=1}^{r-1} dT_j$$
$$= \prod_{i=1}^{r} \Gamma_m(a_i,\kappa_i)[\Gamma_m(\sum_{i=1}^{r} a_i,\phi)]^{-1} \theta_\phi^{\kappa_1,\ldots,\kappa_r} C_\phi(I) . \tag{3.26}$$

$$\underset{0<\sum_{i=1}^{r} T_i<I}{\int_0^I \cdots \int_0^I} \prod_{i=1}^{r} |T_i|^{a_i-p} |I - \sum_{i=1}^{r} T_i|^{b-p} C_\phi^{\kappa_1,\ldots,\kappa_r}(T_1,\ldots,T_r) \prod_{j=1}^{r} dT_j$$
$$= \prod_{i=1}^{r} \Gamma_m(a_i,\kappa_i)\Gamma_m(b)[\Gamma_m(\sum_{i=1}^{r} a_i+b,\phi)]^{-1} \theta_\phi^{\kappa_1,\ldots,\kappa_r} C_\phi(I) . \tag{3.27}$$

$$\int_0^I |T|^{a-p}|I-T|^{b-p} C_\phi^{\kappa_1,\ldots,\kappa_r}(A_1T,\ldots,A_rT)dT$$
$$= \Gamma_m(a,\phi)\Gamma_m(b)[\Gamma_m(a+b,\phi)]^{-1} C_\phi^{\kappa_1,\ldots,\kappa_r}(A_1,\ldots,A_r) . \tag{3.28}$$

$$\int_0^I |T|^{a-p}|I-T|^{b-p} C_\phi^{\kappa_1,\ldots,\kappa_r}(A_1TA_1',A_2,\ldots,A_r)dT$$
$$= \Gamma_m(a,\kappa_1)\Gamma_m(b)[\Gamma_m(a+b,\kappa_1)]^{-1}C_\phi^{\kappa_1,\ldots,\kappa_r}(A_1A_1',A_2,\ldots,A_r) . \tag{3.29}$$

$$\int_0^I |T|^{a-p}|I-T|^{b-p} C_\phi^{\kappa_1,\ldots,\kappa_r}(A_1T^{-1},\ldots,A_rT^{-1})dT$$
$$= \Gamma_m(a,-\phi)\Gamma_m(b)[\Gamma_m(a+b,-\phi)]^{-1} C_\phi^{\kappa_1,\ldots,\kappa_r}(A_1,\ldots,A_r) . \tag{3.30}$$

$$\int_0^I |T|^{a-p}|I-T|^{b-p} C_\phi^{\kappa_1,\ldots,\kappa_r}(A_1T^{-1}A_1',A_2,\ldots,A_r)dT$$
$$= \Gamma_m(a,-\kappa_1)\Gamma_m(b)[\Gamma_m(a+b,-\kappa_1)]^{-1} C_\phi^{\kappa_1,\ldots,\kappa_r}(A_1A_1',A_2,\ldots,A_r) . \tag{3.31}$$

$$\int_0^X |T|^{a-p} C_\phi^{\kappa_1,\ldots,\kappa_r}(A_1T,\ldots,A_rT)dT$$
$$= \Gamma_m(p)\Gamma_m(a,\phi)[\Gamma_m(a+p,\phi)]^{-1}|X|^a C_\phi^{\kappa_1,\ldots,\kappa_r}(A_1X,\ldots,A_rX) . \tag{3.32}$$

$$\int_0^X |T|^{a-p} C_\phi^{\kappa_1,\ldots,\kappa_r}(A_1TA_1',A_2,\ldots,A_r)dT$$
$$= \Gamma_m(p)\Gamma_m(a,\phi_1)[\Gamma_m(a+p,\phi_1)]^{-1}|X|^a C_\phi^{\kappa_1,\ldots,\kappa_r}(A_1XA_1',A_2,\ldots,A_r) . \tag{3.33}$$

Incomplete gamma and beta functions are expanded by utilizing (3.32) and (3.33).

Generalized Laguerre Polynomials:

Define

$$L^{u_1,\ldots,u_q}_{\kappa_1,\ldots,\kappa_q,\kappa_{q+1},\ldots,\kappa_r;\phi}(X_1,\ldots,X_q,X_{q+1},\ldots,X_r) = \operatorname{etr}(\sum_{i=1}^q X_i)$$
$$\int_{R_1>0}\cdots\int_{R_q>0} \prod_{i=1}^q [\operatorname{etr}(-R_i)|R_i|^{u_i} A_t(X_iR_i)]C_\phi^{\kappa_1,\ldots,\kappa_r}(R_1,\ldots,R_q,X_{q+1},\ldots,X_r)\prod_{i=1}^q dR_i , \tag{3.34}$$

where A_t is the Bessel function of Herz [8].

(Laplace transform)

$$\int_{R_1>0}\cdots\int_{R_q>0} \prod_{i=1}^{q} [\mathrm{etr}(-X_iR_i)|R_i|^{u_i}]$$

$$L^{u_1,\ldots,u_q}_{\kappa_1,\ldots,\kappa_q,\kappa_{q+1},\ldots,\kappa_r;\phi}(R_1,\ldots,R_q,X_{q+1},\ldots,X_r) \prod_{i=1}^{q} dR_i$$

$$= \prod_{i=1}^{q} [\Gamma_m(u_i+p,\kappa_i)|X_i|^{-u_i-p}] C_\phi^{\kappa_1,\ldots,\kappa_r}(I-X_1^{-1},\ldots,I-X_q^{-1},X_{q+1},\ldots,X_r) . \tag{3.35}$$

(Serial expression)

$$L^{u_1,\ldots,u_q}_{\kappa_1,\ldots,\kappa_q,\kappa_{q+1},\ldots,\kappa_r;\phi}(X_1,\ldots,X_r) = \prod_{i=1}^{q} (u_i+p)_{\kappa_i} C_\phi(I)$$

$$\sum_{r_1=0}^{k_1}\cdots\sum_{r_q=0}^{k_q} \sum_{\substack{\rho_1,\ldots,\rho_q \\ (\zeta\in\rho_1\cdot\ldots\cdot\rho_q\cdot\kappa_{q+1}\cdot\ldots\cdot\kappa_r)}} (-1)^{\sum_{i=1}^{q} r_i} b^{\kappa_1,\ldots,\kappa_q,\kappa_{q+1},\ldots,\kappa_r;\phi}_{\rho_1,\ldots,\rho_q,\kappa_{q+1},\ldots,\kappa_r;\zeta}$$

$$C_\zeta^{\rho_1,\ldots,\rho_q,\kappa_{q+1},\ldots,\kappa_r}(X_1,\ldots,X_r)/\prod_{i=1}^{q} (u_i+p)_{\kappa_i} C_\zeta(I) , \tag{3.36}$$

where the b are given by (3.20).

(Generating function)

$$\prod_{i=1}^{q} |I-A_i|^{-u_i-p} \mathrm{etr}(\sum_{i=q+1}^{r} A_i)\int_{O(m)} \mathrm{etr}[-\sum_{i=1}^{q} X_iH'A_i(I-A_i)^{-1}H]dH$$

$$= \sum^{\infty}_{\kappa_1,\ldots,\kappa_r;\phi} L^{u_1,\ldots,u_q}_{\kappa_1,\ldots,\kappa_q,\kappa_{q+1},\ldots,\kappa_r;\phi}(X_1,\ldots,X_r)$$

$$C_\phi^{\kappa_1,\ldots,\kappa_r}(A_1,\ldots,A_q,I,\ldots,I)/\prod_{i=1}^{r} k_i!\ C_\phi(I) . \tag{3.37}$$

(A relation with Khatris' polynomial [11])

$$\int_{O(m)} \prod_{i=1}^{q} L^{u_i}_{\kappa_i}(HA_iH',X_i) \prod_{j=q+1}^{r} C_{\kappa_j}(HA_jH'X_j)dH$$

$$= \sum_{\phi\in\kappa_1\cdot\ldots\cdot\kappa_r} C_\phi^{\kappa_1,\ldots,\kappa_r}(X_1,\ldots,X_r)L^{u_1,\ldots,u_q}_{\kappa_1,\ldots,\kappa_q,\kappa_{q+1},\ldots,\kappa_r;\phi}(A_1,\ldots,A_r)/C_\phi(I) . \tag{3.38}$$

Expansions like in Section 6 of Chikuse [3] can be also derived for the case of general r matrix arguments. Especially the equation (6.3.4) in the same article is extended as a generating function of our generalized Laguerre polynomials as follows: Generalizing (6.1) of Chikuse [3], we have

$$\operatorname{etr}\Big(-\sum_{i=1}^{q} X_i\Big)\sum^{\infty}_{\kappa_1,\ldots,\kappa_r;\phi} C_{\phi}^{\kappa_1,\ldots,\kappa_r}(A_1,\ldots,A_r)C_{\phi}^{\kappa_1,\ldots,\kappa_r}(X_1,\ldots,X_r)\Big/\prod_{i=1}^{r} k_i!C_{\phi}(I)$$

$$= \sum^{\infty}_{\kappa_1,\ldots,\kappa_r;\phi} C_{\phi}^{\kappa_1,\ldots,\kappa_r}(A_1-I,\ldots,A_q-I,A_{q+1},\ldots,A_r)$$

$$C_{\phi}^{\kappa_1,\ldots,\kappa_r}(X_1,\ldots,X_r)\Big/\prod_{i=1}^{r} k_i!C_{\phi}(I) . \tag{3.39}$$

Replacing $A_1,\ldots,A_q$ by $A_1^{-1},\ldots,A_q^{-1}$, multiplying by $\prod_{i=1}^{q} |A_i|^{-u_i}$ and inverting the Laplace transform give

$$\operatorname{etr}\Big(-\sum_{i=1}^{q} X_i\Big)\int_{O(m)}\prod_{i=1}^{q} {}_{\theta}F_1(u_i,A_iHX_iH')\operatorname{etr}\Big(\sum_{j=q+1}^{r} A_jHX_jH'\Big)dH$$

$$= \sum^{\infty}_{\kappa_1,\ldots,\kappa_r;\phi}(-1)^{\sum_{i=1}^{q} k_i} C_{\phi}^{\kappa_1,\ldots,\kappa_r}(X_1,\ldots,X_r)$$

$$L^{u_1,\ldots,u_q}_{\kappa_1,\ldots,\kappa_q,\kappa_{q+1},\ldots,\kappa_r;\phi}(A_1,\ldots,A_r)\Big/\prod_{i=1}^{q}(u_i)_{\kappa_i}\prod_{j=1}^{r} k_j!\ C_{\phi}(I) . \tag{3.40}$$

<u>Differential Identities</u>:

Differential identities in Chikuse [3, Section 7] are generalized as, for $1 \le q \le s \le r$,

$$\prod_{i=1}^{q} a_1(\kappa_i)\prod_{j=q+1}^{s}\{3a_1(\kappa_j)^2 - a_2(\kappa_j) + k_j\}C_{\phi}^{\kappa_1,\ldots,\kappa_r}(X_1,\ldots,X_r)$$

$$= \prod_{i=1}^{q}\operatorname{tr}(X_i\partial_i)^2\prod_{j=q+1}^{s}\{3(\operatorname{tr}(X_j\partial_j)^2)^2 + 8\operatorname{tr}(X_j\partial_j)^3\}$$

$$C_{\phi}^{\kappa_1,\ldots,\kappa_r}(\Sigma_1,\ldots,\Sigma_s,X_{s+1},\ldots,X_r)\Big|_{\Sigma_i=X_i,\ i=1,\ldots,s} . \tag{3.41}$$

where

$$a_1(\kappa) = \sum_{j=1}^{m} k_j(k_j-j) ,$$

$$a_2(\kappa) = \sum_{j=1}^{m} k_j(4k_j^2-6jk_j+3j^2) \text{ for the partition } \kappa = (k_1,\ldots,k_m) ,$$

and $\partial_i = (\partial_{j\ell}^{(i)})$, with $\partial_{j\ell}^{(i)} = \frac{1}{2}(1+\delta_{j\ell})\partial/\partial\sigma_{j\ell}^{(i)}$, $\Sigma_i = (\sigma_{j\ell}^{(i)})$, $i = 1,\ldots,s$, is the matrix of differential operators.

4. APPLICATIONS

In this section we shall consider some applications of the polynomials $C_\phi^{\kappa_1,\ldots,\kappa_r}$ developed in the previous sections in multivariate normal distribution theory.

4.1. Sums of (Noncentral) Wishart Matrices. The distributions of the sums of independent central or noncentral Wishart matrices may be useful in covariance structure analysis, multivariate linear hypothesis testing or Behrens-Fisher problems. Let S_i be independently distributed as, in general, noncentral Wishart $W_m(n_i,\Sigma_i,\Omega_i)$, $i = 1,\ldots,r$. There is no loss of generality in assuming that $\Omega_i \neq 0$, $i = 1,\ldots,q$ and $\Omega_i = 0$, $i = q+1,\ldots,r$. Making the transformations

$$U = \sum_{i=1}^{r} S_i \ , \ T_i = U^{-\frac{1}{2}}S_iU^{-\frac{1}{2}} \ , \ T_i \to H'T_iH \ , \ H \in O(m) \ , \ \text{for} \ \ i=1,\ldots,r-1 \ , \qquad (4.1)$$

and averaging over $O(m)$ in the joint density function of S_i, $i = 1,\ldots,r$, leads to the density function of U in the form

$$f(U) = c_1 \operatorname{etr}(-\tfrac{1}{2}\Sigma_r^{-1}U)|U|^{\frac{1}{2}\sum_{i=1}^{r} n_i - p} \sum_{\kappa_1,\ldots,\kappa_{r-1},\lambda_1,\ldots,\lambda_q;\phi}^{\infty}$$

$$C_\phi^{\kappa_1,\ldots,\kappa_{r-1},\lambda_1,\ldots,\lambda_q}(\Lambda_1U,\ldots,\Lambda_{r-1}U,\Gamma_1U,\ldots,\Gamma_qU)\int_0^I \cdots \int_{\substack{0 \\ 0<\sum_{i=1}^{r-1} T_i<I}}^I \prod_{i=1}^{r-1} |T_i|^{\frac{1}{2}n_i-p}$$

$$|I - \sum_{i=1}^{r-1} T_i|^{\frac{1}{2}n_r-p} C_\phi^{\kappa_1,\ldots,\kappa_{r-1},\lambda_1,\ldots,\lambda_q}(T_1,\ldots,T_{r-1},T_1,\ldots,T_q) \prod_{j=1}^{r-1} dT_j$$

$$/ \prod_{i=1}^{r-1} k_i! \prod_{j=1}^{q} [\ell_j!(\tfrac{1}{2}n_j)_{\lambda_j}]C_\phi(I)$$

$$= c_1 \operatorname{etr}(-\tfrac{1}{2}\Sigma_r^{-1}U)|U|^{\frac{1}{2}\sum_{i=1}^{r} n_i - p} \sum_{\kappa_1,\ldots,\kappa_{r-1},\lambda_1,\ldots,\lambda_q;\phi}^{\infty}$$

$$C_\phi^{\kappa_1,\ldots,\kappa_{r-1},\lambda_1,\ldots,\lambda}(\Lambda_1U,\ldots,\Lambda_{r-1}U,\Gamma_1U,\ldots,\Gamma_qU)$$

$$\sum_{\sigma_1\in\kappa_1\cdot\lambda_1,\ldots,\sigma_q\in\kappa_q\cdot\lambda_q}\beta_{\sigma_1}^{\kappa_1,\lambda_1;\ldots;\kappa_q,\lambda_q,(\kappa_{q+1}),\ldots,(\kappa_{r-1});}{}_{\sigma_q}$$

$$\int_0^I\cdots\int_0^I\prod_{i=1}^{r-1}|T_i|^{\frac{1}{2}n_i-p}\left|I-\sum_{i=1}^{r-1}T_i\right|^{\frac{1}{2}n_r-p}C_\phi^{\sigma_1,\ldots,\sigma_q,\kappa_{q+1},\ldots,\kappa_{r-1}}(T_1,\ldots,T_{r-1})$$
$$0<\sum_{i=1}^{r-1}T_i<I$$

$$\prod_{j=1}^{r-1}dT_j/\prod_{i=1}^{r-1}k_i!\prod_{j=1}^q[\ell_j!(\tfrac{1}{2}n_j)_{\lambda_j}]C_\phi(I)\quad(\because (3.7))$$

$$=c_2\text{etr}(-\tfrac{1}{2}\Sigma_r^{-1}U)|U|^{\frac{1}{2}\sum_{i=1}^r n_i-p}\sum_{\kappa_1,\ldots,\kappa_{r-1},\lambda_1,\ldots,\lambda_q;\phi}^{\infty}\sum_{\sigma_1\in\kappa_1\cdot\lambda_1,\ldots,\sigma_q\in\kappa_q\cdot\lambda_q}$$

$$\beta_\sigma^{\kappa_1,\lambda_1;\ldots;\kappa_q,\lambda_q,(\kappa_{q+1}),\ldots,(\kappa_{r-1});\phi}{}_{\sigma_q}\prod_{i=1}^q(\tfrac{1}{2}n_i)_{\sigma_i}\prod_{j=q+1}^{r-1}(\tfrac{1}{2}n_j)_{\kappa_j}$$

$$\theta_\phi^{\sigma_1,\ldots,\sigma_q,\kappa_{q+1},\ldots,\kappa_{r-1}}C_\phi^{\kappa_1,\ldots,\kappa_{r-1},\lambda_1,\ldots,\lambda_q}(\Lambda_1 U,\ldots,\Lambda_{r-1}U,\Gamma_1 U,\ldots,\Gamma_q U)$$

$$/\prod_{i=1}^{r-1}k_i!\prod_{j=1}^q[\ell_j!(\tfrac{1}{2}n_j)_{\lambda_j}](\tfrac{1}{2}\sum_{i=1}^r n_i)_\phi\ ,\quad(\because (3.27))\tag{4.2}$$

where

$$c_1=\text{etr}(-\tfrac{1}{2}\sum_{i=1}^q\Omega_i)/\prod_{i=1}^r[\Gamma_m(\tfrac{1}{2}n_i)|2\Sigma_i|^{\frac{1}{2}n_i}]\ ,$$

$$c_2=\text{etr}(-\tfrac{1}{2}\sum_{i=1}^q\Omega_i)/\Gamma_m(\tfrac{1}{2}\sum_{i=1}^r n_i)\prod_{i=1}^r|2\Sigma_i|^{\frac{1}{2}n_i}\ ,$$

$$\Lambda_i=-\tfrac{1}{2}(\Sigma_i^{-1}-\Sigma_r^{-1})\ ,\quad i=1,\ldots,r-1\ ,$$

$$\Gamma_i=\tfrac{1}{4}\Sigma_i^{-\frac{1}{2}}\Omega_i\Sigma_i^{-\frac{1}{2}}\ ,\quad i=1,\ldots,q\ .$$

If we seek a symmetric form of the density function of U for the case $\Omega_i=0$, $i=1,\ldots,r$, making the transformations (4.1) leads to the density function of U in the form

$$f(U)=c_1\text{etr}(-\Delta_0 U)|U|^{\frac{1}{2}\sum_{i=1}^r n_i-p}\sum_{\kappa_1,\ldots,\kappa_r;\phi}^{\infty}C_\phi^{\kappa_1,\ldots,\kappa_r}(\Delta_1 U,\ldots,\Delta_r U)$$

$$\int_0^I\cdots\int_0^I\prod_{i=1}^{r-1}|T_i|^{\frac{1}{2}n_i-p}\left|I-\sum_{i=1}^{r-1}T_i\right|^{\frac{1}{2}n_r-p}C_\phi^{\kappa_1,\ldots,\kappa_r}(T_1,\ldots,T_{r-1},I-\sum_{i=1}^{r-1}T_i)$$
$$0<\sum_{i=1}^{r-1}T_i<I$$

$$\prod_{j=1}^{r-1} dT_j / \prod_{i=1}^{r} k_i! C_\phi(I)$$

$$= c_3 \operatorname{etr}(-\Delta_0 U)|U|^{\frac{1}{2}\sum_{i=1}^{r} n_i - p} \Sigma^\infty_{\kappa_1,\ldots,\kappa_r;\phi} \prod_{i=1}^{r} (\tfrac{1}{2}n_i)_{\kappa_i} \theta_\phi^{\kappa_1,\ldots,\kappa_r}$$

$$C_\phi^{\kappa_1,\ldots,\kappa_r}(\Delta_1 U,\ldots,\Delta_r U) / \prod_{i=1}^{r} k_i! (\tfrac{1}{2}\sum_{i=1}^{r} n_i)_\phi \ , \quad (\because (3.26)) \tag{4.4}$$

where

$$c_3 = [\Gamma_m(\tfrac{1}{2}\sum_{i=1}^{r} n_i) \prod_{i=1}^{r} |2\Sigma_i|^{\frac{1}{2}n_i}]^{-1}$$

$$\Delta_0 = \tfrac{1}{4}\sum_{i=1}^{r} \Sigma_i^{-1} \ ,$$

$$\Delta_i = -\tfrac{1}{4}(\Sigma_i^{-1} - \sum_{j=1, j\neq i}^{r} \Sigma_j^{-1}) \ , \ i = 1,\ldots,r \ .$$

These are generalizations of the results of Chikuse [2, Section 3].

4.2. Doubly Noncentral F with Unequal Covariance Matrices. Letting S_i be independently distributed as noncentral Wishart $W_m(n_i,\Sigma_i,\Omega_i)$, $i = 1,2$, we consider the latent roots of $F = S_2^{-\frac{1}{2}} S_1 S_2^{-\frac{1}{2}}$. There is no loss of generality in assuming $\Sigma_2 = I$. F has the same roots as $\tilde{F} = \tilde{S}_2^{-\frac{1}{2}}\tilde{S}_1\tilde{S}_2^{-\frac{1}{2}}$, where $\tilde{S}_i = H'S_iH$, $H \in O(m)$, $i = 1,2$. Averaging over $O(m)$ gives the joint density function of $\tilde{S}_1$ and $\tilde{S}_2$ in the form

$$f(\tilde{S}_1,\tilde{S}_2) = c_1' \operatorname{etr}(-\tilde{S}_2) \prod_{i=1}^{2} |\tilde{S}_i|^{\frac{1}{2}n_i - p} \Sigma_{\kappa,\lambda,\nu;\phi} \Sigma_{\sigma\in\kappa\cdot\lambda} \beta_\sigma^{\kappa,\lambda,(\nu);\phi}$$

$$C_\phi^{\kappa,\lambda,\nu}(-\tfrac{1}{2}\Sigma_1^{-1}, \tfrac{1}{4}\Sigma_1^{-\frac{1}{2}}\Omega_1\Sigma_1^{-\frac{1}{2}}, \tfrac{1}{4}\Omega_2) C_\phi^{\sigma,\nu}(\tilde{S}_1,\tilde{S}_2)/k!\ell!n!(\tfrac{1}{2}n_1)_\lambda(\tfrac{1}{2}n_2)_\nu C_\phi(I) \ , \tag{4.5}$$

where

$$c_1' = \operatorname{etr}(-\tfrac{1}{2}(\Omega_1+\Omega_2)) / \prod_{i=1}^{2} \Gamma_m(\tfrac{1}{2}n_i) 2^{\frac{1}{2}m(n_1+n_2)} |\Sigma_1|^{\frac{1}{2}n_1} \ .$$

Hence we have the density function of $\tilde{F}$

$$f(\tilde{F}) = c_4|\tilde{F}|^{\frac{1}{2}n_1 - p} \Sigma^\infty_{\kappa,\lambda,\nu;\phi} \Sigma_{\sigma\in\kappa\cdot\lambda} \beta_\sigma^{\kappa,\lambda,(\nu);\phi} (\tfrac{1}{2}(n_1+n_2))_\phi \theta_\phi^{\sigma,\nu}$$

$$C_\phi^{\kappa,\lambda,\nu}(-\Sigma_1^{-1}, \tfrac{1}{2}\Sigma_1^{-\frac{1}{2}}\Omega_1\Sigma_1^{-\frac{1}{2}}, \tfrac{1}{2}\Omega_2) C_\sigma(\tilde{F})/k!\ell!n!(\tfrac{1}{2}n_1)_\lambda(\tfrac{1}{2}n_2)_\nu C_\phi(I) \ , \tag{4.6}$$

where

$$c_4 = \mathrm{etr}(-\tfrac{1}{2}(\Omega_1+\Omega_2))\Gamma_m(\tfrac{1}{2}(n_1+n_2))/\prod_{i=1}^{2}\Gamma_m(\tfrac{1}{2}n_i)|\Sigma_1|^{\frac{1}{2}n_1} .$$

From (4.6) the standard methods yield the joint density function of the roots of F and the distribution function of the largest root of F. The density function of $T = \mathrm{tr}\, F$ is obtained by a similar method to that in Davis [7, Section 8] as

$$f(T) = c_5 T^{\frac{1}{2}mn_1-1} \sum_{\kappa,\lambda,\nu;\phi}^{\infty} \sum_{\sigma\in\kappa\cdot\lambda} \beta_{\sigma}^{\kappa,\lambda,(\nu);\phi} (\tfrac{1}{2}(n_1+n_2))_{\phi}(\tfrac{1}{2}n_1)_{\sigma}\, \theta_{\phi}^{\sigma,\nu}$$
$$C_{\phi}^{\kappa,\lambda,\nu}(-\Sigma_1^{-1},\tfrac{1}{2}\Sigma_1^{-\frac{1}{2}}\Omega_1\Sigma_1^{-\frac{1}{2}},\tfrac{1}{2}\Omega_2)T^{k+\ell}/k!\ell!n!(\tfrac{1}{2}n_1)_{\lambda}(\tfrac{1}{2}n_2)_{\nu}(\tfrac{1}{2}mn_1)_{k+\ell} , \qquad (4.7)$$

where $c_5 = c_4\Gamma_m(\frac{1}{2}n_1)/\Gamma(\frac{1}{2}mn_1)$, and (4.7) is a generalization of (9.8) in Davis [7] for $\Sigma_1 = \Sigma_2 = I$.

4.3. <u>Extended Noncentral MANOVA W</u> . Let S_i be independently distributed as Wishart $W_m(n_i,\Sigma_i,\Omega_i)$, $i = 0,1,2,$ and assume $\Omega_i = 0$, $i = 1,2$. The 'extended' noncentral MANOVA matrix $W = (S_1+S_2)^{-\frac{1}{2}}S_0(S_1+S_2)^{-\frac{1}{2}}$ may be interesting in connection with the multivariate Behrens-Fisher discriminant problem. We have already derived in Chikuse [2, Sections 4 and 5] the explicit forms of the distributions, conditional and unconditional, of the roots, the largest root and the trace of W (conditional on those of a preliminary test matrix $B = (S_1+S_2)^{-\frac{1}{2}}S_1(S_1+S_2)^{-\frac{1}{2}}$) . It is shown here that the results obtained in Chikuse [2] can be extended for our general noncentral case $\Omega_0 \neq 0$.

Starting from the joint density function of S_0 , S_1 and S_2 we have the joint density function of $\tilde{W} = H'WH$ and $\tilde{B} = H'BH$, $H \in O(m)$, in the form

$$f(\tilde{W},\tilde{B}) = c_6|\tilde{W}|^{\frac{1}{2}n_0-p}|\tilde{B}|^{\frac{1}{2}n_1-p}|I-\tilde{B}|^{\frac{1}{2}n_2-p} \sum_{\kappa,\lambda,\nu;\phi}\sum_{\sigma\in\kappa\cdot\lambda}(-1)^{\ell+n}$$
$$\beta_{\sigma}^{\kappa,\lambda,(\nu);\phi}(\tfrac{1}{2}\sum_{i=0}^{2} n_i)_{\phi}C_{\phi}^{\kappa,\lambda,\nu}(\tfrac{1}{2}\Sigma_0^{-\frac{1}{2}}\Omega_0\Sigma_0^{-\frac{1}{2}}\Sigma_2,\Sigma_0^{-1}\Sigma_2,\Sigma_1^{-1}\Sigma_2-I)C_{\phi}^{\sigma,\nu}(\tilde{W},\tilde{B})$$
$$/k!\ell!n!(\tfrac{1}{2}n_0)_{\kappa}C_{\phi}(I) , \qquad (4.8)$$

where

$$c_6 = \Gamma_m(\tfrac{1}{2}\sum_{i=0}^{2} n_i)\mathrm{etr}(-\tfrac{1}{2}\Omega_0)[\prod_{i=0}^{2}\Gamma_m(\tfrac{1}{2}n_i)\prod_{j=0}^{1}|\Sigma_j\Sigma_2^{-1}|^{\frac{1}{2}n_j}]^{-1} .$$

This involves only the invariant polynomials $C_{\phi}^{\sigma,\nu}(\tilde{W},\tilde{B})$ with two matrix arguments; hence, all of the distributions considered in Sections 4 and 5 of Chikuse [2] can be similarly obtained for $\Omega_0 \neq 0$. Especially the density function of $V = \mathrm{tr}\, W$ is given by

$$f(V) = c_7 V^{\frac{1}{2}mn_0 - 1} \Sigma^{\infty}_{\kappa,\lambda,\nu;\phi} \Sigma_{\sigma\in\kappa\cdot\lambda} (-1)^{\ell+n} \beta_{\sigma}^{\kappa,\lambda,(\nu);\phi} (\tfrac{1}{2}n_0)_{\nu} (\tfrac{1}{2}n_1)_{\nu} (\tfrac{1}{2} \sum_{i=0}^{2} n_i)_{\phi}$$

$$\theta_{\phi}^{\sigma,\nu} C_{\phi}^{\kappa,\lambda,\nu} (\tfrac{1}{2}\Sigma_0^{-\frac{1}{2}} \Omega_0 \Sigma_0^{-\frac{1}{2}} \Sigma_2 , \Sigma_0^{-1}\Sigma_2, \Sigma_1^{-1}\Sigma_2 - I) V^{k+\ell}$$

$$/k!\ell!n!(\tfrac{1}{2}n_0)_{\kappa} (\tfrac{1}{2}(n_1+n_2))_{\nu} (\tfrac{1}{2}mn_0)_{k+\ell} , \tag{4.9}$$

where

$$c_7 = \Gamma_m(\tfrac{1}{2} \sum_{i=0}^{2} n_i) \operatorname{etr}(-\tfrac{1}{2}\Omega_0) [\Gamma(\tfrac{1}{2}mn_0) \Gamma_m(\tfrac{1}{2}(n_1+n_2)) \prod_{j=0}^{1} |\Sigma_j \Sigma_2^{-1}|^{\frac{1}{2}n_j}]^{-1} .$$

(4.9) is an extension of the result for Hotelling's generalized T_0^2 given by Constantine [5].

ACKNOWLEDGEMENT

The author would like to thank Dr. A.W. Davis at the Division of Mathematics and Statistics, C.S.I.R.O., Australia, for having had access to his invariant polynomials with two matrix arguments, which led to developing the results for the general case of this paper.

REFERENCES

[1] Boerner, H. (1963). Representations of Groups. North-Holland, Amsterdam.

[2] Chikuse, Y. (1979). Distributions of some matrix variates and latent roots in multivariate Behrens-Fisher discriminant analysis. Submitted for publication.

[3] Chikuse, Y. (1979). Invariant polynomials with three matrix arguments, extending the polynomials with lower numbers of matrix arguments. Submitted for publication.

[4] Constantine, A.G. (1963). Some non-central distribution problems in multivariate analysis. Ann. Math. Statist. 34, 1270-1285.

[5] Constantine, A.G. (1966). The distribution of Hotelling's generalized T_0^2. Ann. Math. Statist. 37, 215-225.

[6] Davis, A.W. (1978). Invariant polynomials with two matrix arguments, extending the zonal polynomials. Submitted for publication.

[7] Davis, A.W. (1978). Invariant polynomials with two matrix arguments extending the zonal polynomials: applications to multivariate distribution theory. Submitted for publication.

[8] Herz, C.S. (1955). Bessel functions of matrix argument. Ann. Math. 61, 474-523.

[9] James, A.T. (1960). The distribution of the latent roots of the covariance matrix. Ann. Math. Statist. 31, 151-158.

[10] James, A.T. (1964). Distributions of Matrix variates and latent roots derived from normal samples. Ann. Math. Statist. 35, 475-501.

[11] Khatri, C.G. (1977). Distribution of a quadratic form in noncentral normal vectors using generalized Laguerre polynomials. South African Statist. J. 11, 167-179.

[12] Robinson, G. de B. (1961). Representation Theory of the Symmetric Group. Edinburgh University Press, Edinburgh.

[13] Saw, J.G. (1977). Zonal polynomials: an alternative approach. J. Multivar. Anal. 7 461-467.

Multivariate Statistical Analysis
R.P. Gupta (ed.)

A NONPARAMETRIC METHOD TO DISCRIMINATE TWO POPULATIONS*

G. Gabor
Mathematics Department
Dalhousie University
Halifax, N.S. B3H 4H8
Canada

The purpose of this paper is to give a nonparametric easy-to-compute procedure for approximation of the Bayesian separation surface of two populations if a well classified sample is given.

INTRODUCTION

Consider the well-known discrimination problem for two populations. Let an independent and well classified sample (x_1,θ_1) , $(x_2,\theta_2),\dots,(x_m,\theta_m)$, $x_i \in R^n$, $\theta_i = 1$ or 2 , $i = 1,2,\dots,m$ be given. We are looking for a nonparametric procedure to classify a nonlabelled observation $x \in R^n$ into one of the two populations, using only the given labelled sample. If a metric is given on the sample space, the Nearest Neighbor procedure (NN) is appropriate [1], i.e. x is classified according to the nearest one in the sample. Many mofidications of this simple procedure and its asymptotic properties have been analyzed by Wagner [2] and Fritz [3]. Despite the good asymptotic behaviour of the NN, good results can be achieved only with large samples. Moreover the necessity to store and "move" the whole sample for every single classification could make the NN very difficult to apply particularly in higher dimensions. A sequential procedure to reduce the sample size for NN classification purposes has been proposed by Gabor [4]. In this we preserve only those units that carry important information in the NN sense, i.e., if in a certain sphere around an observation one of the populations is dominating enough then this sphere can be represented by the centre with the label of the dominating population. Any future classification within this sphere will be made by the centre only. In this way a large amount of misclassifications can be avoided that show the theoretical error ratio only in the long run. The theoretical and computer results have shown this method surprisingly good. The procedure is able to reduce the classification error and the sample size at the same time. The centre selection, however, has not been solved in a sphere which does not have a dominating population. In such cases both populations have almost the same probability. Assuming the continuity of the probabilities $P(\theta|x)$ on X we have $P(\theta|S) \simeq P(\theta|x)$, $x \in S$ in a sufficiently small sphere $S \subset X$.

* This research was supported in part by NRC of Canada, A3123.

It implies that in a small sphere S in which $P(\theta=1|S) \simeq P(\theta=2|S)$, i.e. in a sphere without dominating population, the posterior probabilities $P(\theta=1|x)$ and $P(\theta=2|x)$, $x \in S$ are almost equal. Thus the centre of such a sphere is close to the Bayesian separation surface and even the optimal classification is not better than a guessing. In this case if we do not preserve this point we will not lose much information in the NN sense. Curiously enough there is a completely reverse reasoning. We preserve only those centres of spheres without dominating population since they are close to the separation surface; hence, we can approximate the surface in the sense that we have points close to it. Any smooth surface that fits to these observations can be used as an estimation of the Bayesian separation surface. There is no contradiction between the two reasonings since the resulting selections are designed for different classification methods. Note that no knowledge about the priors is required.

THE ALGORITHM

Step 1. Select an observation x_i from the sample.
Find m_1^i, m_2^i the number of observations from the two populations in $S_\epsilon(x_i)$, where $\epsilon>0$ is a given number and $S_\epsilon(x_i)$ is a sphere with centre x_i and radius ϵ.

Step 2. Test the hypothesis H_0^i: $P(\theta-1|S_\epsilon(x_i)) = P(\theta-2|S_\epsilon(x_i)) = \frac{1}{2}$ using $\frac{m_1^i}{m_1^i+m_2^i} = \hat{P}(\theta=1|S_\epsilon(x_i))$ and $\frac{m_2^i}{m_1^i+m_2^i} = \hat{P}(\theta=2|S_\epsilon(x_i))$.

(See Lehman [5].)

If H_0^i is accepted then qualify x_i as an observation close to the Bayesian separation surface and call it x_i^q. In case of rejection repeat Step 1 and Step 2.

Repeating the procedure for all observations yields the qualified subset $Q = \{x_1^q, x_2^q, \ldots, x_\ell^q\}, \ell < m$ of the sample. Given the ϵ and the size of the test α this subset is uniquely defined.

There is a common tradeoff between the accuracy of the procedure ϵ and the reliability of the text. On one hand, the smaller the ϵ the closer we can get to the separation surface. On the other hand, the more observations a sphere contains the better for the test. As usual these two criterions work against each other. A good choice of ϵ requires some knowledge of the particular problem.

COMPUTER RESULTS

300 independent observations were drawn from a mixture of two bivariate normal distributions with various parameters and equal priors. $\epsilon = 1.5$ and $\alpha = 0.1$ were used in all cases. As it is shown on the computer graphics, a linear or a

quadratic curve was fit to the qualified subset. These curves were used for the classifications. The probability of error was estimated by the number of misclassified observations divided by 300.

Fig. 1-5 display the result of one run out of the 50 runs on each case. The stars on the lower pictures indicate only the location of the observations in Q , for the qualified observations have no population characteristic. The parameters of the distributions, the Bayes errors (when the covariance matrices are equal) and the estimated error probabilities were as follows:

Fig. 1 $0 \sim N[(0,0);\ \begin{pmatrix}1&0\\0&1\end{pmatrix}],\ \star \sim N[(0,3);\ \begin{pmatrix}1&0\\0&1\end{pmatrix}),\quad P_B = 0.067,\ \hat{P}_{50} = 0.069$

Fig. 2 $0 \sim N[(0,2);\ \begin{pmatrix}4&0\\0&1\end{pmatrix}],\ \star \sim N[(0,0);\ \begin{pmatrix}4&0\\0&1\end{pmatrix}],\quad P_B = 0.159,\ \hat{P}_{50} = 0.161$

Fig. 3 $0 \sim N[(2,2);\ \begin{pmatrix}4&0\\0&1\end{pmatrix}],\ \star \sim N[(0,0);\ \begin{pmatrix}4&0\\0&1\end{pmatrix}],\quad P_B = 0.131,\ \hat{P}_{50} = 0.134$

Fig. 4 $0 \sim N[(0,2);\ \begin{pmatrix}1&0\\0&1\end{pmatrix}],\ \star \sim N[(0,0);\ \begin{pmatrix}4&0\\0&1\end{pmatrix}],\quad \hat{P}_{50} = 0.141$

Fig. 5 $0 \sim N[(4,4);\ \begin{pmatrix}1&0\\0&4\end{pmatrix}],\ \star \sim N[(0,0);\ \begin{pmatrix}4&0\\0&1\end{pmatrix}],\quad \hat{P}_{50} = 0.033$

REFERENCES

[1] Cover, M. & Hart, P.E. (1967). The Nearest Neighbor Pattern Classification. IEEE IT.-13, 21-27.

[2] Wagner, T.J. (1971). Covergence of the Nearest Neighbor Rule. IEEE IT.-17, 566-571.

[3] Fritz, F. (1975). Distribution-Free Exponential Error Bound for Nearest Neighbor Pattern Classification. IEEE IT-21, 552-557.

[4] Gabor, G. (1977). The ϵ-NN Method. A Sequential Feature Selection for Nearest Neighbor Decision Rule. in I. Csiszar ed. Topics in Information Theory, North Holland.

[5] Lehmann, E. (1959). Testing Statistical Hypotheses. John Wiley and Sons, Inc., p. 128.

Acknowledgement

The author thanks the referee for the valuable comments.

Multivariate Statistical Analysis
R.P. Gupta (ed.)

SOME METHODS OF SEARCHING FOR OUTLIERS

Jane F. Gentleman

University of Waterloo
Waterloo, Ontario
Canada

This paper considers the detection of outliers in data, which may be described by linear model. Using a 7x7 two-way table from Daniel (1978), a comparison of five approached is given.

INTRODUCTION

The detection of outliers will be considered in the general situation where multiple outliers may be present in data described by the usual linear model. The following section describes the data and the model. The remaining sections consider and compare five approaches to outlier detection, using as an example a 7 by 7 two-way table from Daniel (1978).

Figs. 1-4 are normal probability plots of residuals from the example. To enable the reader to judge them independently and objectively, it is suggested that the reader look at these figures before reading further, and form an opinion as to whether any of the points on the plots may represent outliers.

DATA AND MODEL

Assume that y is an n-vector of data described by the usual linear model $y = X\beta+\varepsilon$, where the design matrix X is n by p , the coefficients β are p by 1 , and the $N(0,\sigma^2)$ errors are n by 1 . As a specific example, a two-way table model will be used throughout this paper. The original r by c matrix-structured doubly-subscripted data array is described by the usual model: $y_{ij} = \mu + \alpha_i + \gamma_j + \varepsilon_{ij}$ $(i=1,r;\ j=1,c)$, where μ is the usual overall constant, α_i is the i-th row constant, γ_j is the j-th column constant, and the ε_{ij}'s are normal errors. For purposes of generality and for computational purposes, it is convenient to retain the notation of the general linear model. In it, the y_{ij}'s are assumed to be ordered in y by rows, i.e. $y = [y_{11},y_{12},\ldots,y_{1c},y_{21},\ldots,y_{rc}]'$. The design matrix X which corresponds to the two-way table model is a set of columns of indicator variables which can be defined in a number of ways. For example, for a 3 by 3 table, let $\beta = [\mu,\alpha_2,\alpha_3,\gamma_2,\gamma_3]'$ (α_1 and γ_1 being omitted because of the usual linear constraints $\Sigma\alpha_i = \Sigma\gamma_j = 0$). Then X is n by p , where $n = rc = 9$ and

$p = 1 + (r-1) + (c-1) = 5$:

$$X = \begin{bmatrix} 1 & -1 & -1 & -1 & -1 \\ 1 & -1 & -1 & 1 & 0 \\ 1 & -1 & -1 & 0 & 1 \\ 1 & 1 & 0 & -1 & -1 \\ 1 & 1 & 0 & 1 & 0 \\ 1 & 1 & 0 & 0 & 1 \\ 1 & 0 & 1 & -1 & -1 \\ 1 & 0 & 1 & 1 & 0 \\ 1 & 0 & 1 & 0 & 1 \end{bmatrix} .$$

Let $\hat{\varepsilon}$ be the n-vector of residuals. Then for the general linear model, $\hat{\varepsilon} = Ty$, where T is the usual projection matrix $I - X(X'X)^{-1}X'$. Also, $\sigma^2 T$ is the n covariance matrix of $\hat{\varepsilon}$. For the two-way table model, the elements of T can be obtained as follows: Each row of T corresponds to one of the cells of the original two-way table, in the order $(1,1),(1,2),\ldots,(1,c),(2,1),\ldots,(r,c)$. The same is true for each column of T . Thus, a given element t_{ab} of T is associated with two pairs of subscripts: (i_a,j_a) corresponding to the row a , and (i_b,j_b) corresponding to the column b . Then $t_{ab} = AB/rc$, where $A = -(r-1)$ if $i_a = i_b$, and $A = 1$ otherwise, and $B = -(c-1)$ if $j_a = j_b$, and $B = 1$ otherwise. (See Gentleman and Wilk, 1975b.)

DIRECT ANALYSIS OF LEAST SQUARES RESIDUALS

It is now fairly well accepted that the fitting and analysis of a linear model should necessarily include the calculation of residuals. These residuals can then be plotted and otherwise analyzed in a number of ways, providing insight as to the adequacy of the usual assumptions. A single outlier can often be detected via the direct analysis of residuals, but the presence of multiple outliers may perturb the fitted model and residuals in such a way that mutual concealment of outliers occurs. (See Gentleman and Wilk, 1975a, 1975b.)

The data set used to illustrate the five methods described here is a 7 by 7 two-way table from Daniel (1978). Fig. 1 is a normal probability plot of the residuals from this example. The three most extreme positive residuals on the plot are from cells (2,3), (1,3), and (7,3), and they ranked 4, 5, and 6 in magnitude among the 49 residuals. The three most extreme negative residuals on the plot are from cells (5,3), (4,3), and (3,2). They ranked 1, 2, and 3 in magnitude. The plot suggests that (5,3), (4,3), and (3,2) may be outliers.

ITERATIVE ANALYSIS OF LEAST SQUARES RESIDUALS

A natural reaction to the inspection of residuals in Fig. 1 is to remove the three suspected outliers from the data and then refit the model without them. Fig. 2 shows a normal probability plot of the resulting 46 new residuals. (The residuals in Figs. 1-3 have been standardized; residuals from a complete two-way table all have the same variance, but this property is lost if observations are missing.) The two most extreme positive residuals in Fig. 2 are from cells (6,2) and (5,7), ranking 3 and 4 in magnitude (both before and after standardizing). The two most extreme negative residuals are from cells (6,3) and (3,1), ranking 1 and 2 in magnitude. The latter two cells appear to be detached from the rest of the points, although they are not greatly out of line with the other points, considering the usual raggedness of probability plots. One might or might not be tempted to continue the iteration and/or to investigate these two cells further. Looking back at Fig. 1, however, one is startled to see how well concealed these same two cells were among the original 49 residuals; cell (3,1), although ranking fourth among the most extreme positive residuals, ranked tenth in magnitude, and it is not at all out of line among the points on the plot. Cell (6,3) is not out of line and is nowhere near the extremes; it is 10th among the most extreme positive residuals and 23rd in magnitude. Note also that the ordering between residuals (3,1) and (6,3) is reversed between Fig. 1 and Fig. 2.

Fig. 3 shows a normal probability plot of 44 standardized residuals obtained by refitting the model without the five possibly suspect observations. At the lower end, the configuration curves slightly upward, a frequently-encountered phenomenon in probability plots of residuals, probably due to their non-independence and to the fact here that observations causing large residuals have been removed. The upper end shows two detached residuals, from cells (5,7) and (6,2) , but they are not excessively large. The largest and smallest residuals on the plot do not differ greatly in magnitude, whereas the configurations of Figs. 1 and 2 were more assymetrically positioned about zero. The iterative analysis has to stop somewhere; perhaps this is an appropriate place.

DANIEL'S METHOD OF LOOKING FOR PATTERNS IN RESIDUALS

To detect outliers in two-way table residuals, Daniel looks for patterns in them which resemble those of the expected values of these residuals when outliers are present. He then isolates a sub-table of possible outliers and tests each cell in the sub-table as a possible outlier. Using this method, Daniel found a sub-table of 12 cells, from which he selected as outliers the same five cells pegged as possible outliers by the iterative method.

GENTLEMAN AND WILK'S Q_K CRITERION

Gentleman and Wilk (1975b) define the K most likely outliers in linear model data to be those K observations whose removal from the data, followed by refitting, most reduces the sum of squared residuals. Having identified these K observations, one can then test them to see if they are outliers. The tests performed by Daniel (1978), for example, are equivalent to the usual t-tests of hypotheses that certain coefficients in the model are zero, where the model being fitted has an additional coefficient (additive constant for the size of the perturbation) for each cell suspected of being an outlier.

Refitting a model and recomputing residuals and their sums of squares can be very arduous if $\binom{n}{K}$, the number of sets of K possible outliers, is large. Instead, the reduction Q_K in the sum of squared residuals caused by removing a specific set y_* of K cells can be computed directly, as follows (Gentleman and Wilk, 1975b):

Let $\hat{\varepsilon}_*$ be the K residuals for those cells. Let $\sigma^2 T_*$ be the K by K covariance matrix of $\hat{\varepsilon}_*$. T_* is a principal submatrix of T and can be calculated as described in Section 2. Then $Q_K = \hat{\varepsilon}_*' T_*^{-1} \hat{\varepsilon}_*$. The K simultaneous missing value estimates for the removed observations y_* are $y_* - T_*^{-1}\hat{\varepsilon}_*$, and $T_*^{-1}\hat{\varepsilon}_*$ estimates the amount of perturbation in each of the K cells.

Results from applying Gentleman and Wilk's method to the 7 by 7 table were as follows:

1 most likely outlier: (5,3).

(This is always the cell with the largest absolute residual.)

2 most likely outliers: (5,3), (4,3).

3 most likely outliers: (5,3), (4,3), (3,2).

4 most likely outliers: (5,3), (4,3), (3,2), (6,3).

5 most likely outliers: (5,3), (4,3), (3,2), (6,3), (3,1).

(These are the same as the 5 cells considered in Section 4 and picked by Daniel.)

The $K-1$ most likely outliers are not necessarily a subset of the K most likely outliers, as happened here, although when outliers of appreciable size are really present, experience has shown this to be the case. The ordering suggested by the nesting here is the same as the ordering of the absolute estimated perturbations when $K = 5$, calculated as described above.

LEAST ABSOLUTE RESIDUALS INSTEAD OF LEAST SQUARED RESIDUALS

As a final example of methods for detecting multiple outliers, a different fitting criterion is used. Instead of minimizing the sum of squared residuals in order to fit the model, the sum of absolute residuals is minimized. This is one of the simpler methods of performing a robust regression. The program used to perform these calculations was provided by Bartels and Conn (1980).

Fig. 4 shows a normal probability plot of 36 residuals resulting from the least absolute residuals fit to the 7 by 7 data. (Parameter values used to initialize the algorithm were the least squares estimates.) 13 residuals are not plotted because they are identically zero, due to the constraints imposed by this fitting method; the fitted surface necessarily passes through $p = 13$ points. While these residuals are not the usual normally distributed linear combinations of observations, the plot is nevertheless provided for purposes of comparison with the other figures. The five most extreme positive residuals are from cells (5,3), (4,3), (3,2), (3,1), and (6,3), the five suspect cells considered above, with the ordering between (3,1) and (6,3) now reversed. The main portion of the configuration is surprisingly straight. Cells (3,1) and (6,3) are not greatly out of line, but they have moved much closer to the extremes than they were in the original set of least squares residuals (see Fig. 1). Once again, an iterative application of this fitting method seems to be a natural next step.

CONCLUSION

The five approaches considered above (direct analysis of least squares residuals, iterative analysis of least squares residuals, Daniel's method, Gentleman and Wilk's criterion, and least absolute residuals) are not intended to be presented as competitive. On the contrary, it is useful in the exploration of data to apply several techniques, especially when each technique has its limitations. The direct analysis of residuals is useful for detecting outliers, but more can be done. Iterative techniques can help ferret out multiple outliers, but mutual concealment may still occur. A few objective iterative tests have been proposed (see, for example, Anscombe (1960), Daniel (1978), Grubbs (1950), John and Draper (1978), Pearson and Sekar (1936), but these can be used for only a few models and/or with small numbers of outliers. Daniel's method, like the probability plot analyses above, is subjective and probably requires an experienced data analyst. Also, his significance tests, as do some of the other tests, ignore the fact that an ordering, or pre-selecting of the data, has occurred. Gentleman and Wilk's method, like many others, leaves to the user the choice of a value for K. The use of least absolute residuals can yield non-unique answers (a trivial example being the fitting of the model $y_i = \mu + \varepsilon_i$ to an even-sized single sample $y_1, y_2, \ldots y_n$, where the least absolute residuals estimate of μ is the non-unique sample median). Also, this method would appear to force abandonment of the normal theory with which we are so comfortable, although the surprisingly good results exemplified by Fig. 4 suggest that perhaps further research could provide something similarly useful.

More development of outlier tests is needed. Meanwhile, in seeking out the causes of any disagreements among these various methods, we can discover the sensitive aspects of the data.

REFERENCES

[1] Anscombe, F.J. (1960). Rejection of outliers. Technometrics, 2, 123-147.

[2] Bartels, R.J. and Conn, A.R. (1980). To appear in Trans. of Math. Software.

[3] Daniel, C.D. (1978). Patterns in residuals in the two-way layout. Techno-

metrics, 20, 385-395.

[4] Gentleman, J.F. and Wilk , M.B. (1975a). Detecting outliers in a two-way table: I. Statistical behavior of residuals. Technometrics, 17, 1-14.

[5] Gentleman, J.F. and Wilk , M.B. (1975b). Detecting outliers. II. Supplementing the direct analysis of residuals. Biometrics, 31, 387-410.

[6] Grubbs, F.E. (1950). Sample criteria for testing outlying observations. Ann. Math. Statist., 21, 27-58.

[7] John, J.A. and Draper, N.F. (1978). On testing for two outliers or one outlier in two-way tables. Technometrics, 20, 69-78.

[8] Pearson, E.S. and Sekar, C.C. (1936). The efficiency of statistical tools and a criterion for the rejection of outlying observations. Biometrika, 28, 308-319.

FIG. 1. NORMAL PROBABILITY PLOT OF 49 STANDARDIZED LEAST SQUARES RESIDUALS FROM 7x7 TWO-WAY TABLE.

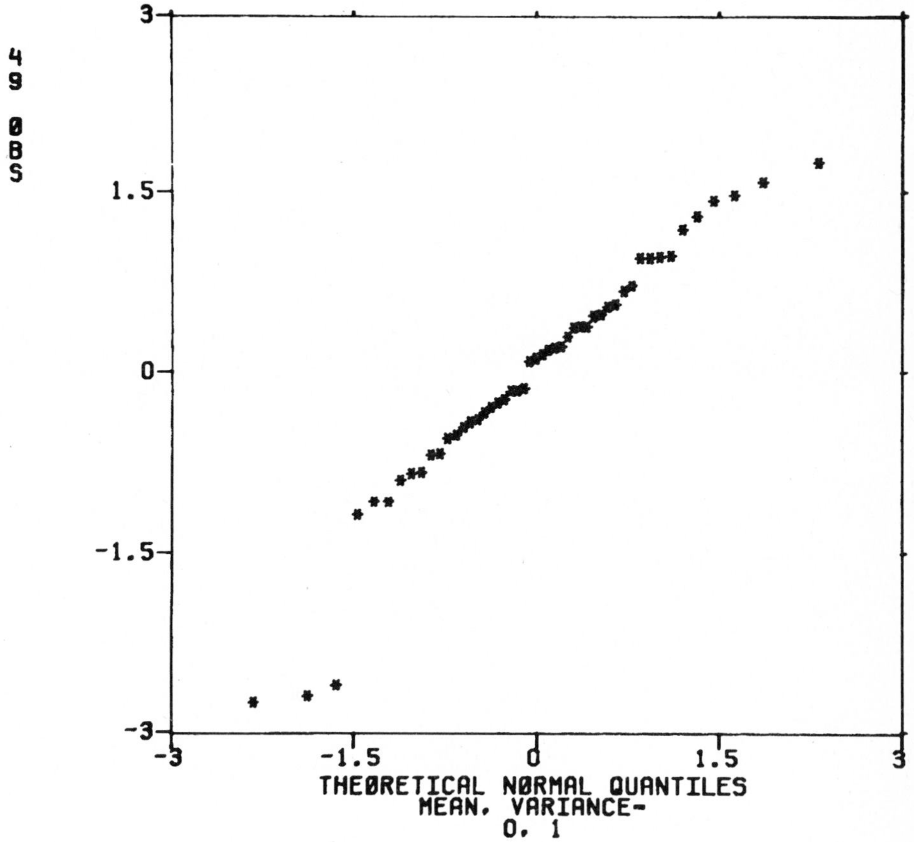

FIG. 2. NORMAL PROBABILITY PLOT OF 46 STANDARDIZED LEAST SQUARES RESIDUALS FROM 7x7 TWO-WAY TABLE WITH 3 CELLS OMITTED.

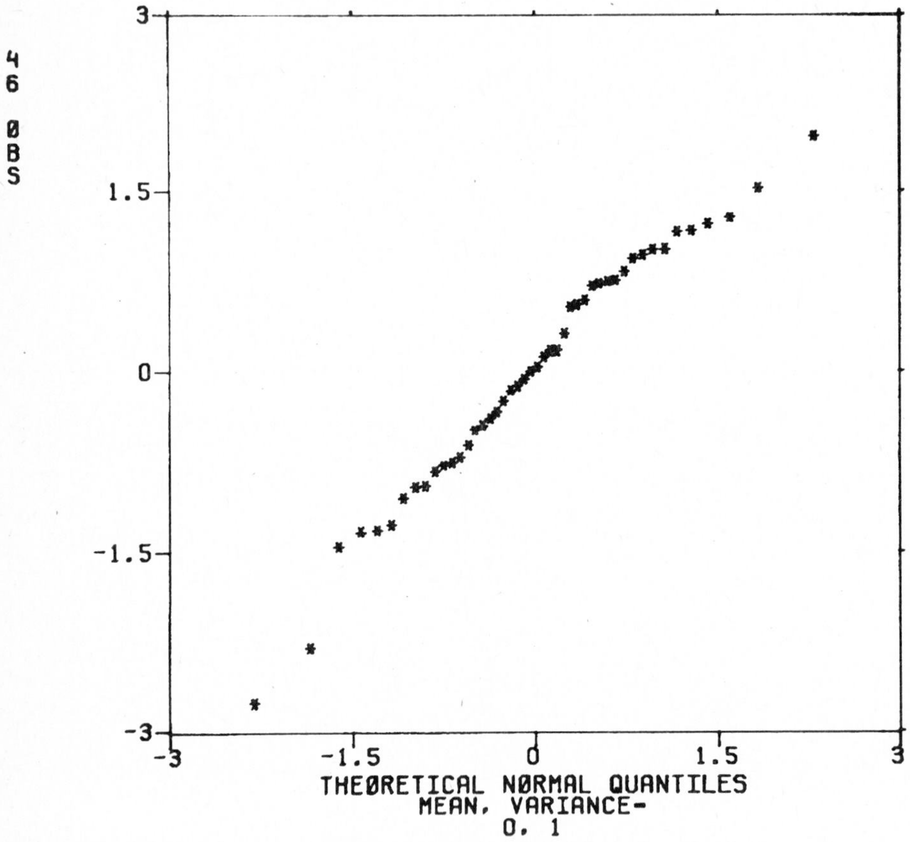

FIG. 3. NORMAL PROBABILITY PLOT OF 44 STANDARDIZED LEAST SQUARES RESIDUALS FROM 7x7 TWO-WAY TABLE WITH 5 CELLS OMITTED.

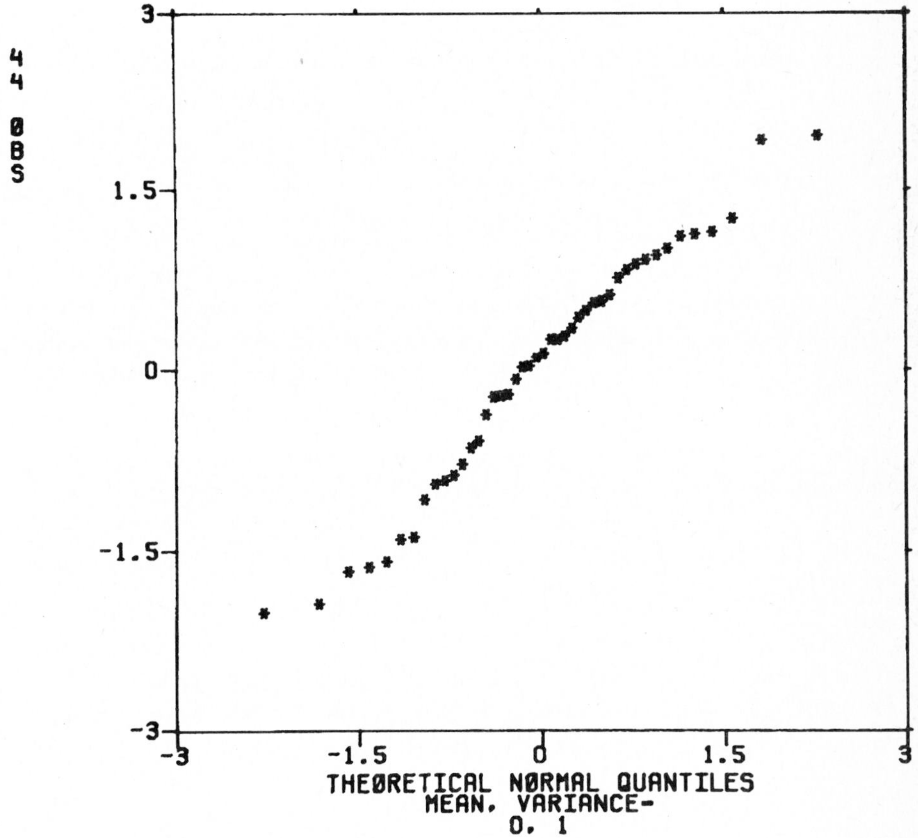

FIG. 4. NORMAL PROBABILITY PLOT OF 36 LEAST ABSOLUTE RESIDUALS FROM 7x7 TWO-WAY TABLE.

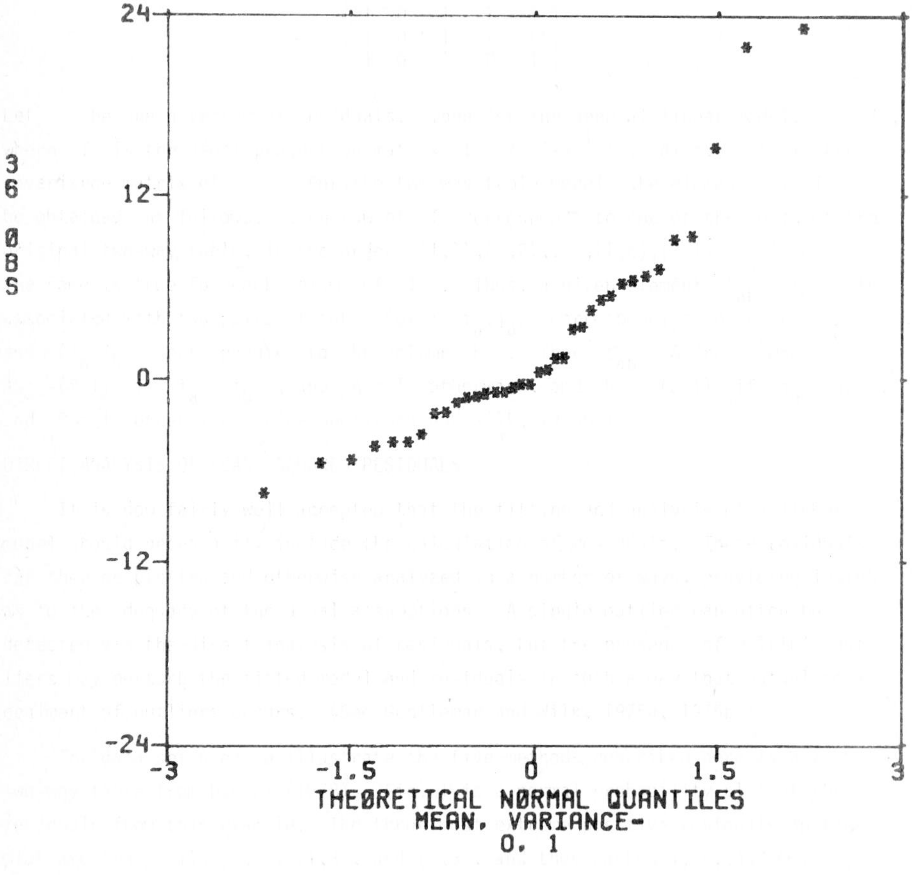

Multivariate Statistical Analysis
R.P. Gupta (ed.)

ON A MULTIVARIATE STATISTICAL CLASSIFICATION MODEL*

A. K. Gupta
Department of Mathematics and Statistics
Bowling Green State University
Bowling Green, Ohio 43403, U.S.A.

Classification procedures, based on the maximum-likelihood criterion, for classification into one of two multivariate populations when multiple observations are available on the same variable for each individual, have been studied in the present paper. The distribution of the classification statistics is derived and formulae for the probabilities of misclassification, exact and approximate, are given when the parameters are known. These procedures are then extended to more than two populations.

1. INTRODUCTION AND FORMULATION OF THE PROBLEM.

Different authors study different formulations of the classification problems and the overwhelming majority manage without a formulation and therefore work on the solution of a problem which is not formulated or is, at most vaguely formulated. There are even half a dozen or more different names for such studies, e.g., diagnosis problem, pattern recognition problem, discernment theory, discriminant analysis, etc. However, at the present time most authors agree that a priori restrictions substantially narrowing the class of solutions must be present in every classification problem. More or less universal methods may exist only for the case of very small dimension. Restrictions on the class of solutions are determined both by the characteristics of the individuals to be classified and by the technical possibilities of the realizations of the solution. The consideration of technical restrictions in the general formulation of the problem should be overcome. It is necessary to consider the individuals to be classified as the basis of the mathematical model. Such a model should describe the variation in the characteristics of the individuals to be classified from group to group. Having such a model, it is then possible to deduce classification rules which are, in some sense, optimal. Below we study such a classification model in the diagnostic context or in the context of pattern recognition.

Let $p \times 1$ vector Y_{ijk} denote the kth measurement on the jth individual in the ith population (pattern) π_i $(i = 1,\ldots,m)$. The basic model for the population π_i will be represented by

*Research partially supported by a Faculty Research Fellowship of Bowling Green State University.

$$\underset{p\times 1}{Y_{ijk}} = \underset{p\times 1}{\mu_i} + \underset{p\times 1}{I_{ij}} + \underset{p\times 1}{E_{ijk}}\ , \quad j = 1,\ldots,n_i\ , \quad k = 1,\ldots,n, \tag{1.1}$$

i.e., the observation of the jth individual in the ith population is made up of a fixed constant portion vector $\mu_i(p \times 1)$, common to all observations, a contribution vector $I_{ij}(p \times 1)$, common to all measurements on the jth individual and a deviation, vector $E_{ijk}(p \times 1)$, particular to the kth measurement of the jth individual.

The classification problem may now be described as how to ascribe an individual with observations $X(n \times p) = (X_1,\ldots,X_n)'$ into one of the groups π_i when the basic model is described by (1.1).

First let us note that there is vast literature on classification and discrimination (see Anderson et al. [2]). Fisher [7] suggested the use of discriminant function as a basis for classification decisions. Other bases for classification include likelihood ratio tests (e.g., see Anderson [1], Basu and Gupta [4], [5], Gupta [10]); information theory (Kullback [17]), Bayesian techniques (Geisser [8]), nonparametric techniques (Gupta and Kim [13], [14], Govindarajulu and Gupta [9]) and heuristic criteria (Gupta and Govindarajulu [11], [12]), to mention but a few. However, as pointed out by Choi [6], no results appear to be available when each individual observation is made in replication although one often encounters such situations in biomedical applications.

In this paper, section 2 sets forth a decision-theoretic model of the classification problem, assuming there are losses, a priori probabilities for each possible population, and known probability distributions for the random variables in each of the populations. In section 3, the distribution of the classification statistic is considered and the probability of misclassification is computed. In section 4 the classification problem is extended to $m(> 2)$ populations.

2. BAYES SOLUTION OF CLASSIFICATION PROBLEM

Here we consider the classification procedure (recognition criterion) for ascribing the multiple vector observations $x_1,\ldots,x_n$ to one of the specified populations (classes) which ensures the least possible value of the risk or of the mathematical expectation of loss on classification (recognition). First we will make the following assumptions.

Assumptions. (i) The number of patterns to be recognized is two (i.e., $m = 2$).

(ii) The random vectors I_{ij} $(p \times 1)$ and E_{ijk} $(p \times 1)$ are distributed as $N(0,\Omega_i)$ and $N(0,\Sigma_i)$ respectively with real, symmetric and positive-definite covariance matrices Ω_i and Σ_i.

(iii) I_{ij} and E_{ijk} are independent.

(iv) The arriving sequence of pattern realizations is stationary, that is, the parameters of the populations are stable.

The possibility of weakening of some of these restrictions will be considered later. The covariance matrices Ω_i and Σ_i can be interpreted as the dispersions due to individual differences and that due to different observations respectively. The model (1.1) is that of a two factor mixed hierarchical design and was also considered by Choi [6]. In the case of two populations the best regions S_i of classification for population π_i $(i = 1,2)$ are given by

$$S_1: \frac{p_1(X)}{p_2(X)} \geq \frac{c(1/2)q_2}{c(2/1)q_1} \qquad\qquad (2.1)$$
$$S_2: \qquad\qquad <$$

where $c(i/j)$ is the cost (loss or penalty) of ascribing to the class i a pattern belonging to class j and q_i is the apriori probability that an individual in question has been extracted from class π_i with density function $p_i(X)$, $i = 1,2$ $(q_1 + q_2 = 1)$. Further if $P\{(p_1(X)/p_2(X)) = c(1/2)q_2/c(2/1)q_1 \mid \pi_i\} = 0$, $(i = 1,2)$, then the procedure is unique except for sets of probability zero (see [2], p. 131).

Now consider the problem of classifying an individual, under the assumptions (i)-(iv), with observations $X = (X_1,\ldots,X_n)'$ into one of the two classes π_i , $i = 1,2$, when the basic model is given by (1.1). Denote the density of the class π_i by $p(x: \mu_i,\Sigma_i,\Omega_i) = p_i(x)$, $i = 1,2$, which can be shown to be

$$p_i(x) = [(2\pi)^{pn/2}|\Sigma_i|^{n/2}|n\Omega_i\Sigma_i^{-1} + I|^{1/2}]^{-1}\exp(-Q_i) \quad , \qquad (2.2)$$

where

$$Q_i = (\tfrac{n-1}{2})\mathrm{tr}S\Sigma_i^{-1} + \tfrac{n}{2}(\bar{x} - \mu_i)'(\Sigma_i + n\Omega_i)^{-1}(\bar{x} - \mu_i) \quad , \qquad (2.3)$$

$\bar{X} = \frac{1}{n}\sum_{i=1}^{n} X_i$ is the sample mean vector $(p \times 1)$ and

$S = \frac{1}{n-1}\sum_{i=1}^{n} (X_i - \bar{X})(X_i - \bar{X})'$ is the sample covariance matrix $(p \times p)$.

The following theorem gives the best classification regions and directly follows from (2.1).

Theorem 2.1. For the model (1.1), the best regions of classification, under the assumptions (i)-(iv), are given by

$$S_1:\ U(X) \geq c$$
$$S_2:\ U(X) < c \tag{2.4}$$

where

$$U(X) = (n-1)\mathrm{tr}S(\Sigma_2^{-1}-\Sigma_1^{-1}) - n[(\bar{X}-\mu_1)'\Gamma_1^{-1}(\bar{X}-\mu_1) - (\bar{X}-\mu_2)'\Gamma_2^{-1}(\bar{X}-\mu_2)], \tag{2.5}$$

$$\Gamma_i = \frac{1}{n}\Sigma_i + \Omega_i\ ,\quad i = 1,2, \tag{2.6}$$

$$c = (n-1)\ \ln|\Sigma_1\Sigma_2^{-1}| + \ln|\Gamma_1\Gamma_2^{-1}| + 2\ \ln\ k\ , \tag{2.7}$$

and

$$k = c(1/2)q_2/c(2/1)q_1\ . \tag{2.8}$$

Hence the Bayes classification rule for our problem can be stated as

$$R:\ \text{classify}\ X\ \text{to}\ \pi_1\ \text{iff}\ U(X) \geq c\ . \tag{2.9}$$

It may be noted that the classification regions defined by Choi [6] are in error. It has been arbitrarily decided to include the boundary $U(X) = c$ in π_1, however one may like to make a randomized decision, by tossing a coin or one could delete measurements on a characteristic chosen at random and assign X on the basis of the remaining $(p - 1)$ characteristics.

3. DISTRIBUTION OF THE CLASSIFICATION STATISTICS AND THE PROBABILITY OF MISCLASSIFICATION

One of the immediate problems arising in investigations into recognition systems is the problem of evaluating the probability of erroneous recognition. For this purpose we shall derive the distribution of the classification statistic $U(X)$. However we first state the following lemma which will be useful in the sequel.

Lemma 3.1. Given that the observations $X(n \times p)$ are from the population π_i, under the model (1.1), we have

(i) $\bar{X}, S$ are independently distributed and form a set of minimal sufficient and jointly completes statistics for μ_i, Σ_i and Ω_i,

(ii) $\bar{X} \sim N(\mu_i, \frac{1}{n}\Sigma_i + \Omega_i)$,

(iii) $S \sim W(\Sigma_i, n-1)$,

where $W(\Sigma, n)$ denotes the Wishart distribution with covariance matrix Σ and degrees of freedom n .

The proof of the lemma follows from multivariate normal theory and is omitted.

Now for the computation of the probability of misclassification we shall consider several different cases.

Case (i): $\Omega_1 = \Omega_2 = \Omega$ and $\Sigma_1 = \Sigma_2 = \Sigma$

It is noticable that the sample covariance matrix S has no value in classification when $\Sigma_1 = \Sigma_2$, however $\bar{X}$ would even if $\mu_1 = \mu_2$. The classification statistic in this case reduces to the linear discriminant function and we get

$$R_1: \text{classify } X \text{ to } \pi_1 \text{ iff } U_1(X) \geq \frac{1}{n} \ln k \tag{3.1}$$

where

$$U_1(X) = (\mu_1-\mu_2)'\Gamma^{-1}\bar{X} - \frac{1}{2}(\mu_1-\mu_2)'\Gamma^{-1}(\mu_1+\mu_2) \quad , \tag{3.2}$$

and

$$\Gamma = \frac{1}{n}\Sigma + \Omega \quad .$$

It is interesting to note that the operation of matrix inversion can be avoided by using the statistic

$$U_1(X) = \tfrac{1}{2}[|\Gamma + (\bar{X}-\mu_1)(\bar{X}-\mu_1)'| - |\Gamma + (\bar{X}-\mu_2)(\bar{X}-\mu_2)'|]|\Gamma|^{-1}$$

or equivalently

$$U_1(X) = -|\Gamma|^{-1} \begin{vmatrix} \gamma_{11} & \cdots\gamma_{1p} & \bar{X}_1 - \frac{1}{2}[\mu_{11}+\mu_{21}] \\ \cdots & \cdots & \cdots \\ \gamma_{p1} & \cdots\gamma_{pp} & \bar{X}_p - \frac{1}{2}[\mu_{1p}+\mu_{2p}] \\ \mu_{12}-\mu_{21} & \cdots\mu_{1p}-\mu_{2p} & 0 \end{vmatrix} = -|\Gamma|^{-1} \begin{vmatrix} \Gamma & \bar{X}-\frac{1}{2}(\mu_1+\mu_2) \\ (\mu_1-\mu_2)' & 0 \end{vmatrix}$$

where we write $\Gamma = (\gamma_{ij})$, $i,j = 1,2,\ldots,p$; $\mu_i' = (\mu_{i1}\cdots\mu_{ip})$, $i = 1,2,$ and $\bar{x}' = (\bar{x}_1,\ldots,\bar{x}_p)$.

Since $U_1(X) \mid \pi_1 \sim N[\alpha,\alpha]$ and $U_1(X) \mid \pi_2 \sim N(-\alpha,\alpha)$ where $\alpha = (\mu_1-\mu_2)'\Gamma^{-1}(\mu_1-\mu_2)$, it is easy to compute the probabilities of misclassification .

$$P(1/2;R_1) = 1 - \Phi\left(\frac{\frac{1}{n}\ \ell n\ k + \frac{\alpha}{2}}{\sqrt{\alpha}}\right) \tag{3.3}$$

$$P(2/1;R_1) = \Phi\left(\frac{\frac{1}{n}\ \ell n\ k - \frac{\alpha}{2}}{\sqrt{\alpha}}\right) \tag{3.4}$$

In case $c(1/2)q_2 = c(2/1)q_1$ both these probabilities are simply $\Phi(-\sqrt{\alpha}/2)$ and are monotonic decreasing functions of α which may be taken as the square of the "distance" between the two populations based on p characteristics. If apriori probabilities are not known we may select the threshold value $\frac{1}{n}\ln\ k = k_1$, say, on the bias of making the expected losses due to misclassification equal. The minimax solution is obtained from

$$c(1/2)[1 - \Phi(\frac{k_1 + \frac{\alpha}{2}}{\sqrt{\alpha}})] = c(2/1)\Phi(\frac{k_1 - \frac{\alpha}{2}}{\sqrt{\alpha}}) \tag{3.5}$$

Further, if $c(1/2) = c(2/1)$, k_1 could be determined to sufficient accuracy by a trial-and-error method with the normal tables.

Case (ii): $\Sigma_1 = \Sigma_2 = \Sigma$

Here $\Gamma_i = \frac{1}{n}\Sigma + \Omega_i$, $i = 1,2$. The classification statistic is a quadratic discriminant function and the rule becomes

R_2: classify X to π_1 iff

$$U_{12}(X) < -\frac{1}{n}\{\ell n|\Gamma_1\Gamma_2^{-1}| + 2\ \ell n\ k\} = k_2(\text{say}), \tag{3.6}$$

where

$$U_{12}(X) = (\bar{X}-\mu_1)'\Gamma_1^{-1}(\bar{X}-\mu_1) - (\bar{X}-\mu_1)'\Gamma_2^{-1}(\bar{X}-\mu_2)\ . \tag{3.7}$$

Special cases of the quadratic discriminant function $U_{12}(X)$ have been studied by Bartlett and Please [3], Hahn [15], [16]. As in case (i) , we can represent $U_{12}(X)$ so that the matrix inversion is avoided.

$$U_{12}(X) = |\Gamma_1|^{-1}|\Gamma_1 + (\bar{X}-\mu_1)(\bar{X}-\mu_1)'| - |\Gamma_2|^{-1}|\Gamma_2 + (\bar{X}-\mu_2)(\bar{X}-\mu_2)'|$$

$$= |\Gamma_1|^{-1} \begin{vmatrix} \gamma_{111} & \cdots\gamma_{11p} & \cdots\bar{X}_1-\mu_{11} \\ \cdots & \cdots & \cdots \\ \gamma_{1p1} & \cdots\gamma_{1pp} & \cdots\bar{X}_p-\mu_{1p} \\ \mu_{11}-\bar{X}_1 & \cdots\mu_{1p}-\bar{X}_p & 1 \end{vmatrix}$$

$$- |\Gamma_2|^{-1} \begin{vmatrix} \gamma_{211} & \cdots\gamma_{21p} & \bar{X}_1-\mu_{21} \\ \cdots & \cdots & \cdots \\ \gamma_{2p1} & \cdots\gamma_{2pp} & \bar{X}_p-\mu_{2p} \\ \mu_{21}-\bar{X}_1 & \cdots\mu_{2p}-\bar{X}_p & 1 \end{vmatrix} ,$$

where now $\Gamma_i = (\gamma_{ijk})$, $j,k = 1,\ldots,p$; $i = 1,2$ and $\bar{X}_i$ and μ_i are as in case (i).

To compute the probability of misclassification, we need to know the distribution of $U_{12}(X)$. We note that $U_{12}(X) = -U_{21}(X)$. Write $U_{12}(X) = \bar{X}'A_1\bar{X} + 2A_2\bar{X} + A_3$, where

$$A_1 = \Gamma_1^{-1} - \Gamma_2^{-1},\quad A_2 = \mu_2'\Gamma_2^{-1}-\mu_1'\Gamma_1^{-1} \text{ and } A_3 = \mu_1'\Gamma_1^{-1}\mu_1-\mu_2'\Gamma_2^{-1}\mu_2 .$$

Then transforming $Y = \bar{X} + A_4$ and requiring the coefficient of Y to be zero, we obtain

$$U_{12}(X) = Y'(\Gamma_1^{-1}-\Gamma_2^{-1})Y + A_5$$

where

$$\begin{aligned} A_5 &= A_3 - A_2A_4 \\ &= \mu_1'\Gamma_1^{-1}\mu_1 - \mu_2'\Gamma_2^{-1}\mu_2 + (\mu_2'\Gamma_2^{-1}-\mu_1'\Gamma_2^{-1})(\Gamma_1^{-1}-\Gamma_2^{-1})^{-1}(\Gamma_2^{-1}\mu_2-\Gamma_1^{-1}\mu_1) \end{aligned} ,$$

and the distribution of Y, given X is from π_i, is $N(\mu_i+A_1^{-1}A_2',\Gamma_i)$, $i = 1,2$.

The conditional (given $X \in \pi_i$) distribution of $U_{12}(X)$ now can be approximated by a Gram-Charlier series of Type A. Let

$$U_{12}^*(X) = \frac{U_{12}(X) - E(U_{12}(X))}{\sqrt{Var(U_{12}(X))}} ,$$

then the approximate p.d.f. of $U^*_{12}(X)$ given $X \in \pi_i$ is given by

$$g_{U^*_{12}(X)}(x \mid \pi_i) \simeq \phi(x)[1 + \frac{1}{6}\sqrt{\beta_{1i}}\, H_3(x) + \frac{1}{24}(\beta_{2i}-3)H_4(x)] \tag{3.8}$$

where $H_r(x)$ are the Hermite polynomials of order r and $\sqrt{\beta_1}$ and β_2 are the coefficients of skewness and Kurtosis respectively and are given in terms of the cumulants of $U_{12}(X)$ by $\beta_{1i} = \frac{\kappa_{3i}^2}{\kappa_{2i}^3}$ and $\beta_{2i} - 3 = \frac{\kappa_{4i}}{\kappa_{2i}^2}$, and

$$\kappa_{ri} = \kappa_r(U_{12}(X) \mid \pi_i)$$

$$= 2^{r-1}(r-1)![tr(A_1\Gamma_i)^r + r(\mu_i + A_1^{-1}A_2')'\, A_1(\Gamma_iA_1)^{r-1}(\mu_i + A_1^{-1}A_2')] \tag{3.9}$$

It may be noted that terms in (3.8) are not necessarily in decreasing order of importance, and a different ordering is sometimes used.

Hence the probabilities of misclassification can be approximated by

$$P(1/2\colon R_2) = \int_{k_{22}}^{\infty} g_{U^*_{12}(X)}(x \mid \pi_2)dx$$

and

$$P(2/1\colon R_2) = \int_{-\infty}^{k_{21}} g_{U^*_{12}(X)}(x \mid \pi_1)dx$$

and

$$P(mc) = q_1P(2/1\colon R_2) + q_2P(1/2\colon R_2) \quad ,$$

where $k_{2i} = [k_2 - E[U_{12}(x) \mid \pi_i]/\sqrt{Var(U_{12}(X) \mid \pi_i)}$, $i = 1,2$.

Case (iii): General case

In the general case the classification rule is given by

$$R\colon \text{ classify } X \text{ to } \pi_1 \text{ iff } U(x) \geq c$$

where

$$U(X) = (n-1)trS(\Sigma_2^{-1} - \Sigma_1^{-1}) - nU_{12}(X)$$

and $U_{12}(X)$ is given by (3.7). The conditional distribution of $U(X)$, given

$X \in \pi_i$, can similarly be approximated as follows. Let

$$U^*(X) = \frac{U(X) - EU(X)}{\sqrt{\mathrm{Var}\ U(X)}}$$

Then $g_{U^*(X)}(x \mid \pi_i)$ is given by (3.8) where the cumulants of $U(X)$ can be found as follows. According to Lemma 1, $\mathrm{tr}S$ is distributed independently of $U_{12}(X)$. Also $\mathrm{tr}S(\Sigma_2^{-1}-\Sigma_1^{-1})$, given $X \in \pi_i$, is distributed as $\sum_1^p \lambda_j Y_j$ where Y_j are distributed as $\chi^2(n-1)$ and λ_j's are the eigenvalues of $\Sigma_i(\Sigma_2^{-1}-\Sigma_1^{-1})$. Hence the cumulants of $U(X)$ are given by

$$\begin{aligned} \kappa^*_{ri} &= \kappa^*_r(U(X) \mid \pi_i) \\ &= 2^{r-1}(n-1)^{r+1}(r-1)! \sum_{j=1}^{p} \lambda_j^r + (-n)^r \kappa_{ri} \end{aligned}$$

where κ_{ri} is given by (3.9). In this case

$$P(1/2\colon R) = \int_{c_2}^{\infty} g_{U^*(X)}(x \mid \pi_2)dx$$

and

$$P(2/1\colon R) = \int_{-\infty}^{c_1} g_{U^*(X)}(x \mid \pi_1)dx \quad ,$$

where $c_i = \dfrac{c - E(U(X) \mid \pi_i)}{\sqrt{\mathrm{Var}(U(X) \mid \pi_i)}}$, $i = 1,2.$

4. CLASSIFICATION INTO ONE OF SEVERAL POPULATIONS

The classification procedure of section 2 is extended to $m\ (> 2)$ populations π_i, $i = 1,2,\ldots,m$. The basic model remains (1.1) where now $i = 1,2,\ldots,m$. To classify an individual X to π_i , the best region of classification (corresponding to the minimum expected loss), S_i $(i = 1,2,\ldots,m)$ is given by (Anderson [1])

$$S_i\colon U_{ij}(X) \geq c_{ij} \quad , \qquad j = 1,2,\ldots,m \quad j \neq i$$

where

$$U_{ij}(X) = (n-1)\text{tr}S(\Sigma_j^{-1}-\Sigma_i^{-1}) - n[(\bar{X}-\mu_i)'\Gamma_i^{-1}(\bar{X}-\mu_i) - (\bar{X}-\mu_j)'\Gamma_j^{-1}(\bar{X}-\mu_j)]$$

$$\Gamma = \frac{1}{n}\Sigma_i + \Omega_i \quad ,$$

$$c_{ij} = (n-1)\ \ln|\Sigma_i\Sigma_j^{-1}| + \ln|\Gamma_i\Gamma_j^{-1}| + 2\ \ln k_{ij} \quad ,$$

and

$$k_{ij} = c(i/j)q_j/c(j/i)q_i; \quad \sum_1^m q_i = 1 \quad .$$

Hence the Bayes classification rule is given by

R_3: classify X to π_i iff

$$U_{ij}(X) - c_{ij} \geq 0 \quad , \quad j = 1,\ldots,m \quad j \neq i$$

The probability of misclassification in this case is given by

$$P(mc \mid R_3) = \sum_{i=1}^{m} \sum_{\substack{j=1 \\ j\neq i}}^{m} q_i P[U_{ij}(X)-c_{ij} \geq 0,\ j = 1\ldots m,\ j \neq i \mid \pi_i]$$

and can be computed using the results of the previous section as the case may be.

5. CONCLUDING REMARKS

Classification problem has received considerable attention in recent years. There are three main aspects of this problem, the engineering, the artificial intelligence and analytical aspects. The analytical aspect is concerned with the mathematical techniques of decision, estimation and optimization under uncertainty. Here we have been interested in this later aspect, especially in the derivation and estimation of a discriminant function which is optimum in the sense of Bayes rule, which minimizes the probability of misclassification. We have also estimated the probabilities of erroneous classification. However, it should be mentioned that any classification procedure needs updating from time to time and that even then they are only probabilistically correct. In case parameters are unknown, the unbiased estimation of the probability of misclassification is under investigation. Also the asymptotic properties of this classification procedure are being investigated; and will be reported elsewhere.

REFERENCES

[1] Anderson, T.W., An Introduction to Multivariate Statistical Analysis, New York: John Wiley and Sons, Inc., 1958.

[2] Anderson, T.W., Das Gupta, S., and Styan, G.P.H., A Bibliography of Multivariate Statistical Analysis, Edinburgh: Oliver and Boyd, 1972.

[3] Bartlett, M.S., and Please, N.W., "Discrimination in the case of zero means difference," Biometrika, 50(1963), 17-21.

[4] Basu, A.P., and Gupta, A.K., "Classification rules for exponential populations," Reliability and Biometry, Proceedings of the Conference held at Florida State Univ., SIAM Publications (1974), 637-650.

[5] Basu, A.P., and Gupta, A.K., "Classification rules for exponential pop ulations: Two parameter cases," Theory and Applications of Reliability, Vol. I, Proceedings of the Conference held at the Univ. of South Florida, Academic Press (1977), 507-525.

[6] Choi, S.C., "Classification of multiply observed data," Biom. Z., 14(1972), 8-11.

[7] Fisher, R.A., "Use of multiple measurements in taxonomic problems," Ann. Eug., 7(1936), 179-188.

[8] Geisser, S., "Posterior odds for multivariate normal classifications," J. Roy. Statist. Soc., B, 26(1964), 69-76.

[9] Govindarajulu, Z., and Gupta, A.K., "Certain nonparametric classification rules and their asymptotic efficiencies," Canad. J. Statist., 5(1977), 167-178.

[10] Gupta, A.K., "On the equivalence of two classification rules," Biom. Z., 19(1977), 365-367.

[11] Gupta, A.K., and Govindarajulu, Z., "Some new classification rules for c univariate normal populations," Canad. J. Statist., 1(1973), 139-157.

[12] Gupta, A.K., and Givindarajulu, Z., "Distribution of the quotient of two Hotelling's-T^2 variates," Comm. Statist., 4(1975), 449-453.

[13] Gupta, A.K., and Kim, B.K., "A classification rule based on permutations of multivariate observations," Bull. Internat. Statist. Inst., 46, no. 3(1975), 327-329.

[14] Gupta, A.K., and Kim, B.K., "On a distribution-free discriminant analysis," Biom. Z., 20(1978), 729-736.

[15] Han, C.P., "A note on discrimination in the case of unequal covariance matrices," Biometrika, 55(1968), 586-587.

[16] Han, C.P., "Distribution of discriminant function when covariance matrices are proportional," Ann. Math. Statist., 40(1969), 979-985.

[17] Kulback, S., Information Theory and Statistics, New York: John Wiley and Sons, Inc., 1959.

Multivariate Statistical Analysis
R.P. Gupta (ed.)

SOME APPLICATIONS OF CONDITIONAL EXPECTATION MINIMIZATION THEORY TO PSYCHOLOGICAL TESTS

R.P. Gupta

Department of Mathematics
Dalhousie University
Halifax, N.S.
Canada, B3H 4H8

Several authors, see e.g. Darroch (1965), Fortier (1966) Maxwell (1971), Jackson and Novick (1970), implicitly use conditional expectation minimization theory (CEMT) to estimate (or predict) a random vector by a linear function of another correlated random vector. Our purpose here is to make an explicit statement of this implicitly used CEMT.

1. INTRODUCTION

Given a bivariate random variable (α,β) the correlation between α and $f(\beta)$, a function of β, is maximized when $E(\alpha-f(\beta))^2$ is minimized. The value $f(\beta)$ which gives the required minimum is $f(\beta) = E(\alpha|\beta)$, i.e. $f(\beta)$ estimates $E(\alpha|\beta)$, see Wilks (1962,p.84). If β is a complete and sufficient estimator for $E(\alpha)$, then $f(\beta) = E(\alpha|\beta)$ is the minimum variance unbiased estimator for $E(\alpha)$, which gives an indication for choosing β.

Let y be a p component column vector $E(y) = 0$, $E(yy') = \Sigma$, and x another k component random vector with $E(x) = 0$, $E(xx') = \Delta$, $E(yx') = B$, B $p\times k$ and of rank k, then the following results are obvious.

$$E(y-Ax)'(y-Ax) = tr[\Sigma-B\Delta^{-1}B' + (A-B\Delta^{-1})\Delta(A-B\Delta^{-1})] . \qquad (1)$$

is minimized with respect to (w.r.t) A and x, when

$$AE(xx') = E(yx') \text{ , i.e., } A\Delta = B \text{ , } A \ p\times k \text{ of rank } k \text{ ,} \qquad (2)$$

and

$$\hat{x} = E(x|y) = E(xy')(E(yy'))^{-1}y = \Delta^{-1}B'\Sigma^{-1}y \quad . \qquad (3)$$

It follows that

$$A\hat{x} = E(y|x) = B\Delta^{-1}x \quad , \qquad (4)$$

and, from (3), that $t'x$ is estimated by $t'\hat{x}$ for any arbitrary t. A future

x_0 may also be predicted from current values of y. If $E(x_0x_0') = \Delta_0$, $E(yx_0') = B_0$, then $\hat{x}_0 = \Delta_0^{-1}B_0'\Sigma^{-1}y$. If the distribution of $y - A\hat{x}_0$ is known, then prediction intervals for $\hat{x}_0$ may be obtained.

Since from (3) $\hat{x}$ is a linear function of y, the generalized variance of $(y',\hat{x}')'$ vanishes, i.e.

$$|\Sigma - B\Delta^{-1}B'| = |\Sigma - A\Delta A'| = 0 \quad , \tag{5}$$

which shows that

$$A\Delta A' = \theta_1\alpha_1\alpha_1' + \ldots + \theta_k\alpha_k\alpha_k' \quad , \tag{6}$$

and

$$\Sigma - A\Delta A' = \theta_{k+1}\alpha_{k+1}\alpha_{k+1}' + \ldots + \theta_p\alpha_p\alpha_p' \quad , \tag{7}$$

where $\theta_1 >\ldots> \theta_p$ are roots of Σ and

$$\Sigma = \theta_1\alpha_1\alpha_1' + \ldots + \theta_p\alpha_p\alpha_p' \quad . \tag{8}$$

Obviously, from (6) $A\hat{x}$ is a better estimator for $E(y)$ than y itself. If $\hat{x}$ is complete and sufficient, then $A\hat{x}$ is unique.

It follows that

$$\text{Min } tr(\Sigma - B\Delta^{-1}B') = \theta_{k+1} + \ldots + \theta_p \quad . \tag{9}$$

If we write

$$y = Ax + e \ , \ E(e) = 0 \ , \ E(ee') = \sigma^2 I \quad , \tag{10}$$

then, from linear least squares regression theory, it is known that

$$E[(y-A\hat{x})'(y-A\hat{x})|\hat{x}] = (p-k)\sigma^2 \tag{11}$$

so that

$$E(y-A\hat{x})'(y-A\hat{x}) = \text{Min } tr(\Sigma - B\Delta^{-1}B') = (p-k)\sigma^2 \tag{12}$$

The results (1) to (12) easily generalize to the matrix variate case. Let Y be a $p\times(N+k)$ matrix, F a $p\times k$ matrix, such that the $p\times(N-k)$ component column vector that is obtained, in the usual way, from the elements of (Y,F) has the covariance matrix

$$\Sigma \otimes \begin{pmatrix} \psi & B \\ B' & \Delta \end{pmatrix}, \quad \psi \quad N\times N \quad , B \quad N\times k \quad , \tag{13}$$

then minimizing

$$\mathrm{tr}\; E(Y-FA')(Y-FA')' = (\mathrm{tr}\Sigma)\mathrm{tr}(\psi-B\Delta^{-1}B'+(A-B\Delta^{-1})\Delta(A-B\Delta^{-1})') \tag{14}$$

With respect to A and F we find that,

$$A\Delta = B \;, \quad A \text{ and } B \text{ both of rank } k \;, \tag{15}$$

and that

$$\hat{F} = E(F|Y) = Y\psi^{-1}B \tag{16}$$

so that

$$\hat{F}A' = E(Y|F) = F\Delta^{-1}B' \quad . \tag{17}$$

It follows that

$$\mathrm{Min}\; \mathrm{tr}(\psi-B\Delta^{-1}B') = \sum_{i=k+1}^{N} \lambda_i t_i t_e' \quad , \tag{18}$$

where $\lambda_1 > \dots > \lambda_N$ are roots of ψ' and t's are corresponding latent vectors of ψ . We may write

$$Y = FA' + E \tag{19}$$

where the pN component column vector that is obtained form the elements of E has mean vector zero and covariance matrix $\sigma^2(\Sigma \otimes I)$, I is $N\times N$. The prediction of $\hat{x}$ puts k linear restriction on y and hence only $p-k$ components of e of (10) can be optimally estimated, or that only $p(N-k)$ components of E can be optimally estimated from linear functions of Y . In case $E(ee')$ of (10) has an arbitrary covariance matrix Ω , then the estimation problem of Ω and the allied distribution problems of the estimator are not fully solved, as yet, in statistics, and hence we assume $\Omega = \sigma^2 I$.

Now we show that the results of the above mentioned authors are in the framework of CEMT.

Although, sometimes, the same symbol denotes different quantities in this paper, its meaning made explicit in the context.

2. SOME APPLICATIONS

Darroch's results: Following Hotelling (1933) we take the principal components of y as $u_1 = \alpha_1' y_1, \ldots, u_p = \alpha_p' y$ and $E(xx') = I$, i.e. $\Delta = I$. Then from (5) we note that

$$BB'\Sigma^{-1} = \alpha \begin{pmatrix} I & 0 \\ 0 & 0 \end{pmatrix} \alpha' \quad \text{, where} \quad \alpha\Sigma\alpha' = \text{diag}(\theta_1, \ldots, \theta_p) \tag{20}$$

Now from (2) it follows that $A = B$, and from (3) we have $\hat{x} = B'\Sigma^{-1}y$. Thus we deduce that

$$A\hat{x} = AB'\Sigma^{-1}y = BB'\Sigma^{-1}y = \alpha \begin{pmatrix} I & 0 \\ 0 & 0 \end{pmatrix} \alpha' y = (\alpha_1\alpha_1' + \ldots + \alpha_k\alpha_k')y$$

$$= \alpha_1 u_1 + \ldots + \alpha_k u_k \quad . \tag{21}$$

Darroch characterizes the first k $(k \leq p)$ principal components $u_1, \ldots, u_k$ by the property (21) .

<u>Fortier's results</u>: Let D be a $p \times s$ matrix and G an $s \times k$ matrix, both of full ranks, then Fortier minimizes

$$E(y-DGx)'(y-DGx) = \text{tr}[\Sigma - DGB' - BG'D' + DG\Delta G'D']$$

$$= \text{tr}[\Sigma - BG'\Phi^{-1}GB + (D - BG'\Phi^{-1})\Phi(D - BG'\Phi^{-1})'] \tag{22}$$

w.r.t. D and G , where $\phi = G\Delta G'$. We note that the condition $D = BG'\Phi^{-1}$ is necessary for our minimization problem. Actually, $D = BG'\Phi^{-1}$ is a fact of CEMT. Thus it remains to minimize $\text{tr}[\Sigma - BG'\Phi^{-1}GB']$ over all G , i.e. we have to maximize $\text{tr } BG'\Phi^{-1}GB$ over all G . However, it is known that the maximum of $BG'\Phi^{-1}GB = \delta_1 + \ldots + \delta_s$ where $\delta_1 > \ldots > \delta_s > \ldots > \delta_k$ are roots of $B\Delta^{-1}B$. The value of G which gives us this maximum is the $s \times k$ matrix of the first s latent vectors of $B\Delta^{-1}B'$. If this matrix is p , then $D = Bp'\delta^{-1}$, where $\delta = \text{diag}(\delta_1, \ldots, \delta_s)$. Fortier (sections 2.2 and 2.3, pp. 370-372) solves this minimum value problem in stages by using

$$E(y-ba'x)(y-ba'x)' = \text{tr}[\Sigma - \frac{Baa'B'}{t} + (b - \frac{Ba}{t})t(b - \frac{Ba'}{t})'] \tag{23}$$

is minimum w.r.t b and a , when $tb = Ba$, $t = a'\Delta a$, and a is the first latent vector of $B\Delta^{-1}B'$ corresponding to the first largest root. Thus he first predicts the first residual $y - b_2a_2'x$ and after having found b_1 and a_1 from (23) he predicts the second residual $(y-b_1a_1'x)$, and find b_2 and a_2 and so on.

Maxwell's results: The general problem mentioned by Maxwell (section 3, pp. 196-197) is the problem of multivariate analysis of variance component estimation problem of statistics. We repeat it here for convenience.

We consider a one way classification model with k blocks of n plots each, $N = nk$, and

$$y_{ij} = \mu + \rho_i + e_{ij} \ , \ i = 1,\dots,k \quad j = 1,\dots,n \ , \tag{24}$$

where each p component e_{ij} is normal mean zero and covariance $\sigma_\varepsilon^2\Sigma$, $E(\rho_i) = 0$, $E(\rho_i\rho_i) = \sigma_0^2\Sigma$, ρ_i and e_{ij} are uncorrelated for each i and j . The total model (24) is

$$Y = \mu J + FA' + E \ , \tag{25}$$

where $Y = (y_{11}, y_{12},\dots,y_{1n}, y_{21},\dots,y_{kn})$, $F = (\rho_1,\dots,\rho_k)$, $A' = \text{diag}(d',d',\dots,d')$, where d is an n component column vector of unities and J is an N component row vector of unities. However, the problem of F estimation is already considered in section 1. Here

$$\psi = \text{diag}(\sigma_\varepsilon^2 J + \sigma_0^2 dd' \ , \ \sigma_\varepsilon^2 I + \sigma_0^2 dd' \ , \ \dots, \ \sigma_\varepsilon^2 + \sigma_0^2 dd') \ . \tag{26}$$

It follows from section 1, that $\hat{F}' = \Delta A'\psi^{-1}(Y-\bar{y}J)'$, where $\Delta = \sigma_0^2 I$, and $N\bar{y} = JY'$, i.e., $\bar{y}$ is the grand mean vector. We note that ψ has only two distinct roots, the first is $n\sigma_0^2 + \sigma_\varepsilon^2$ of multiplicity k , and second is σ_ε^2 of multiplicity $k(n-1)$. We denote the roots of ψ by $\psi_1 \geq \psi_2 \geq \dots \geq \psi_N$, and by β the $N \times N$ matrix of the latent vectors of ψ , then

$$\beta\psi\beta' = \psi_1\beta_1\beta_1' + \dots + \psi_N\beta_N\beta_N' \ . \tag{27}$$

It follows that

$$(I - J'J/N) = \beta_2\beta_2' + \dots + \beta_N\beta_N' \ , \tag{28}$$

and hence from (6), (7) and (18)

$$\mathrm{tr}(I - J'J/N)A\Delta A' = \psi_2 + \ldots + \psi_k = (k-1)(n\sigma_0^2+\sigma_\varepsilon^2) \quad , \tag{29}$$

and that

$$\mathrm{tr}(I - J'J/N)(\psi - A\Delta A') = k(n-1)\sigma_\varepsilon^2 \quad . \tag{30}$$

Again obviously we have that

$$E(\hat{F}\Delta^{-1}\hat{F}') = (\mathrm{tr}(I - J'J/N)A\Delta A')\Sigma = (k-1)(n\sigma_0^2 + \sigma_\varepsilon^2)\Sigma \tag{31}$$

and that

$$E(YY'-N\bar{y}\bar{y}' - \hat{F}\Delta^{-1}\hat{F}') = (\mathrm{tr}(I - J'J/N)(\psi - A\Delta A'))\Sigma$$

$$= k(n-1)\sigma_\varepsilon^2\Sigma \tag{32}$$

However, (32) is the expected value of the residual sum of products matrix with N-k degrees of freedom (d.f.), and (31) is the expected value of blocks sum of products matrix with (k-1) d.f. Thus by using (31) and (32) we estimate the variance components $\sigma_0^2\Sigma$ and $\sigma_\varepsilon^2\Sigma$.

Maxwell takes (25) as a factor analysis model and considers some special cases of (25). His equation (12) on p.99 is the same as our equation (10) except that he assumes $\theta_{k+1} = \theta_{k+2} = \ldots = \theta_p = \sigma^2$ and that $\Delta = I$. Maxwell's equation (27) on p. 201, with $\Delta = I$, is our equation (3). In section 1, we have shown that $E(\hat{x}\hat{x}') = E(xx') = \Delta$, $E(y\hat{x}') = E(yx') = B$, and these results are given by Maxwell in section 7 of his paper.

Sometimes it may be necessary or desirable to minimize (1) w.r.t. A, when A is subjected to certain linear restrictions. A problem of this type is considered by Jackson and Novick.

Jackson and Novick's results: Here the model of equation (1) is taken as

$$\psi(t) = D_t T + e(t) \quad , \tag{33}$$

where $\psi(t)$ is an N component column vector, $D_t = \mathrm{diag}(t_1,\ldots,t_N)$, T is an N component column vector with $E(T) = 0$, $E(TT') = G$, $E(e(t)) = 0$, $E(e(t)e(t)') = AD_t$, where now A is a diagonal matrix $A = \mathrm{diag}(a_1^2,\ldots,a_N^2)$, T and e(t) are uncorrelated. We wish to predict η , a single future component of T , by a linear function of 4(t) , say d'4(t) , where now d is any arbitrary vector, where $E(4(t)\eta) = D_t\delta$, δ specified. The prediction is to be carried on subject to the restriction $u'D_tJ = w$, where u is any known vector and J is an N component column vector of unities everywhere, w is known. This involves the problem

$$\min_{\beta} \min_{D_t} E[\eta-\beta d'4(t)]^2 \tag{34}$$

$$= \min_{\beta} \min_{D_t} [\beta^2 d'D_tGD_td + \beta^2 d'AD_td - 2\beta\delta D'_td + \sigma^2(\eta)]$$

$$= \min_{\beta} \min_{D_t} [\beta^2(D_td - G^{-1}\alpha)'G(D_td - G^{-1}\alpha) - \beta^2\alpha'G^{-1}\alpha + \sigma^2(\eta)] ,$$

subject to $u'D_tJ = w$, where $\alpha = (\delta\beta^{-1} - AD/2)$. Now we set $D_td = y$ and observe that (34) involves

$$\min_{y} \beta^2(y-G^{-1}\alpha)'G(y-G^{-1}\alpha), \text{ subject to } h'y = w , \tag{35}$$

where $h' = (u_1/d_1, u_2/d_2, \ldots, u_N/d_N)$, $u' = (u_1, \ldots, u_N)$, $d' = (d_1, \ldots, d_N)$. However, the minimum value problem (35) is a known minimum value problem, i.e.,

$$\min_{y}(y-\mu)'\psi(y-\mu) , \text{ subject to } Hy = v , \tag{36}$$

is given by

$$(v - H\mu)'(H\psi^{-1}H')^{-1}(v-H\mu) , \tag{37}$$

and occurs at

$$y = \mu + \psi^{-1}H'(H\psi^{-1}H')^{-1}(v-H\mu) , \tag{38}$$

where ψ $N\times N$ is symmetric positive definite, H is $q\times N$ and of rank $q<N$, μ is an N component column vector, and v is a q component vector. From (38), a solution of (35) is

$$y = D_t\hat{d} = y(\beta) = G^{-1}\alpha + G^{-1}h(w-h'G^{-1}\alpha)/h'G^{-1}h , \tag{39}$$

a result given by Jackson and Novick, who now substitute $y(\beta)$ for D_td in (34) and then minimize (34) w.r.t. β. If $\hat{\beta}$ and $\hat{d}$ are solutions of (34), then $\hat{\eta} = \hat{\beta}\hat{d}'4(t)$ is the optimal estimator whose variance can be easily found.

Thus in psychological testing theory the techniques of CEMT are used to find an estimate of Ax and its covariance matrix, by using the model (10). The Psychologist called x as the vector of true scores. In statistics it is simply the vector of variable effects.

Now we proceed with the inferential part of our paper. In case $E(x) = \beta$ in section 1, then $E(y) = A\beta$ and the equation (3) changes to

$$\hat{x} = E(x) + \Delta^{-1}B'\Sigma^{-1}(y-E(y)) \ . \tag{40}$$

If β and Σ are unknown, then it may be required to test the hypothesis that $\beta = \beta_0$, β_0 specified. For this purpose we rewrite our model as

$$y = A\beta + A(x-\beta) + \varepsilon = A\beta + u \ , \tag{41}$$

and assume u to be normal with $E(u) = 0$, $E(uu') = \Sigma$, which shows that (41) is the covariance regression model considered by Gleser and Olkin (1966), who provide a maximum likelihood ratio test criterion for testing that $\beta = \beta_0$, β_0 specified. They assume a sample of size N , Y $p\times N$, is available on y , and by $\bar{y}$ and $s = YY' - N\bar{y}\bar{y}'$ the sample mean vector and the sample dispersion matrix, and show $\bar{y}$ and s are sufficient for β and Σ . The maximum likelihood estimator $\hat{\beta}$ of β is $\hat{\beta} = (A's^{-1}A)^{-1}A's^{-1}\bar{y}$, and the test criterion is

$$\Lambda = (\hat{\beta}-\beta_0)'A^{-1}s^{-1}A(\hat{\beta}-\beta_0)[1 + N(\bar{y}-A\beta_0)'s^{-1}(\bar{y}-A\beta_0)]^{-1} \tag{42}$$

which has (i.e. $(N-k)\Lambda/k$ has) an F distribution with k and $N-k$ degrees of freedom. Anderson (1958) provides test for testing that Σ is a specified matrix, and Anderson (1969) provides (asymptotic) tests for testing that $\Sigma = A\Delta A' + \sigma^2 I$. The estimation of unknown A , Δ , σ^2 amounts to choosing A, Δ , σ^2 so that s is a close to Σ as possible in a certain sense. However, no satisfactory theory is as yet available for testing and estimating A , Δ , σ^2 .

Again in (19) we assume $E(F) = \bar{F}$ and write (16) as

$$\hat{F} = E(F) + (Y - (E(Y)))\psi^{-1}B \ , \tag{43}$$

and (19) as

$$Y = \bar{F}A' + (F-\bar{F})A' + \varepsilon = \bar{F}A' + U \ , \tag{44}$$

where $E(U) = 0$, and the usual pN component vector that is obtained from the elements of U has a covariance matrix $\Sigma \otimes \psi^{-1}$, ψ known. Under the assumption of normality of U, (44) is the multivariate normal regression model and Anderson (1958) provides tests for testing that $\bar{F} = \bar{F}_0$, $\bar{F}_0$ specified, Σ unknown.

REFERENCES

[1] Anderson, T.W. (1959). Statistical Inference for Covariance Matrices with Linear Structure. Multivariate Analysis - II, [Proc. 2nd Internat. Symp. Krishnaiah, ed.] 55-66.

[2] Anderson, T.W. (1958), An Introduction to Multivariate Statistical Analysis, John Wiley and Sons, N.Y.

[3] Darroch, J.N. (1965), An optimal property of principal components. Ann. Math. Statist. 36, 1574-1582.

[4] Fortier, J.J. (1966) Simultaneous linear prediction. Psychometrika 31, 369-381.

[5] Hotelling, H. (1933). Analysis of a complex of statistical variable into principal components. J. Educ. Psychology, 24, 147-444.

[6] Jackson, F. and Novick, M.R. (1970). Maximizing the validity of a unit weight composite as a function of relative weight component length with a fixed total testing time. Psychometrika, 35, 333-348.

[7] Maxwell, A.E. (1971). Estimating true scores and their reliabilities in the case of composite psychological tests. Br. J. Math. and Statist. Psychol. 24, 195-204.

[8] Wilks, S.S. (1962), Mathematical Statistics, John Wiley, New York.

Multivariate Statistical Analysis
R.P. Gupta (ed.)

AN APPLICATION OF THE SINGULAR NORMAL DISTRIBUTION IN LINEAR MODELS

Anant M. Kshirsagar and Noel Wheeler

Department of Biostatistics
University of Michigan
Ann Arbor, Michigan 48109
U.S.A.

The regression of $\underset{\sim}{y}$ on $\underset{\sim}{z}$ (where $\underset{\sim}{y}$ and $\underset{\sim}{z}$ have a singular joint multivariate normal distribution) is used to prove that, in a mixed model, estimates of functions of fixed effects are still unbiased as long as estimates of the variance components are of a particular type. A simple expression is given for the increase in the variance due to the estimation of these variance components.

INTRODUCTION

If $\underset{\sim}{y}$ and $\underset{\sim}{z}$ are two vectors having a singular multivariate normal distribution with means $\underset{\sim}{\mu}$ and $\underset{\sim}{\nu}$ and variance-covariance matrix

$$\begin{bmatrix} \Sigma_{11} & \Sigma_{12} \\ \Sigma_{21} & \Sigma_{22} \end{bmatrix}$$

then the regression of $\underset{\sim}{y}$ and $\underset{\sim}{z}$ is

$$\underset{\sim}{\mu} + \Sigma_{12}\Sigma_{22}^{-}(\underset{\sim}{z}-\underset{\sim}{\nu})$$

where Σ_{22}^{-} is any generalized inverse of Σ_{22} . This result is proved in Rao (1973). It is proposed to give an application of this result in linear models.

Consider the model

$$\underset{\sim}{y} = X\underset{\sim}{\beta} + Z\underset{\sim}{\phi} + \underset{\sim}{\varepsilon}$$

where $\underset{\sim}{\varepsilon}$ and $\underset{\sim}{\phi}$ are independent random vectors having multivariate normal distributions with means zero and variance-covariance matrices $\sigma^2 I$ and $\sigma^2 \mathrm{diag}(\lambda_1 I_{m_1}, \lambda_2 I_{m_2}, \ldots, \lambda_q I_{m_q})$ respectively, and $\underset{\sim}{\beta}$ is a vector of fixed effects. Such a model arises in design of experiments when inter-block information (in one-way designs) or inter-row and inter-column information (in two-way designs) is to be utilized in estimating the treatment effects. In this case $\underset{\sim}{\beta}$ is the vector of treatment effects and $\underset{\sim}{\phi}$ represents the vector of block or row and column effects. Usually the variance components are unknown and it is common practice to substitute for those estimates obtained from $\underset{\sim}{y}$ itself. This note provides a proof that if estimates of a particular type are used, the resulting estimates of functions of treatment effects are still unbiased. This result was first proved in its general form by Khatri and Shah (1977), but their proof is involved. The regression result quoted earlier provides an alternative simple derivation. Also the increase in the variance due to the use of the estimates of $\sigma^2\lambda_1,\ldots,\sigma^2\lambda_q$ is expressed in a much simpler form using Stein's Theorem (1950).

ESTIMATION OF LINEAR FUNCTIONS OF $\underset{\sim}{\beta}$

From the model it is easy to see that

$$E(\underset{\sim}{y}) = X\underset{\sim}{\beta}$$

and

$$V(\underset{\sim}{y}) = \sigma^2[I + Z\,\mathrm{diag}(\lambda_1 I_{m_1},\ldots,\lambda_q I_{m_q})Z'] = \Sigma$$

By the Gauss-Markov Theorem the best linear unbiased estimate of $\underset{\sim}{h}'\underset{\sim}{\beta}$, an estimable function of $\underset{\sim}{\beta}$, is $\underset{\sim}{h}'\underset{\sim}{\tilde{\beta}}$ where $\underset{\sim}{\tilde{\beta}}$ is any solution to

$$X'\Sigma^{-1}X\underset{\sim}{\tilde{\beta}} = X'\Sigma^{-1}\underset{\sim}{y} \quad .$$

If $\hat{\Sigma}$ is of the particular type described by Khatri and Shah (1977) then the estimate of each of $\sigma^2,\sigma^2\lambda_1,\ldots,\sigma^2\lambda_q$ is obtained as the maximum of zero and a quadratic form in the residual vector $\underset{\sim}{u} = T\underset{\sim}{y}$. It is intuitively obvious that the estimate of Σ must be obtained from $\underset{\sim}{u} = T\underset{\sim}{y}$ only, where T satisfies $TX = 0$. This condition is essential as it ensures that $E(\underset{\sim}{u})$ does not involve the fixed effects and that $E(\underset{\sim}{u}) = \underset{\sim}{0}$. Alternatively this means that $\underset{\sim}{u}$ belongs to the error space. This particular type of $\hat{\Sigma}$ is positive definite and the estimate of $\underset{\sim}{h}'\underset{\sim}{\beta}$ is taken to be $\underset{\sim}{h}'\underset{\sim}{\hat{\beta}}$ where $\underset{\sim}{\hat{\beta}}$ is any solution to

$$X'\hat{\Sigma}^{-1}X\underset{\sim}{\tilde{\beta}} = X'\hat{\Sigma}^{-1}\underset{\sim}{y} \quad .$$

Then

$$\underset{\sim}{h}'\underset{\sim}{\hat{\beta}} = \underset{\sim}{h}'(X'\hat{\Sigma}^{-1}X)^{-}X'\hat{\Sigma}^{-1}\underset{\sim}{y}$$

and

$$E(\underset{\sim}{h}'\underset{\sim}{\hat{\beta}}) = E\{E(\underset{\sim}{h}'\underset{\sim}{\hat{\beta}}|\hat{\Sigma}^{-1})\} = E\{E[\underset{\sim}{h}'(x'\hat{\Sigma}^{-1}X)^{-}X'\hat{\Sigma}^{-1}\underset{\sim}{y}|T\underset{\sim}{y}]\}$$

Now consider the vector $\begin{bmatrix} y \\ Ty \end{bmatrix}$ having a singular normal distribution we mean $\begin{bmatrix} X\beta \\ 0 \end{bmatrix}$ and variance-covariance matrix $\begin{bmatrix} \Sigma & \Sigma T' \\ T\Sigma & T\Sigma T' \end{bmatrix}$. Using Rao's results for the regression of y on Ty

$$E[h'(X'\hat{\Sigma}^{-1}X)^{-}X'\hat{\Sigma}^{-1}y|Ty] = h'(X'\hat{\Sigma}^{-1}X)^{-}X'\hat{\Sigma}^{-1}[X\beta + \Sigma T'\ (T\Sigma T')^{-}Ty] \ .$$

But $X(X'\hat{\Sigma}^{-1}X)^{-}X'\hat{\Sigma}^{-1}X = X$. Also the fact that $h'\beta$ is estimable implies that $h' = d'X$ for some vector d , so that

$$E[h'(X'\hat{\Sigma}^{-1}X)^{-}X'\hat{\Sigma}^{-1}y|Ty] = h'\beta + h'(X'\hat{\Sigma}^{-1}X)^{-}X'\hat{\Sigma}^{-1}\Sigma T'(T\Sigma T')^{-}Ty.$$

Then

$$E(h'\hat{\beta}) = h'\beta + E\{h'(X'\hat{\Sigma}^{-1}X)^{-}X'\hat{\Sigma}^{-1}\Sigma T'(T\Sigma T')^{-}Ty\}$$

and, in Khatri and Shah's notation, this becomes

$$E(h'\hat{\beta}) = h'\beta + E\{f'G_1u\}$$

where $f' = h'(X'\hat{\Sigma}^{-1}X)^{-}X'\hat{\Sigma}^{-1}$

and $G_1 = \Sigma T/(T\Sigma T')^{-}$.

With f an even function of u , a normal random vector with mean zero, it follows that $E(f'G_1u) = 0$ and hence

$$E(h'\hat{\beta}) = h'\beta \ .$$

Now, for the model with $V(y) = \Sigma$, $h'\tilde{\beta}$ is the minimum variance unbiased estimator of $h'\beta$. Since $h'\hat{\beta}$ is another unbiased estimator, by result (i) in 5a.2 on page 317 of Rao (1973) and sometimes called Stein's Theorem (1950), $h'\tilde{\beta}$ is uncorrelated with $h'\hat{\beta} - h'\tilde{\beta}$. Then $V(h'\hat{\beta}) = V(h'\tilde{\beta}) + V(h'\hat{\beta} - h'\tilde{\beta})$. The increase in variance due to using $\hat{\Sigma}$ is seen to be $V(h'\hat{\beta} - h'\tilde{\beta})$. Khatri and Shah (1977) write this increase as $E[(v + w)^2$ where v and w are complicated expressions defined in their paper. By some matrix algebra it can be shown that $v + w = h'\hat{\beta} - h'\tilde{\beta}$.

REFERENCES

[1] Khatri, C.G. and Shah, K.R. (1977). On the estimation of fixed effects in a mixed model. Proceedings of the 41st Session of the International Statistical

Institute, 284-287.

[2] Rao, C.R. (1973). Linear Statistical Inference and its Applications, Second Edition. Wiley and Sons, New York.

[3] Stein, C. (1950). Unbiased estimates with minimum variance. Ann. Math. Statist. 21, 406-415.

Multivariate Statistical Analysis
R.P. Gupta (ed.)

ASYMPTOTIC DISTRIBUTION OF QUANTILES FROM A MULTIVARIATE DISTRIBUTION*

K.S. Kuan
Department of Mathematics
University of Science of Malaysia

Mir M. Ali
Department of Mathematics
University of Western Ontario
London, Ontario

The asymptotic distribution of several quantiles from several components of a multivariate continuous population is derived.

1. INTRODUCTION

Consider a k-dimensional random variable and a sample of size N. Also consider r_i sample quantiles for the ith component, $r_i \geq 0$ for $1 = 1,\ldots,k$. We are primarily concerned with the joint asymptotic distribution of $r_1 + \ldots + r_k$ sample quantiles.

The asymptotic distribution of sample quantiles from univariate population has received a great deal of attention in the literature. The limiting distribution of a single quantile and also that of two sample quantiles are dealt with by Cramer (1946), while the case of several quantiles from the univariate population was obtained by Mosteller (1946). Several authors, among them we mention Mosteller (1946), Ogawa (1951) and Sarhan & Greenberg (1962), have used these results in the estimation of location and scale parameters.

The limiting distribution of sample quantiles from multivariate population has received relatively less attention. Mood (1941) gives the joint distribution of medians from bivariate population; Siddiqui (1960) derived the asymptotic distribution of quantiles one from each component of a bivariate population while Weiss (1960) obtained the limiting distribution of sample quantiles one from each component of a multivariate population.

The present work may therefore be viewed as an extension of the above works and brings to a reasonable conclusion a study of quantile distributions extending over thirty-five years.

2. SOME PRELIMINARIES

2.1. SAMPLE QUANTILES. Let $(X_1,X_2,\ldots,X_m)$ be a continuous m-variate random variable $(m \geq 2)$ with strictly increasing cumulative distribution function $F(x_1,x_2,\ldots,x_m)$ and p.d.f. $f(x_1,x_2,\ldots,x_m)$. Let $F_i(x_i)$ and $f_i(x_i)$ denote

*This work was partially supported by a grant from the National Research Council of Canada.

respectively the marginal c.d.f. and p.d.f. of X_i, for $i = 1,2,\ldots,m$. The equation $F_i(x_i) = \beta$ for $0 < \beta < 1$, has a unique solution in x_i say ξ_β. Then ξ_β is called the β-quantile of x_i.

Consider a sample of size N from the m-variate variable $(X_1,X_2,\ldots,X_m)$, say $(X_{1j},X_{2j},\ldots,X_{mj})$ for $j = 1,2,\ldots,N$. The order statistics of the ith component is obtained by arranging $X_{i1},X_{i2},\ldots,X_{iN}$ in ascending order of magnitude and is denoted by

$$X_{(i,1)} < X_{(i,2)} < \ldots < X_{(i,N)}, \text{ for } i = 1,2,\ldots,m .$$

For any β such that $0 < \beta < 1$, the sample β-quantile of the ith component X_i is defined by $X_{(i,[N\beta]+1)}$ where $[a]$ denotes the largest integer in a.

Let α_{ij}, $j = 1,2,\ldots,r_i$; $i = 1,2,\ldots,m$ be a set of real numbers such that

$$0 < \alpha_{ir_i} < \alpha_{i(r_i-1)} < \ldots < \alpha_{i1} < 1 , \; i = 1,2,\ldots,m .$$

Corresponding to these real numbers, denote the r_i population quantiles of X_i by $\xi_{ir_i},\xi_{i(r_i-1)},\ldots,\xi_{i1}$ with

$$\xi_{ir_i} < \xi_{i(r_i-1)} < \ldots < \xi_{i1} , \; i = 1,2,\ldots,m .$$

The corresponding sample quantiles of X_i are $Z_{ir_i},Z_{i(r_i-1)},\ldots,Z_{i1}$ with

$$Z_{ir_i} < Z_{i(r_i-1)} < \ldots < Z_{i1} , \; i = 1,2,\ldots,m .$$

For the case $r_1 = r_2 = r_3 = \ldots = r_m = 1$ there is one quantile from each component. We will call $(Z_{11},Z_{21},\ldots,Z_{m1})$ the sample quantiles of order $(\alpha_{11},\alpha_{21},\ldots,\alpha_{m1})$ or simply the sample $(\alpha_{11},\alpha_{21},\ldots,\alpha_{m1})$-quantiles, and we write $(Z_{11},Z_{21},\ldots,Z_{m1}) = (Z_1,Z_2,\ldots,Z_m)$.

2.2. As a simple notation, the expression $u = v(1+O(1/\sqrt{N}))$ will be abbreviated to read $u = \cdot v$; the symbol $O(1/\sqrt{N})$ represents any function such that $\sqrt{N} \times O(1/\sqrt{N})$ remains bounded as N tends to infinity.

2.3. We shall use the following well-known normal approximation to the multinomial distribution (see Cramer (1946))

$$[N!/(n_1!n_2!\ldots n_r!)]p_1^{n_1}p_2^{n_2}\ldots p_r^{n_r} = \cdot[|A|/(2\pi)^{r-1}]^{1/2}\exp\{-\tfrac{1}{2}\Sigma A_{ij}t_it_j\}\prod_{i=1}^{r-1} dt_i$$

where $t_i = (n_i - Np_i)/\sqrt{N}$, $i = 1,2,\ldots,r-1$,

and A is the matrix $A = (A_{ij})$ with

$$A_{ii} = 1/p_i + 1/p_r \ , \quad i = 1,2,\ldots,r-1 \ , \quad A_{ij} = 1/p_r \ , \quad i \neq j \ .$$

The matrix has determinant value $\prod_{i=1}^{r} (1/p_i)$.

We note that the multinomial probability converges uniformly to the multinormal density.

2.4. A CONVERGENCE THEOREM IN DISTRIBUTION

The following lemma is well-known; see for example, Tucker (1967).

Lemma. Let $X, X^{(1)}, X^{(2)}, \ldots$ be k-dimensional random variables and $X^{(n)}$ converge in distribution to X . Let $\phi_1(X), \phi_2(X), \ldots, \phi_m(X)$, $m \le k$ be real continuous function on E^k then $[\phi_1(X^{(n)}), \phi_2(X^{(n)}), \ldots, \phi_m(X^{(n)})]$, converges in distribution to $[\phi_1(X), \phi_2(X), \ldots, \phi_m(X)]$.

Corollary. Let $X, X^{(1)}, X^{(2)}, \ldots$ be k-dimensional random variables and $X^{(n)}$ converge in distribution to X , then $(X_i^{(n)}, X_j^{(n)})'$ converges in distribution to $(X_i, X_j)'$ where $X_h^{(n)}$ is the hth component of $X^{(n)}$ and X_h is the hth component of X , $h = 1,2,\ldots,k$.

2.5. NOTATION AND ASSUMPTIONS

Throughout the rest of the paper we use the assumptions and notations introduced in section 2.1 and 2.2.

In addition, we also make the further assumption that the marginal p.d.f.'s of the component population random variables satisfy the following condition:

$$f_i(x + \tfrac{1}{N}) = f_i(x) + O(1/N) \quad \text{for} \quad i = 1,2,\ldots,m \ .$$

2.6. SOME IMPORTANT RESULTS

We quote the following well-known theorems since they will be needed in the sequel.

THEOREM 1. With the assumptions and notations stated in section 2.5 consider the sample quantiles $Z_{i1}, Z_{iz}, \ldots, Z_{ir_i}$ of order $(\alpha_{i1}, \alpha_{iz}, \ldots, \alpha_{ir_i})$ belonging to the ith component and let $W_{ij} = \sqrt{N}\, f_i(\xi_{ij})(Z_{ij} - \xi_{ij})$. Then the joint distribution of $W_{i1}, W_{i2}, \ldots, W_{ir_i}$, as $N \to \infty$, tends to a multivariate normal distribution of r_i dimensions with mean $(0,0,\ldots,0)$ and variance-covariance matrix given by:

$$\mathrm{Var}(W_{ij}) = \alpha_{ij}(1-\alpha_{ij}) \quad \text{for} \quad j = 1,2,\ldots,r_i$$

and $\mathrm{Cov}(W_{ij},W_{ik}) = \alpha_{ij}(1-\alpha_{ik})$ for $\alpha_{ij} < \alpha_{ik}$.

Proof: The proof follows from Mosteller (1946).

THEOREM 2. With the assumptions and notations introduced in section 2.5 let $r_i = 1$ for $i = 1,\ldots,m$. Consider the sample quantiles $(Z_1,Z_2,\ldots,Z_m)$ of order $(\alpha_{i1},\alpha_{z1},\ldots,\alpha_{m1})$ as defined in section 2.2. Define

$$W_i = \sqrt{N}\, fi(\xi_{i1})(Z_i-\xi_{i1}) .$$

Then the joint distribution of $W_1,W_2,\ldots,W_m$, as $N \to \infty$, tends to a multivariate distribution of m-dimensions having mean $(0,0,\ldots,0)$ and variance-covariance matrix given by

$$\mathrm{Var}(W_i) = \alpha_{i1}(1-\alpha_{i1}) \quad \text{for} \quad i = 1,2,\ldots,m$$

and $\mathrm{Cov}(W_i,W_j) = F_{ij}(\xi_{i1},\xi_{j1}) - \alpha_{i1}\alpha_{j1}$ for $i \neq j$.

Proof: This theorem has been proved by Weiss (1960). We note that if we have only two quantiles Z_{i1} and Z_{j1} one from the ith component and one from the jth component then the limiting distribution of W_i and W_j is a bivariate normal distribution with mean $(0,0)$ and variance-covariance matrix of the above form.

3. THE RESULT

3.1. We will establish the following theorem:

THEOREM 3. With the assumption and notation stated in section 2.5 consider the sample quantiles $(Z_{ij} | j = 1,2,\ldots,r_i \,,\; i = 1,2,\ldots,m)$ of order (α_{ij}) . Then the joint distribution of

$$W_{ij} = \sqrt{N}\, f_i(\xi_{ij})(Z_{ij}-\xi_{ij}) .$$

for $j = 1,2,\ldots,r_i$; $i = 1,2,\ldots,m$; as $N \to \infty$, tends to a $(r_1+r_2+\ldots+r_m)$ dimensional normal distribution with mean $(0,0,\ldots,0)$ and variance and covariances

$$\mathrm{Var}(W_{ij}) = \alpha_{ij}(1-\alpha_{ij}) \quad \text{for} \quad j = 1,2,\ldots,r_i \quad \text{and} \quad i = 1,2,\ldots,m ;$$

$$\mathrm{Cov}(W_{ij},W_{ks}) = F_{ik}(\xi_{ij},\xi_{ks}) - \alpha_{ij}\alpha_{ks} \; ; \; i \neq K ;$$

$$\mathrm{Cov}(W_{ij},W_{is}) = \alpha_{ij}(1-\alpha_{is}) \quad \text{for} \quad \alpha_{ij} < \alpha_{is}$$

where F_{ik} denotes the marginal cumulative distribution function of the population ith and kth component.

The proof will be given in the next section. First, we only show that the limiting form of the joint distribution is multivariate normal but in view of very tedious computations we do not actually compute the means or the variances, covariances of the normal distribution. Then the lemma and its corollary of section 2.4 along with Theorem 1 and Theorem 2 yields the parameters of the limiting distribution. If two quantiles belong to the same component the covariance is given by theorem 1 while theorem 2 gives the covariance when two quantiles belong to two different components.

The technique used is essentially similar to that of Mood (1946). The sample space is divided into appropriate mutually disjoint regions giving rise to a multinormal setup to which the normal approximation is applied. Certain regions, namely those when the sample quantiles are determined by distinct sample observations, are called primary regions and it will be called as regions without differential dimensions. The complement of the primary region will have differential dimensions and terms arising from this case can be neglected in the asymptotic expression as compared to the terms arising from the primary regions. The regions for $m = 2$, $r_1 = r_2 = 1$ is shown in Figure 1, while the regions arising out of the case $m = 3$ with $r_1 = r_2 = r_3 = 1$ is shown in Figure 2.

3.2. THE PROOF. Consider the probability of the following event

$$z_{ij} - \frac{1}{2} dz_{ij} < z_{ij} < z_{ij} + \frac{1}{2} z_i \ , \quad j = 1,2,\ldots,r_i$$
$$i = 1,2,\ldots,m \ .$$

Divide the m-dimensional space into different regions by hyperplanes

$$x_i = z_{ij} - \frac{1}{2} dz_{ij} \ , \quad j = 1,2,\ldots,r_i$$

$$x_i = z_{ij} + \frac{1}{2} dz_{ij} \ , \quad i = 1,2,\ldots,m \ .$$

Let R_i, $i = 1,2,\ldots,\prod_{i=1}^{m} (r_i+1)$, denote the primary regions without differential dimension; let R_i' denote the regions with one differential dimension and let p_i, p_i' be the probabilities that an element falls in R_i and R_i' respectively.

We label the primary regions R_i ad follows: Label the region where $x_i < z_{ir_i}$ for all i, $i = 1,2,\ldots,m$, as $R_{\Pi(r_i+1)}$. On the positive side of $x_1 = z_{11}$, label the region where $x_i < z_{ir_i}$, $i \neq 1$, as R_1, and label the (r_1-1) regions on the negative side of $x_1 = z_{11}$ where $x_i < z_{ir_i}$, $i \neq 1$ as $R_2, R_3, \ldots, R_{r_1}$.

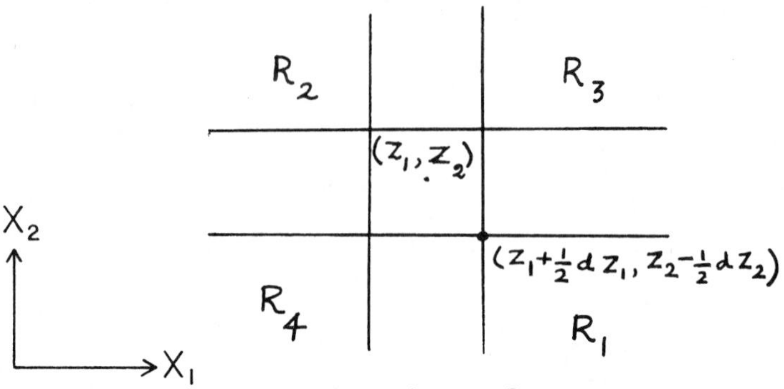

FIG. 1. $m=2$, $r_1=r_2=1$

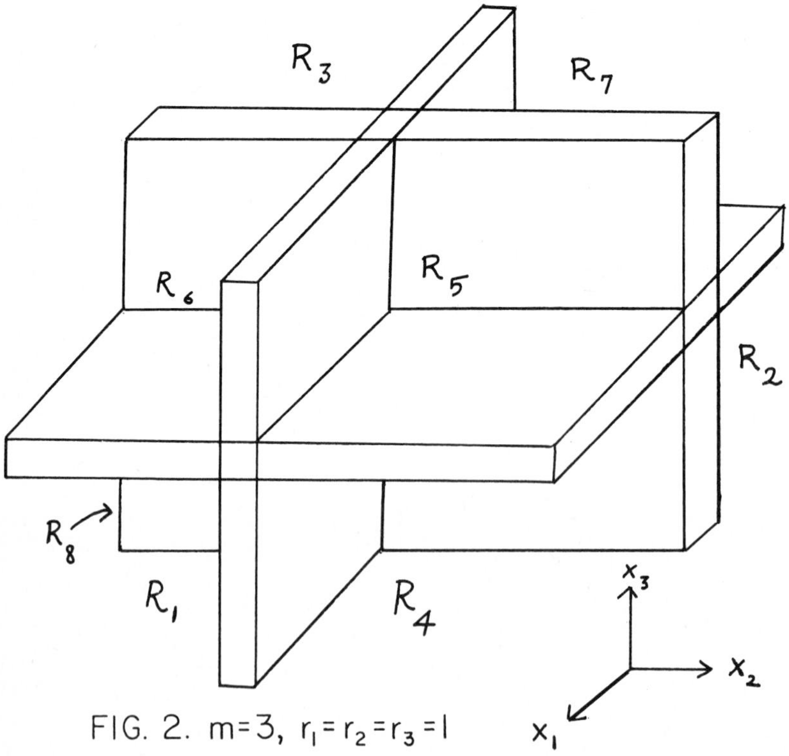

FIG. 2. $m=3$, $r_1=r_2=r_3=1$

On the positive side of $x_2 = z_{21}$ label the region where $x_i < z_{ir_i}$, $i \neq 2$ as R_{r_1+1} and label the (r_2-1) regions on the negative side of $x_2 = z_{21}$ where $x_i < z_{ir_i}$, $i \neq 2$ as $R_{r_1+2}, R_{r_1+3}, \ldots, R_{r_1+r_2}$.

$\vdots$

On the positive side of $x_m = z_{m1}$ label the region where $x_i < z_{ir_i}$, $i \neq m$ as $R_{\sum_{i=1}^{m-1} r_i+1}$ and label the r_m-1 regions on the negative side of $x_m = z_{m1}$ where $x_i < z_{ir_i}$, $i \neq m$ as $R_{\sum_{i=1}^{m-1} r_i+2}, \ldots, R_{\sum_{i=1}^{m} r_i}$. The remaining of the R_i's are arbitrarily labelled. Let

$$p_i = \int_{R_i} f(x_1, x_2, \ldots, x_m) \prod_{i=1}^{m} dx_i$$

$$p_i' = \int_{R_i'} f(x_1, x_2, \ldots, x_m) \prod_{i=1}^{m} dx_i .$$

Neglecting terms involving differentials of higher order it is seen that,

$$p_i = \int_{R_i^*} f(x_1, x_2, \ldots, x_m) \prod_{i=1}^{m} dx_i$$

$$p_i' = \int_{R_i'^*} f(x_1, x_2, \ldots, x_m) \Pi^* dx_i dz_{\beta\gamma}$$

where R_i^* is the region R_i with its possible boundaries $z_{ij} \pm \frac{1}{2} dz_{ij}$ replaced by z_{ij}, and $R_i'^*$ is one-dimensionless region obtained from R_i' by omitting the differential dimension. Π^* indicates that one of the dx_i's is omitted. If the differential dimension is $dz_{\beta\gamma}$, and is parallel to the x_i-axis, then x_i is replaced by $z_{\beta\gamma}$ in $f(x_1, x_2, \ldots, x_m)$.

If $\{Z_{ij} \mid j = 1,2,\ldots,r_i ; i = 1,2,\ldots,m\}$ is determined by less than D $(D = \sum_{i=1}^{m} r_i)$ elements of the sample, it can be shown that the terms arising from this case can be neglected in the asymptotic expression.

We are only concerned with terms which arise from the case where $\{Z_{ij} \mid j = 1,2,\ldots,r_i ; i = 1,2,\ldots,m\}$ is determined by D different elements of the sample. If this is so, then there is one element in each of the r_i slides, $i = 1,2,\ldots,m$

$$z_{ij} - \frac{1}{2} dz_{ij} < x_i < z_{ij} + \frac{1}{2} dz_{ij} , \; j = 1,2,\ldots,r_i .$$

Consider one of these possibilities where one element is in each of those slides $z_{ij} - \frac{1}{2} dz_{ij} < x_i < z_{ij} + \frac{1}{2} dz_{ij}$, with $x_k > z_{k1}$, $k \neq i$; the probability of this occurrence is, with $r = \prod_{i=1}^{m} (r_i+1)$, is given by

$$B = \prod_{\gamma=1}^{D} p'_{i_\gamma} \; \Sigma \frac{N!}{\Pi\, n_i!} \prod_{i=1}^{r} p_i^{n_i} \tag{3.2.1}$$

where n_i are number of elements in R_i and $\sum_{i=1}^{r} n_i = N-D$. If $g(z_{11}, z_{12}, \ldots, z_{mr_m})$ is the density that gives the distribution of $Z_{11}, Z_{12}, \ldots, Z_{m1}, \ldots, Z_{mr_m}$,

$$g(z_{11}, z_{12}, \ldots, z_{mr_m}) \prod_{i=1}^{m} \prod_{j=1}^{r_i} dz_{ij} = \Sigma_1 C + \Sigma_2 B \tag{3.2.2}$$

where Σ_1 means sum of all such C which arises from the case where $\{Z_{ij} \mid j = 1,2,\ldots,r_i ; i = 1,2,\ldots,m\}$ is determined by $D-h$ $(D > h > 0)$ elements of the sample and Σ_2 means sum of all such B which arises from the case that it is determined by D distinct elements. Consider the term B in (3.2.1),

$$B = N(N-1)\ldots(N-D+1) \prod_{\gamma=1}^{D} p'_{i_\gamma} \; \Sigma \frac{(N-D)!}{\Pi\, n_i!} \prod_{i=1}^{r} p_i^{n_i} ,$$

neglecting terms of lower power in N and applying the normal approximation,

$$B = N^D \prod_{\gamma=1}^{D} p'_{i_\gamma} \; \Sigma[|A|/(2\pi)^{r-1}]^{1/2} \exp\{-\tfrac{1}{2} \Sigma A_{ij} t_i t_j\} \prod_{i=1}^{r-1} dt_i \tag{3.2.3}$$

where

$$t_i = \frac{n_i - Np_i}{\sqrt{N}} , \quad i = 1,2,\ldots,r-1 ,$$

and $A = (A_{ij})$ with

$$A_{ii} = \frac{1}{p_i} + \frac{1}{p_r} , \quad i = 1,2,\ldots,r-1 ,$$

and

$$A_{ij} = \frac{1}{p_r} \quad \text{for } i \neq j .$$

Define

$$u_1 = \sqrt{N} \, [\Sigma_1 \frac{n_i}{N} - \Sigma_1 p_i]$$

where Σ_1 indicates sum over the regions R_i on the positive side of $x_1 = z_{11}$,

$$u_2 = \cdots$$

$$\vdots$$

$$u_{r_1} = \sqrt{N}\,[\Sigma_{r_1} \frac{n_i}{N} - \Sigma_{r_1} p_i]$$

where Σ_{r_1} indicates sum over the regions R_i on the positive side of $x_1 = z_{1r_1}$,

$$u_{r_1+1} = \cdots$$

$$\vdots$$

$$u_{r_1+r_2} = \cdots$$

$$\vdots$$

$$u_D = \sqrt{N}\,[\Sigma_D \frac{n_i}{N} - \Sigma_D p_i]$$

where Σ_D indicates sum over the regions R_i on the positive side of $x_m = z_{mr_m}$. We see that

$$t_1 = u_1 - \Sigma_{-1} t_j ,$$

$$t_2 = u_2 - \Sigma_{-2} t_j ,$$

$$\vdots$$

$$t_D = u_D - \Sigma_{-D} t_j , \tag{3.2.4}$$

where Σ_{-i} indicates sum over the same indices as in Σ_i except the index i.

It is seen that by the way we label the primary regions R_i, $t_1, t_2, \ldots, t_D$ are linear functions of $u_1, u_2, \ldots, u_D, t_{D+1}, \ldots, t_{r-1}$. Since $t_1, \ldots, t_{r-1}$ are joint normal, $u_1, \ldots, u_D, t_{D+1}, \ldots, t_{r-1}$ are joint normal.

Substitute (3.2.4) in (3.2.3); then

$$B = N^D \Pi p_i' \sum_{\gamma} [|A|/(2\pi)^{r-1}]^{1/2} \exp\{-\tfrac{1}{2}\, Q(u_1, \ldots, u_D, t_{D+1}, \ldots, t_{r-1})\} \prod_{i=1}^{r-1} dt_i$$

$$= N^{D/2} \Pi p_i' \sum_{\gamma} [|A|/(2\pi)^{r-1}]^{1/2} \exp\{-\tfrac{1}{2}\, Q(u_1, \ldots, u_D, t_{D+1}, \ldots, t_{r-1})\} \prod_{i=D+1}^{r-1} dt_i .$$

where Q is quadratic form in $u_1, \ldots, u_{D+1}, t_{D+1}, \ldots, t_{r-1}$. Q is used generically to denote quadratic form and is not the same from equation to equation.

In order to get rid of the summation sign, we integrate t_i, $i = D+1, \ldots, r-1$ each from $-\infty$ to ∞. This is equivalent to finding the joint marginal of u_1, $u_2, \ldots, u_D$; we get

$$B = K|A|^{1/2} [\frac{N}{2}]^{D/2} \Pi p_{i_\gamma}' \exp|-\tfrac{1}{2}\, Q(u_1, u_2, \ldots, u_D)\} \tag{3.2.5}$$

where the rest of the constant of integration is absorbed in K . Define

$$q_i = \int_{R_i^{**}} f(x_1,x_2,\ldots,x_m) \prod_{i=1}^{m} dx_i$$

$$q_i' = \int_{R_i'^{**}} f(x_1,x_2,\ldots,x_m)\Pi^* dx_i$$

where R_i^{**} , $R_i'^{**}$ are respectively R_i^* , $R_i'^*$ with possible boundaries z_{ij} replaced by ξ_{ij} .

We see that

$$p_i = q_i ,$$

and

$$\prod_{\gamma=1}^{D} p'_{i_\gamma} = \prod_{\gamma=1}^{D} q'_{i_\gamma} \prod_{i=1}^{m} \prod_{j=1}^{r_i} dz_{ij} \tag{3.2.6}$$

Substitute (3.2.6) in (3.2.5), we get

$$B = K[\frac{N}{2\pi}]^{D/2} \exp\{-\frac{1}{2} Q(u_1,u_2,\ldots,u_D)\} \prod_{i=1}^{m} \prod_{j=1}^{r_i} dz_{ij} \tag{3.2.7}$$

with the rest of the constant absorbed in K . But we have

$$\begin{aligned}
u_1 &= \sqrt{N}[\Sigma_1 \frac{n_i}{N} - \Sigma_1 p_i] \\
&= \sqrt{N}\,[\int_{\xi_{11}} (\int_{-\infty}^{\infty} \cdots \int_{-\infty}^{\infty} f(x_1,\ldots,x_m) \prod_{i=2}^{m} dx_i)dx_1 \\
&\quad - \int_{z_{11}}^{\infty} (\int_{-\infty}^{\infty} \cdots \int_{-\infty}^{\infty} f(x_1,\ldots,x_m) \prod_{i=2}^{m} dx_i)dx_1] \\
&= \sqrt{N}\,[\int_{\xi_{11}}^{\infty} f_1(x_1)dx_1 - \int_{z_{11}}^{\infty} f_1(x_1)dx_1] \\
&= \sqrt{N}(z_{11}-\xi_{11})f_1(\xi_{11}) \\
&= w_{11}
\end{aligned}$$

similarly,

$$\begin{aligned}
u_2 &= w_{12} \\
&\vdots \\
u_D &= w_{mr_m} .
\end{aligned} \tag{3.2.8}$$

Substitute (3.2.8) in (3.2.7), B becomes

$$B = K[\frac{N}{2\pi}]^{D/2}\exp\{-\frac{1}{2}\, Q(w_{11},w_{12},\ldots,w_{mr_m})\} \prod_{i=1}^{m} \prod_{j=1}^{r_i} dz_{ij} \quad . \tag{3.2.9}$$

Other B's will give rise to identical asymptotic expression as in (3.2.9) except that the factor K will be different; it is clear then that

$$g(z_{11},z_{12},\ldots,mr_m) \prod_{i=1}^{m} \prod_{j=1}^{r_i} dz_{ij}$$

$$= K^*[\frac{N}{2\pi}]^{D/2}\exp\{-\frac{1}{2}\, Q(w_{11},w_{12},\ldots,w_{mr_m})\} \prod_{i=1}^{m} \prod_{j=1}^{r_i} dz_{ij}$$

$$= \frac{K^*}{(2\pi)^{D/2}} \exp\{-\frac{1}{2}\, Q(w_{11},w_{12},\ldots,w_{mr_m})\} \prod_{i=1}^{m} \prod_{j=1}^{r_i} dw_{ij}$$

where $D = \sum_{i=1}^{m} r_i$, the constant K^* can be determined by integrating the right-hand side & equated to one. W_{ij} , $j = 1,2,\ldots,r_i$; $i = 1,2,\ldots,m$ are joint normal since $u_1,u_2,\ldots,u_D$ are joint normal and each w_{ij} is a linear function of the u_i's .

In view of the lemma in section 2.4, to specify the asymptotic distribution of $W_{11},W_{12},\ldots,W_{mr_m}$ only the asymptotic means and variances and covariances between the variables are needed. However, that can be done by considering the bivariate distribution of any two of the W_{ij}'s , say W_{ij} , W_{kc} as in the last chapter if $i \neq k$; if $i = k$, then the sample quantiles comes from the same component and this is well-known. Therefore, the joint distribution of $W_{11},W_{12},\ldots,W_{mr_m}$ tends to a $\sum_{i=1}^{m} r_i$-dimensional normal distribution with means and variances and covariances as mentioned in the theorem. This establishes Theorem 3.

ACKNOWLEDGEMENTS

The authors wish to thank Professor D.A.S. Fraser for some valuable suggestions.

REFERENCES

[1] Cramer, H. (1946). Mathematical Method of Statistics, Princeton University Press.

[2] Mood, A. (1941). On the Joint Distribution of the Median in Samples from Multivariate Population. Annals of Math. Stat. 12, 268-278.

[3] Mosteller, F. (1946). On Some Useful "inefficient' Statistics. Annals of Math. Stat. 17, 377-408.

[4] Ogawa, J. (1951). Contribution to the Theory of Systematic Statistics, I. Osaka Math. J. 3, 175-213.

[5] Sarhan, A.E. and Greenberg, B.G. (1962). Contributions to Order Statistics. John Wiley & Sons Inc.

[6] Siddiqui, M.M. (1960). Distribution of Quantiles in Samples from a Bivariate Population. Journal of Research of the National Bureau of Standards 64B, No. 3, 145-150.

[7] Tucker, H.G. (1967). A Graduate Course in Probatility. Academic Press.

[8] Weiss, L. (1964). On the Asymptotic Joint Normality of Quantiles from a Multivariate Distribution. Journal of Research of the National Bureau of Standards 68B, 65-66.

Multivariate Statistical Analysis
R.P. Gupta (ed.)

DISTRIBUTIONAL PROPERTIES OF CERTAIN TESTS FOR DETECTION OF DISCRETE MASS IN CROSS-SPECTRA OF MULTIVARIATE TIME SERIES*

Ian B. MacNeill

Statistics and Actuarial Science Group
University of Western Ontario
London, Ontario
Canada, N6H 5B9

Tests for discrete mass in the cross-spectra of two sets of time series are proposed. The tests are based on sub-matrices of the matrix of periodograms. Exact tests are developed for the case that the sets of series are independent normal white noise series. Selected quantiles of the distribution of the test statistic are tabulated. Extreme value results are obtained for the case of large sample size. Results are given to show that the normality assumption is not important. In the event of large sample size, the exact tests for the white noise case are extended to the case of linear series with specified spectrum. Also, conditions are given under which spectral estimates can be used in place of specified spectra.

INTRODUCTION AND SUMMARY

No real phenomenon is known to be exactly periodic, nor is it possible to gather enough data to verify precisely the existence of a periodicity. However, the concept of discrete mass in an auto or cross-spectrum is a useful fiction since, for modelling purposes, there is little difference between heavy concentration of mass about a frequency in an absolutely continuous spectrum and discrete mass at this frequency in a mixed spectrum. Each case is treated as a mixed spectrum, usually producing useful models. There are, however, times when it is important to distinguish between this concept of a mixed spectrum and the spectrum of a truly non-periodic phenomenon. The latter spectrum may be characterized in an ad hoc manner by requiring the bandwidth of peaks to exceed a specified minimum. The detection of discrete mass, or a close approximation thereof, in the cross-spectrum of two time series is a signal to search for a fundamental relationship between the series. Unfortunately the sample spectra of short data records from time series possessing periodic properties will contain peaks of bandwidth wide enough to cast doubt upon the discreteness of the spectral mass. Hence it is useful to have tests for the presence of discrete mass in auto and cross-spectra.

Under the hypothesis that the theoretical cross-spectrum between two time series is zero Nicholls [8] has obtained distribution theory for a test for

*This research was supported by grants from the Natural Sciences and Engineering Research Council of Canada.

discrete mass in a cross-spectrum. MacNeill [7] extended this result to the case of non-zero cross-spectrum for two or more series. The test is analogous to those that had been proposed earlier for testing for periodic components in univariate time series. The first several of these tests were proposed by Fisher [1,2] for the null hypothesis of white noise. These tests are exact. For large sample size Whittle [13] showed how these tests could be modified so as to apply to normal linear series with a specified spectrum. Hannan [3] and Nicholls [7] removed the requirement that the spectrum be specified in advance by substituting spectral estimates that essentially involve rough regressions on the various harmonics under consideration. Walker [12] derived extreme value theory for the periodogram and also showed that the normality assumption can be dropped with little effect on the upper tails of the distributions. MacNeill [5] discussed tests for periodic components in multiple time series. These tests are based upon the Euclidean norm of the matrix of periodogram ordinates. MacNeill [6] also proposed and discussed tests for detecting periodic components that are common to several time series.

The tests that are discussed in the sequel are analogous to Nicholls' tests for discrete mass in a cross-spectrum. However they are not restricted to a single cross-spectrum but could involve many. The requirement of zero cross-spectrum for the null hypothesis is also eliminated.

We consider real, discrete time parameter, n-vector time series, $\underset{\sim}{Z}(t)$, $t = 0, \pm 1, \pm 2 \ldots$, where $\underset{\sim}{Z}(t) = \underset{\sim}{\mu} + \underset{\sim}{m}(t) + \underset{\sim}{X}(t)$ with $\underset{\sim}{\mu} + \underset{\sim}{m}(t)$ being deterministic and $\underset{\sim}{X}(\cdot)$ being weakly stationary, zero mean, time series possessing a spectral density matrix $\underset{\sim}{f}^{XX}(\lambda)$ with $\operatorname{tr}\{\underset{\sim}{f}^{XX}(\lambda)\}$ being bounded uniformly in λ by $M < \infty$. We also assume $\underset{\sim}{X}(\cdot)$ to be a linear series such that

$$\underset{\sim}{X}(t) = \sum_{s=-\infty}^{\infty} \underset{\sim}{c}(s)\, \underset{\sim}{\xi}(t-s)$$

where $\underset{\sim}{c}(s)$ is an $n \times p$ matrix, $p \leq n$ and the components of the p-dimensional series $\underset{\sim}{\zeta}(t)$, $t = 0, \pm 1, \pm 2, \ldots,$ are white noise and uncorrelated series. Further assumptions are required concerning the coefficients and elementary variables in the linear series so one can relate the tests for discrete mass in white noise cross-spectra which are developed in the next section to similar tests for linear series. The set of conditions we select are those of Hannan [4, p.248]. They are that:

i) the components of $\underset{\sim}{\zeta}(\cdot)$ are independent series of independent variables,

ii) $E\{|\zeta_j(t)|^{2k}\} < \infty$ for some $k > 0$, and

iii) $\Sigma ||c(t)||\ |t|^{\frac{1}{2}} < \infty$

where

$$||\underset{\sim}{c}||^2 = \sum_{j=1}^{n} \sum_{k=1}^{p} |c_{jk}|^2 .$$

We will assume $k > 2$.

We denote a portion of a realization of the series by $\underset{\sim}{z}(t)$, $t = 1,\ldots,N$, and denote its mean-corrected finite Fourier transform by

$$w_N^Z(\lambda) = \frac{1}{\sqrt{2\pi N}} \sum_{t=1}^{N} e^{it\lambda}\{\underset{\sim}{Z}(t) - \bar{\underset{\sim}{Z}}_N\}$$

where $\bar{\underset{\sim}{Z}}_N = \frac{1}{N} \sum_{t=1}^{N} \underset{\sim}{Z}(t)$. The matrix of periodograms for the process is then given by

$$\underset{\sim}{I}_N^{ZZ}(\lambda) = \underset{\sim}{w}_N^Z(\lambda)\, \underset{\sim}{w}_N^Z(\lambda)^*$$

where $*$, here and in the sequel, denotes complex conjugate transpose.

When discussing the Fourier analysis of the mean value function, $\underset{\sim}{m}(\cdot)$, or the weakly stationary part of the series , $X(\cdot)$, we will use m or x as superscripts in place of Z . When specializing the weakly stationary component to that of independent series of independent variables we will use the notation $\underset{\sim}{w}_N^{\zeta}(\cdot)$ and $\underset{\sim}{I}_N^{\zeta\zeta}(\lambda)$.

Discrete mass appears in a cross spectrum of two series if one or both of the series possesses periodic components. Otherwise the spectrum is absolutely continuous (we ignore the possibility of a singular continuous component). Hence the null hypothesis, H_0 , will be that $\underset{\sim}{Z}(t) = \underset{\sim}{\mu} + \underset{\sim}{X}(t)$, $\underset{\sim}{\mu}$ unknown and the alternative hypothesis, H_A , is that $\underset{\sim}{Z}(t) = \underset{\sim}{\mu} + \underset{\sim}{m}(t) + \underset{\sim}{X}(t)$ where

$$m_j(t) = \sum_{r=1}^{R} \rho_{jr}\cos(t\theta_r + \phi_{jr}) \quad j = 1,2,\ldots,n$$

where $\underset{\sim}{\mu}$, R, ρ_{jr}, ϕ_{jr}, θ_r, $j = 1,\ldots,m$, $r = 1,\ldots,R$, are unknown and at least R of the ρ_{jr}'s are nonzero for some $R > 0$. We also assume that $\theta_r > 0$, $r = 1,2,\ldots,R$. Hence we are assuming under H_A that there are R positive frequencies each of which is present in at least one series.

The test that we propose is based on the cross-periodograms between the first n_1 component series and the remaining $n_2 = n - n_1$ series. If $\underset{\sim}{I}_{12N}^{ZZ}(\lambda)$ is a submatrix of $\underset{\sim}{I}_N^{ZZ}(\lambda)$ formed from these cross-periodograms then, using the methods of MacNeill [5, p.59] it can be shown that $E||\underset{\sim}{I}_{12N}^{ZZ}||^2$ is uniformly bounded at all frequencies under H_0 and has peaks of order $O(N^2)$ at frequencies θ_r, $r = 1,2,\ldots,R$ under H_A . This suggests that a suitable test for periodic components is a one-sided test based on the Euclidean norm of the submatrix $\underset{\sim}{I}_{12N}^{ZZ}(\lambda)$. Such a test statistic will have high power since it is $O(N)$ under H_A , which compares favourably with the usual hypothesis testing situation where the test statistic is $O(N^{\frac{1}{2}})$ under the alternative.

INDEPENDENT WHITE NOISE SERIES

As a starting point we discuss the case where the autospectra of the component series are flat and the cross-spectra are zero. We let $\underset{\sim}{\zeta}'(t) = (\zeta_1(t),\zeta_2(t),\ldots,\zeta_{n_1}(t),\zeta_{n_1+1}(t),\ldots\zeta_n(t))$ where $\underset{\sim}{\zeta}(t) \sim N(0,\sigma^2 I_n)$ and $n \geq 2$. We also let

$$\underset{\sim}{I}^{\zeta\zeta}_N(\lambda) = \begin{array}{c} \\ n_1 \\ n_2 \end{array}\begin{array}{c} \begin{array}{cc} n_1 & n_2 \end{array} \\ \left(\begin{array}{c|c} \underset{\sim}{I}^{\zeta\zeta}_{11N}(\lambda) & \underset{\sim}{I}^{\zeta\zeta}_{12N}(\lambda) \\ \hline \underset{\sim}{I}^{\zeta\zeta}_{21N}(\lambda) & \underset{\sim}{I}^{\zeta\zeta}_{22N}(\lambda) \end{array}\right) \end{array} .$$

The periodogram ordinates are computed at $\lambda_u = 2\pi u/N$, $u = 1,2,\ldots,m = [(N-1)/2]$, where $[\chi]$ is the biggest integer in χ. It was shown by MacNeill [7] that if

$$U(\lambda) = \frac{4\pi^2}{\sigma^4} ||\underset{\sim}{I}^{\zeta\zeta}_{12N}(\lambda)||^2 = \frac{4\pi^2}{\sigma^4} \sum_{j=1}^{n_1} \sum_{k=n_1+1}^{n} |I^{\zeta\zeta}_{jkN}(\lambda)|^2$$

then the probability density function for $U(\lambda_u)$ is

$$f_{n_1n_2}(U) = \frac{2U^{\frac{n}{2}-1}}{\Gamma(n_1)\Gamma(n_2)} K_{n_2-n_1}(2\sqrt{U}) \tag{1}$$

where $K_\nu(\cdot)$ is the modified Bessel function of the second kind of order ν. If $D^{n_1,n_2}_{\zeta,m-r}$ is the $m-r^{th}$ order statistic of $U(\lambda_u)$, $u = 1,\ldots,m$, then

$$P[D^{n_1,n_2}_{\zeta,m-r} \leq \chi] = \sum_{j=0}^{r} \binom{m}{j}\left\{\int_\chi^\infty f_{n_1,n_2}(u)du\right\}^j\left\{\int_0^\chi f_{n_1,n_2}(u)du\right\}^{m-j} . \tag{2}$$

This statistic and its distribution function can be used to test whether or not $r+1$ or more elements of discrete mass are present in the spectra. In particular, with $r = 0$, $\max_{1\leq u<m} U(\lambda_u)$ is an appropriate statistic for testing of the presence of at least one element of discrete mass. The 0.99 and 0.95 quantiles for $r = 0,1,2$, $1 \leq n_1 \leq n_2 \leq 10$ and $m = r+1, r+2, \ldots, 50$ have been computed and can be obtained from the nomograms in Appendix 1.

In the event that m is very large it is not feasible to use (2) to obtain critical points. Consequently, we derive the following extreme value results. We let

$$F_{n_1,n_2}(\chi) = P[2\sqrt{U(\lambda_j)} \leq \chi]$$

and let $V^{n_1,n_2}_{\zeta,m-r}$ be the $m-r^{th}$ order statistic of

$2\sqrt{U(\lambda_j)}$, $j = 1,\ldots,m$. Then we seek a divergent sequence of constants a_m , $m = 1,2,\ldots$, such that $P[V^{n_1,n_2}_{\zeta,m-r} - a_m \leq \chi]$ converges to a bonafide probability distribution. From (2) we have

$$P[V^{n_1n_2}_{\zeta,m-r} - a_m \leq \chi] = F^{m-r}_{n_1,n_2}(\chi+a_m) \sum_{j=0}^{r} \binom{m}{j}\{1-F_{n_1,n_2}(\chi+a_m)\}^j .$$

Now, for large m ,

$$m\{1-F_{n_1,n_2}(\chi+a_m)\} = m\int_{\chi+a_m}^{\infty} \frac{y^{n-1}}{2^{n-2}\Gamma(n_1)\Gamma(n_2)} K_{n_2-n_1}(y)dy$$

$$\simeq \frac{m\sqrt{\pi}\,(\chi+a_m)^{n-3/2}\, e^{-(\chi+a_m)}}{2^{n-3/2}\Gamma(n_1)\,\Gamma(n_2)} . \tag{3}$$

Letting $a_m = \log m + (n-3/2)\log\log m$ we see from (3) that

$$\lim_{m\to\infty} m\{1-F_{n_1,n_2}(\chi+a_m)\} = \frac{\sqrt{\pi}\, e^{-\chi}}{2^{n-3/2}\Gamma(n_1)\Gamma(n_2)} = B_{n_1n_2}\, e^{-\chi} .$$

With the same choice of a_m it can also be shown that

$$\lim_{m\to\infty} F^{n-r}_{n_1,n_2}(\chi+a_m) = \exp\{-B_{n_1n_2}\, e^{-\chi}\} .$$

In consequence we have

$$\lim_{m\to\infty} P\{V^{n_1,n_2}_{\zeta,m-r} - \log m - (n-3/2)\log\log m \leq \chi\}$$

$$= \exp\{-B_{n_1n_2}\, e^{-\chi}\} \sum_{j=0}^{r} \frac{1}{j!}\{B_{n_1n_2}\, e^{-\chi}\}^j \tag{4}$$

If

$$\hat{V}^{n_1n_2}_{\zeta,m-r} = \frac{\max\limits_{1\leq j\leq m} \sqrt{U(\lambda_j)}}{\frac{1}{m\sqrt{n_1n_2}}\sum_{k=1}^{m}\sqrt{U(\lambda_k)}} = \frac{\max\limits_{1\leq j\leq m} ||\underset{\sim}{I}^{\zeta\zeta}_{12N}(\lambda_j)||}{\frac{1}{m\sqrt{n_1n_2}}\sum_{k=1}^{m}||\underset{\sim}{I}^{\zeta\zeta}_{12N}(\lambda_k)||}$$

then (4) holds with V replaced by $\hat{V}$ since $\frac{1}{m\sqrt{n_1n_2}}\sum_{k=1}^{m} 2\pi||\underset{\sim}{I}^{\zeta\zeta}_{12N}(\lambda_k)||$ is a consistent estimator for σ^2 . The advantage of

$\hat{V}$ over V is that it is not necessary to know σ^2 in advance.

The extreme value probabilities given by (4) are adequate approximations for m even moderately larger than 50 provided n_1 and n_2 are small. Where n_1 and/or n_2 are large then the sample size will need to be very large in order that the approximations be adequate.

So far we have assumed the variables of the time series to be normally distributed. Using the method of Walker [12, p.112] as adapted for multiple time series by MacNeill [5, p.63] one can demonstrate the following result. Assume that $E\{|\zeta_j(t)|^{\delta}\}<\infty$ for some $\delta>6$ and for $j = 1,2,\ldots,n$. For $\varepsilon > 0$, N_ε can be chosen such that if $N > N_\varepsilon$ then

$$q-\tfrac{1}{2}q^2-\varepsilon^2 \leq P[V_{\zeta,m}^{n_1 n_2} \geq \log\{\frac{\sqrt{\pi}m(\log m)^{m-3/2}}{q2^{n-3/2}\Gamma(n_1)\Gamma(n_2)}\}] \leq q+\varepsilon \quad . \tag{5}$$

If the series were normal we see from the extreme value result (4) that, for q small and positive,

$$P[V_{\zeta,m}^{n_1 n_2} \geq \log\{\frac{\sqrt{\pi}m(\log m)^{n-3/2}}{q2^{n-3/2}\Gamma(n_1)\Gamma(n_2)}\}] \simeq 1 - e^{-q} \simeq q \quad .$$

Consequently, the normality assumption is seen to be unimportant for tail probabilities.

If ω_u, $u = 1,\ldots,s$, is a finite distinct set of positive frequencies not necessarily of the form $2\pi u/N$ then, for series, possessing zero mean and finite variance, $\underset{\sim}{I}_N(\omega_u)$, $u = 1,\ldots,s$, is an asymptotically independent set, each element of which possesses the complex Wishart distribution. Consequently, the exact result (2) holds asymptotically for this more general case.

LINEAR SERIES WITH SPECIFIED SPECTRUM

It will usually be unrealistic to assume flat autospectra and zero cross-spectra so we will now indicate how the results of the previous section can be applied to the case of linear series with specified spectrum. If $\underset{\sim}{f}^{XX}(\lambda)$ is of constant rank $p \leq n$ a.e. $|\lambda| < \pi$ it can be factorized and represented in the form

$$\underset{\sim}{f}^{XX}(\lambda) = \frac{1}{2\pi}\underset{\sim}{\phi}(\lambda)\underset{\sim}{\phi}(\lambda)^*$$

where $\underset{\sim}{\phi}(\lambda)$ is an $n\times p$ matrix. Then there exists a $p\times n$ matrix $\underset{\sim}{\psi}(\lambda)$ such that, a.e.,

$$\underset{\sim}{\psi}(\lambda)\underset{\sim}{\phi}(\lambda) = \underset{\sim}{I}_p \quad .$$

See Rozanov [11, p.39] for a discussion of the factorization of spectral density matrices.

We partition these matrices as follows:

$$\underset{\sim}{f}^{xx}(\lambda) = \begin{array}{c} \\ n_1 \\ n_2 \end{array}\begin{array}{c} \begin{array}{cc} n_1 & n_2 \end{array} \\ \left(\begin{array}{c|c} \underset{\sim}{f}^{xx}_{11}(\lambda) & \underset{\sim}{f}^{xx}_{12}(\lambda) \\ \hline \underset{\sim}{f}^{xx}_{21}(\lambda) & \underset{\sim}{f}^{xx}_{22}(\lambda) \end{array}\right) \end{array}, \quad \underset{\sim}{\phi}(\lambda) = \begin{array}{c} \\ n_1 \\ n_2 \end{array}\begin{array}{c} \begin{array}{cc} p_1 & p_2 \end{array} \\ \left(\begin{array}{c|c} \underset{\sim}{\phi}_{11}(\lambda) & \underset{\sim}{\phi}_{12}(\lambda) \\ \hline \underset{\sim}{\phi}_{21}(\lambda) & \underset{\sim}{\phi}_{22}(\lambda) \end{array}\right) \end{array}$$

$$\underset{\sim}{\psi}(\lambda) = \begin{array}{c} \\ p_1 \\ p_2 \end{array}\begin{array}{c} \begin{array}{cc} n_1 & n_2 \end{array} \\ \left(\begin{array}{c|c} \underset{\sim}{\psi}_{11}(\lambda) & \underset{\sim}{\psi}_{12}(\lambda) \\ \hline \underset{\sim}{\psi}_{21}(\lambda) & \underset{\sim}{\psi}_{22}(\lambda) \end{array}\right) \end{array}$$

where $p_1 + p_2 = p$. If $\underset{\sim}{\Phi}(\cdot)$ is the n-dimensional random measure of the spectral representation

$$\underset{\sim}{X}(t) = \int_{-\pi}^{\pi} e^{-i\lambda t} \underset{\sim}{\Phi}(d\lambda)$$

then one may define a p-dimensional series $\{\underset{\sim}{\zeta}(t)\}_{t=-\infty}^{\infty}$ by

$$\underset{\sim}{\zeta}(t) = \int_{-\pi}^{\pi} e^{-i\lambda t} \underset{\sim}{\psi}(\lambda) \underset{\sim}{\Phi}(d\lambda)$$

where the component series are white noise, uncorrelated series. If

$$\underset{\sim}{\phi}(\lambda) = \sum_{|s|<\infty} \underset{\sim}{c}(s) e^{i\lambda s}$$

then it can be shown that $X(\cdot)$ has the linear representation

$$\underset{\sim}{X}(t) = \sum_{|s|<\infty} \underset{\sim}{c}(s) \underset{\sim}{\zeta}(t-s) \quad .$$

The assumption of the Hannan conditions allows that

$$\underset{\sim}{T}_N(\lambda) = \underset{\sim}{I}_N^{xx}(\lambda) - \underset{\sim}{\phi}(\lambda) \underset{\sim}{I}_N^{\zeta\zeta}(\lambda) \underset{\sim}{\phi}^*(\lambda)$$

has components whose k^{th} absolute moments are $O(N^{-k/2})$ uniformly in λ and similarly for the components of $\underset{\sim}{\psi}(\lambda)\underset{\sim}{T}_N(\lambda)\underset{\sim}{\psi}^*(\lambda)$ provided $tr\{\underset{\sim}{\psi}(\lambda)\underset{\sim}{\psi}^*(\lambda)\} \leq M < \infty$, $|\lambda| \leq \pi$. Using the argument of MacNeill [5, p.66] one can then show that the asymptotic distributions of the order statistics of

$||\underset{\sim}{\psi}(\lambda_u)\underset{\sim N}{I}^{XX}(\lambda_u)\underset{\sim}{\psi}^*(\lambda_u)||$, $u = 1,\ldots,m$, are the same as those of the order statistics of $||\underset{\sim}{I}_N^{\zeta\zeta}(\lambda_u)||$, $u = 1,\ldots,m$, and that, under H_0 ,

$\sum_{j=1}^{p_1}\sum_{k=p_1+1}^{p}|2\pi(\underset{\sim}{\psi}(\lambda)\underset{\sim N}{I}^{XX}(\lambda)\underset{\sim}{\psi}^*(\lambda))_{jk}|^2$ is distributed as

$$\sum_{j=1}^{p_1}\sum_{k=p_1+1}^{p_2}|I_{jkN}^{\zeta\zeta}(\lambda)|^2 = ||\underset{\sim}{I}_{12N}^{\zeta\zeta}(\lambda)||^2$$, the distribution of which was

discussed in the previous section. As a consequence, the results for $D_{\zeta,m-r}^{p_1,p_2}$ embodied in (2), (4) and (5) also apply when $||\underset{\sim}{I}_{12N}^{\zeta\zeta}(\lambda)||^2$ is replaced by

$$\sum_{j=1}^{p_1}\sum_{k=p_1+1}^{p}|(2\pi\underset{\sim}{\psi}(\lambda)\underset{\sim N}{I}^{XX}(\lambda)\underset{\sim}{\psi}^*(\lambda))_{jk}|^2$$

provided the sample size is large. It can be observed that

$$\sum_{j=1}^{p_1}\sum_{k=p_1+1}^{p}|2\pi(\underset{\sim}{\psi}(\lambda)\underset{\sim N}{I}^{XX}(\lambda)\underset{\sim}{\psi}^*(\lambda))_{jk}|^2$$

$$= \mathrm{tr}\{2\pi\underset{\sim N}{I}^{XX}(\lambda)\begin{pmatrix}\underset{\sim}{\psi}_{11}^*(\lambda)\\ \hline \underset{\sim}{\psi}_{21}^*(\lambda)\end{pmatrix}(\underset{\sim}{\psi}_{11}(\lambda)|\underset{\sim}{\psi}_{12}(\lambda))\}$$

$$\times\ \mathrm{tr}\{2\pi\underset{\sim N}{I}^{XX}(\lambda)\begin{pmatrix}\underset{\sim}{\psi}_{12}^*(\lambda)\\ \hline \underset{\sim}{\psi}_{21}^*(\lambda)\end{pmatrix}(\underset{\sim}{\psi}_{21}(\lambda)|\underset{\sim}{\psi}_{22}(\lambda))\} . \qquad (6)$$

If $\underset{\sim}{\psi}_{12}(\lambda) \equiv 0$ the RHS of (6) reduces to

$$\mathrm{tr}\{\underset{\sim}{I}_{11N}^{XX}(\lambda)\underset{\sim}{f}_{11}^{XX^-}(\lambda)\}\mathrm{tr}\{\underset{\sim}{I}_{22N}^{XX}(\lambda)\underset{\sim}{f}_{22}^{XX}(\lambda)\} \qquad (7)$$

where $2\pi\underset{\sim}{\psi}_{11}^*(\lambda)\underset{\sim}{\psi}_{11}(\lambda) = \underset{\sim}{f}_{11}^{XX^-}(\lambda)$ is generalized inverse of $\underset{\sim}{f}_{11}^{XX}(\lambda)$. Computational methods for producing generalized inverses are discussed and referenced by Rao and Mitra [9].

LINEAR SERIES WITH UNSPECIFIED SPECTRUM

Since one will not often know the spectrum in advance it is important to be able to replace it in (6) and (7) with an appropriate estimator. MacNeill [5,p70] gives conditions on the windows and the truncation points such that the bias and variance of smoothed estimators converge uniformly in λ_u to zero at sufficiently fast rates that the tests using (6) and (7) with spectral estimators instead of specified spectra are consistent. Hannan [3] and Nicholls [7] have proposed mofifications to the typical spectrum estimators that greatly improve the power of

the tests for periodic components. These estimators provide for rough regressions on the frequencies $\{\lambda_u\}_{u=1}^m$. In the event that the cross-spectra between
the two sets of series are assumed,under H_0 , to be zero then $f_{11}^{xx-}(\lambda)$ and
$f_{22}^{xx-}(\lambda)$ in (7) can be replaced with suitable estimators, $f_{\sim 11N}^{xx-}(\lambda)$ and $f_{\sim 22N}^{xx-}(\lambda)$,
still providing consistent tests. In the event that one cannot assume zero cross-
spectra between the two sets of series then $f^{xx-}(\lambda)$, the generalized inverse
of $f^{xx}(\lambda)$ will have to be estimated by $f_{\sim N}^{xx\tilde{-}}(\lambda)$, a generalized inverse of the
spectrum estimator $f_{\sim N}^{xx}(\lambda)$. Since, with real data, it is unlikely that $f_{\sim N}^{xx}(\lambda)$
will be singular, an ordinary inverse is all that is likely to have to be computed.
However (6) calls not for $f_{\sim N}^{xx-}(\lambda)$ as a replacement for $f^{xx-}(\lambda)$ but for two
matrices whose sum is $f_{\sim N}^{xx-}(\lambda)$ and which are defined in terms of submatrices of
$\psi_N(\lambda)$ which $2\pi\psi_{\sim N}^*(\lambda)\psi_N(\lambda) = f_{\sim N}^{xx-}(\lambda)$. Thus it is necessary to factorize
$f_{\sim N}^{xx-}(\lambda)$. Robinson [10], p. 190-200] provides a computer algorithm for such a
factorization.

REFERENCES

[1] R.A. Fisher, "Test of significance in harmonic analysis," Proc. R. Soc. A.,
     Vol. 125 , 54-9, 1929.

[2] R.A. Fisher, "On the similarity of the distribution found for the test of
     significance in harmonic analysis and in Stevens' problems in geometrical
     probability, "Ann. Eugen., Vol. 10, 14-7, 1940.

[3] E.J. Hannan, "Testing for a jump in the spectral function," J.R. Statist.
     Soc. B, Vol. 23, 294-404, 1961.

[4] E.J. Hannan, Multiple Times Series. New York: Wiley, 1970.

[5] I.B. MacNeill, "Tests for periodic components in multiple time series,"
     Biometrika, Vol. 61 , 57-70, 1974.

[6] I.B. MacNeill, "A test to determine whether or not several time series share
     common periodicities," Biometrika, Vol. 64, 495-508, 1977.

[7] I.B. MacNeill, "Detection of discrete mass in cross-spectra," (To appear),
     1980.

[8] D.F. Nicholls, "Estimation of the spectral density function when testing for
     a jump in the spectrum," Aust. J. Statist., Vol. 9 , 103-8, 1967.

[9] D.F. Nichols, "Testing for a jump in co-spectra," Aust. J. Statist. Vol. 11,
     7-13, 1969.

[10] C.R. Rao and S.K. Mitra, Generalized Inverse of Matrices and its Applications.
     New York: Wiley, 1971.

[11] E.A. Robinson, Multi-Channel Time Series Analysis. San Francisco: Holden-Day,
     1967.

[12] Yu, A. Rozanov, Stationary Random Processes. San Francisco: Holden-Day,1967.

[13] A.M. Walker, "Some asymptotic results for the periodogram of a stationary time series," J. Aust. Math. Soc. Vol. 5, 107-28, 1965.

[14] P. Whittle, Hypothesis Testing in Time Series Analysis. Uppsala: Almqvist and Wiksell, 1951.

APPENDIX

In this section we consider $P[D^{n_1,n_2}_{\zeta,m-r} \leq x]$ as given in (2). We consider $r = 0,.,2$, $m = r+1, r+2, \ldots, 50$ and $1 \leq n \leq n_2 \leq 10$. The 0.95 and 0.99 quantiles for each case are evaluated. The quantiles for $(n_1 \leq n_2)$ such that $1 \leq n_1 = n_2 \leq 10$ and $1 \leq n_1$, $n_2 = n_1 + 1 \leq 10$ are obtained from figures 1-6. The quantiles for (n_1, n_2) such that $n_2 > n_1 + 1$ are obtained using Tables 1-6 and Figures 1-6. For example, assume $(n_1, n_2) = (4,4)$ and $m = 50$. Then the 0.99 quantile for the second largest of the 50 variates $(r=1)$ is read from Figure 4 as 74.08. As another example assume $(n_1, n_2) = (2,7)$ and $m = 50$. The 0.99 quantile for $r = 1$ is obtained by using both Table 4 and Figure 4. First obtain the (2,7) line of the A column. The corresponding line of the B column is (4,4). Read from the (4,4) curve of Figure 4 the 0.99 quantile of the second largest of 50 variables. This was found to be 74.08. Now evaluate the quantity in the C column. That is:

$$-\{2.31 - 0.160(50)\} = -1.51$$

The 0.99 quantile for $r = 1$, $m = 50$ is found by adding to the B quantile the quantity C, i.e., 74.08 - 1.51 = 72.57.

If the readings taken from the curves of Figures 1-6 were exact then the quantiles as computed from Table 1 - 6 would be in error by less than 2%. It will not likely be possible to read the quantiles from the curves with sufficient accuracy that this 2% error will be cause for concern.

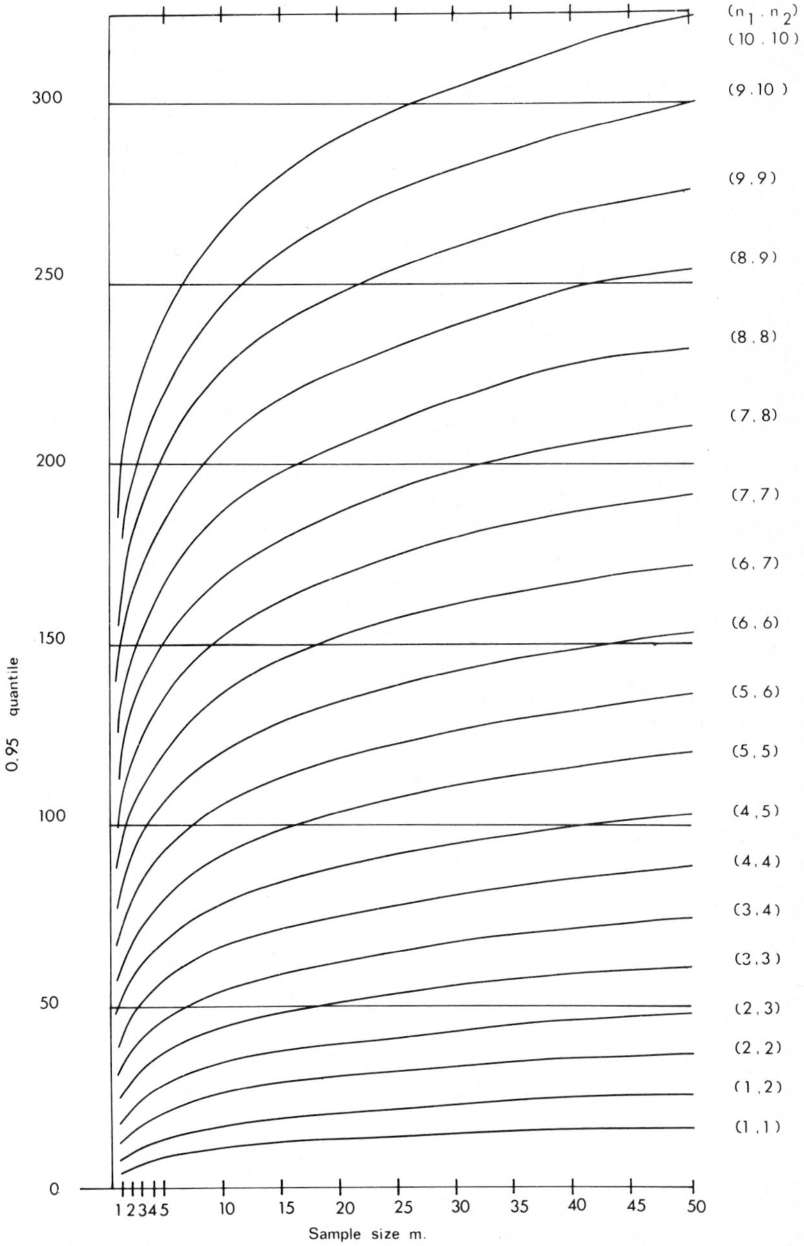

Figure 1. 0.95 quantiles for maximum of products of m gamma variates ($G(n_1) \times G(n_2)$) for selected pairs ($n_1 . n_2$) and samples sizes $1 \leq m \leq 50$.

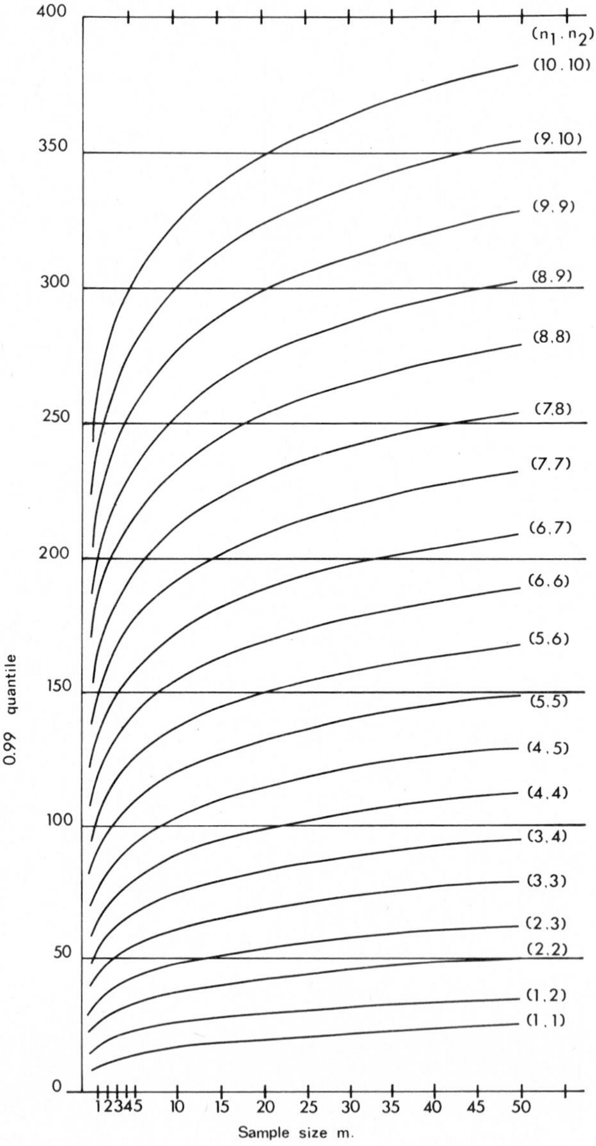

Figure 2. 0.99 quantiles for maximum of m products of gamma variates $(G(n_1) \times G(n_2))$ for selected pairs (n_1, n_2) and sample sizes $1 \leq m \leq 50$.

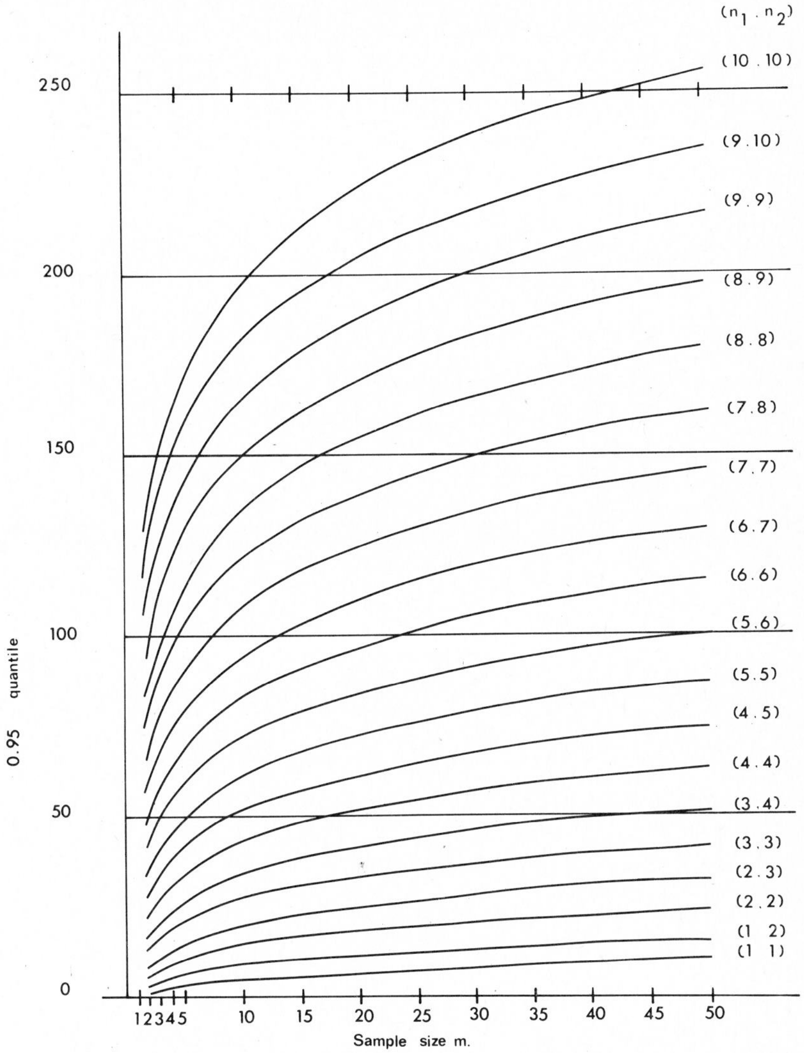

Figure 3. 0.95 quantile for the second largest of m products of gamma variates $(G(n_1) \times G(n_2))$ for selected pairs $(n_1 . n_2)$ and sample sizes $2 \le m \le 50$.

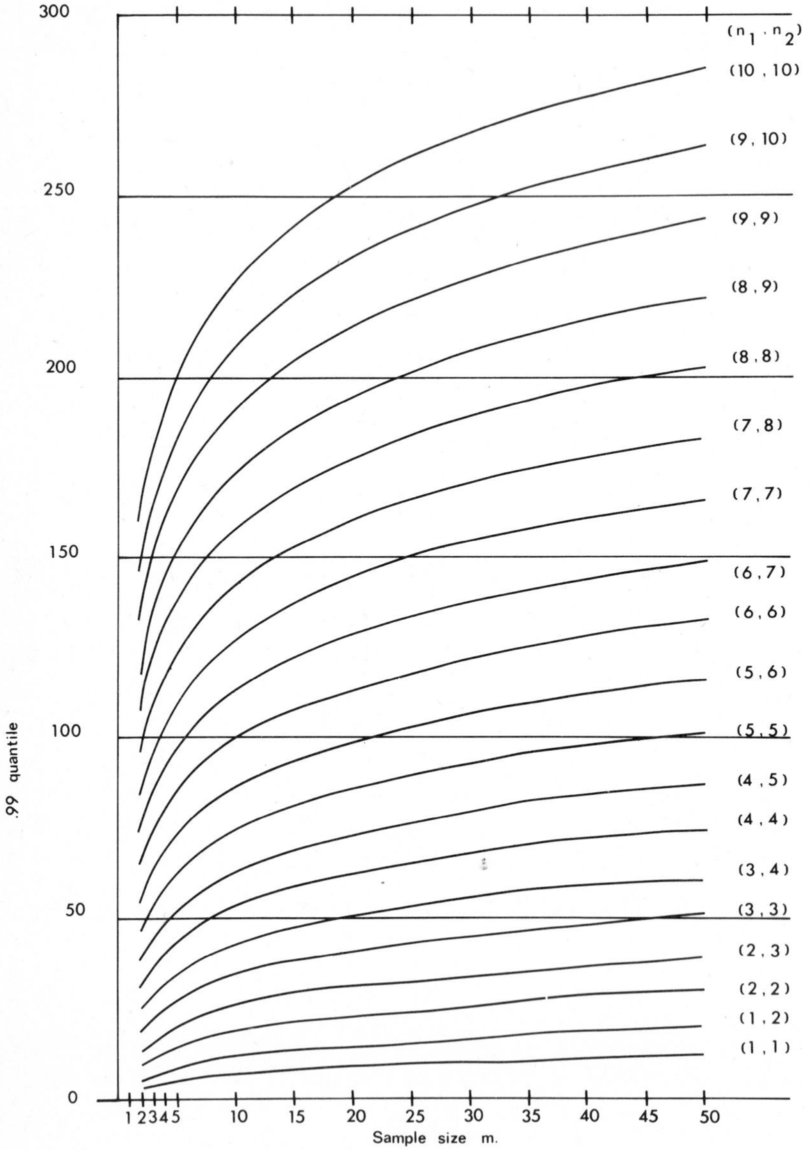

Figure 4. 0.99 quantiles for second largest of m products of gamma variates $\{G(n_1) \times G(n_2)\}$ for selected pairs (n_1, n_2) and sample sizes $2 \leq m \leq 50$.

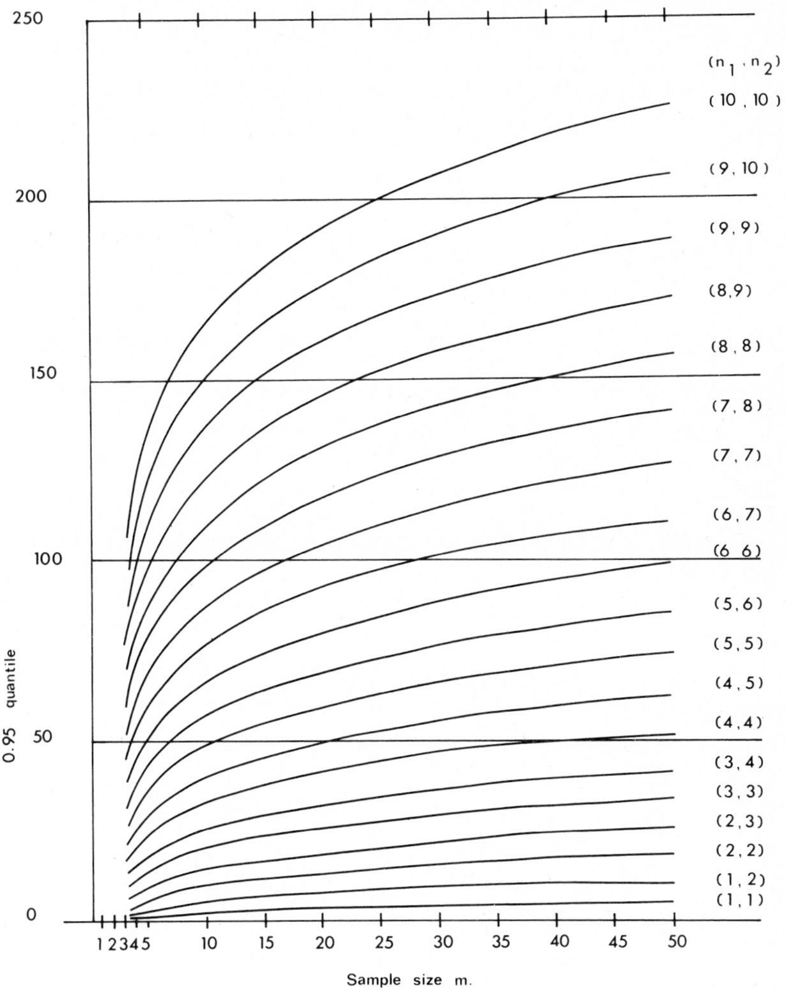

Figure 5. 0.95 quantiles for the third largest of m products of gamma variates $\{G(n_1) \times G(n_2)\}$ for selected pairs (n_1, n_2) and sample sizes $3 \le m \le 50$.

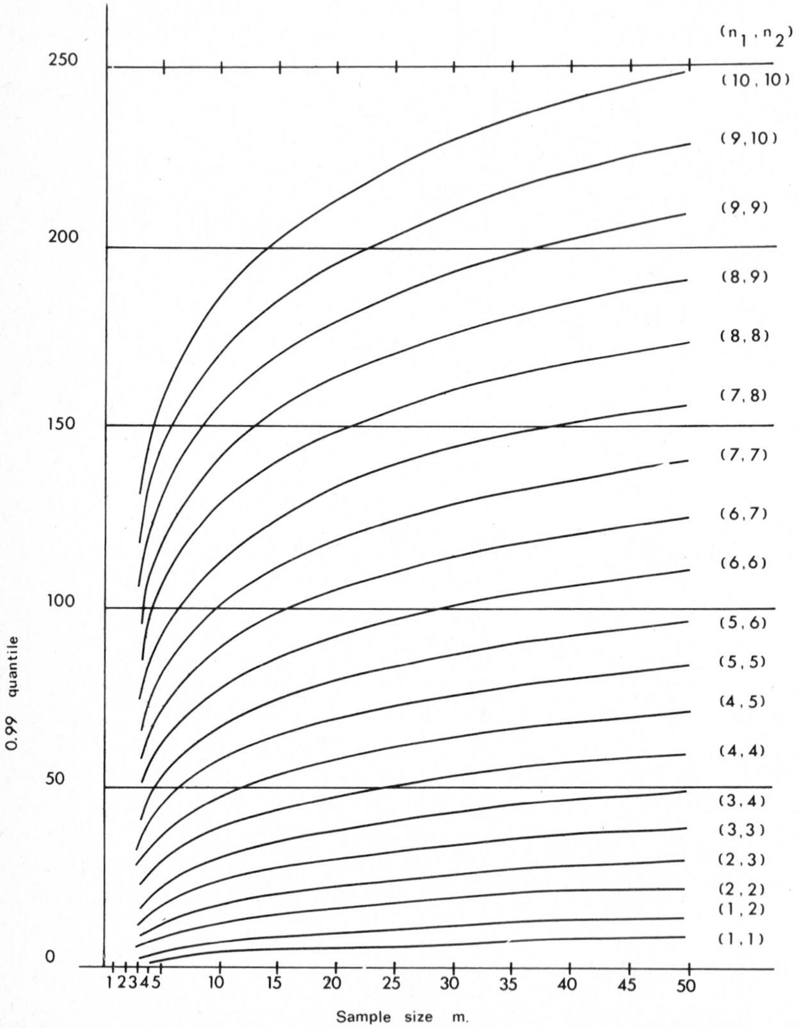

Figure 6. 0.99 quantiles for the third largest of m products of gamma variates $\{G(n_1) \times \{G(n_2)$ for selected pairs (n_1, n_2) and sample sizes $3 \le m \le 50$.

A	B	C
(1,3)	(2,2)	- (2.21 + 0.0350m)
(2.4)	(3,3)	- (1.87 + 0.0235m)
(3,5)	(4,4)	- (1.72 + 0.0180m)
(4,6)	(5,5)	- (1.62 + 0.0150m)
(5,7)	(6,6)	- (1.56 + 0.0130m)
(6,8)	(7,7)	- (1.50 + 0.0120m)
(7,9)	(8,8)	- (1.46 + 0.0110m)
(8,10)	(9,9)	- (1.43 + 0.0100m)
(1,4)	(2,2)	+ (0.84 + 0.0536m)
(2,5)	(3,4)	- (3.62 + 0.0418m)
(3,6)	(4,5)	- (3.35 + 0.0340m)
(4,7)	(5,6)	- (3.18 + 0.0286m)
(5,8)	(6,7)	- (3.06 + 0.0252m)
(6,9)	(7,8)	- (2.98 + 0.0226m)
(7,10)	(8,9)	- (2.91 + 0.0200m)
(1,5)	(2,3)	- (1.15 - 0.0260m)
(2,6)	(3,4)	+ (1.20 + 0.0578m)
(3,7)	(4,5)	+ (3.07 + 0.0766m)
(4,8)	(6,6)	- (6.32 + 0.0338m)
(5,9)	(7,7)	- (6.05 + 0.0418m)
(6,10)	(8,8)	- (5.89 + 0.0432m)
(1,6)	(3,3)	- (4.92 + 0.0202m)
(2,7)	(4,4)	- (2.23 - 0.0218m)
(3,8)	(5,5)	- (0,17 - 0.0460m)
(4,9)	(6,6)	+ (1.62 + 0.0634m)
(5,10)	(7,7)	+ (3.27 + 0.0768m)
(1,7)	(3,3)	- (1.91 - 0.0632m)
(2,8)	(4,5)	- (6.53 - 0.0100m)
(3,9)	(5,6)	- (3.31 - 0.0186m)
(4,10)	(6,7)	- (1.41 - 0.0176m)
(1,8)	(3,4)	- (5.51 - 0.0238m)
(2,9)	(4,5)	- (0.74 - 0.0428m)
(3,10)	(6,6)	- (7.95 + 0.0196m)
(1,9)	(3,4)	- (2.51 - 0.1054m)
(2,10)	(5,5)	- (5.61 - 0.0416m)
(1,10)	(4,4)	- (7.75 - 0.0322m)

Table 1. 0.95 quantiles for maximum of m products of gamma variates $\{G_{n_1} \times G_{n_2}\}$.

The entries in A column are the parameters of the gamma variates, (n_1, n_2).

The entries in B column are the parameters of the gamma variates whose quantiles may be obtained from Figure 1.

The entries in column C indicate what must be added to the B quantile to obtain the A quantile.

A	B	C
(1,3)	(2,2)	- (2.97 + 0.0336m)
(2,4)	(3,3)	- (2.38 + 0.0220m)
(3,5)	(4,4)	- (2.12 + 0.0168m)
(4,6)	(5,5)	- (1.95 + 0.0140m)
(5,7)	(6,6)	- (1.84 + 0.0122m)
(6,8)	(7,7)	- (1.77 + 0.0106m)
(7,9)	(8,8)	- (1.70 + 0.0100m)
(8,10)	(9,9)	- (1.66 + 0.0090m)
(1,4)	(2,2)	+ (0.86 + 0.0794m)
(2,5)	(3,3)	+ (4.60 + 0.0818m)
(3,6)	(4,5)	- (4.10 + 0.0310m)
(4,7)	(5,6)	- (3.82 + 0.0262m)
(4,8)	(6,7)	- (3.63 + 0.0228m)
(5,9)	(7,8)	- (3.48 + 0.0204m)
(6,10)	(8,9)	- (3.37 + 0.0186m)
(1,5)	(2,3)	- (0.73 - 0.0328m)
(2,6)	(3,4)	+ (2.35 + 0.0616m)
(3,7)	(4,5)	+ (4.65 + 0.0780m)
(4,8)	(5,6)	+ (6.63 + 0.0898m)
(5,9)	(6,7)	+ (8.43 + 0.0992m)
(6,10)	(7,8)	+ (10.13 + 0.1070m)
(1,6)	(3,3)	- (5.50 + 0.0102m)
(2,7)	(4,4)	- (1.88 - 0.0278m)
(3,8)	(5,5)	+ (0.73 + 0.0496m)
(4,9)	(6,6)	+ (2.92 + 0.0248m)
(5,10)	(7,7)	+ (4.87 + 0.0768m)
(1,7)	(3,3)	- (0.80 - 0.0752m)
(2,8)	(4,5)	- (5.88 + 0.0014m)
(3,9)	(5,6)	- (3.03 - 0.0244m)
(4,10)	(6,7)	- (0.67 - 0.0424m)
(1,8)	(3,4)	- (5,28 - 0.0390m)
(2,9)	(4,5)	+ (0.90 - 0.0942m)
(3,10)	(5,6)	+ (5.58 + 0.1282m)
(1,9)	(3,4)	- (0.61 - 0.1226m)
(2,10)	(5,5)	- (4.94 - 0.0532m)
(1,10)	(4,4)	- (7.03 - 0.0732m)

Table II. 0.99 quantiles for maximum of m products of gamma variates $\{G_{n_1} \times G_{n_2}\}$.

The entries in A column are the parameters (n_1,n_2) of the gamma variates.

The entries in B column are the parameters of gamma variates whose quantiles may be obtained from Figure 2.

The entries in C column indicate what must be added to the B quantile to obtain the A quantile.

A	B	C
(1,3)	(2,2)	- (1.40 + 0.0344m)
(2,4)	(3,3)	- (1.32 + 0.0232m)
(3,5)	(4,4)	- (1.28 + 0.0182m)
(4,6)	(5,5)	- (1.24 + 0.0158m)
(5,7)	(6,6)	- (1.21 + 0.0138m)
(6,8)	(7,7)	- (1.20 + 0.0122m)
(7,9)	(8,8)	- (1.19 + 0.0114m)
(8,10)	(9,9)	- (1.17 + 0.0106m)
(1,4)	(2,2)	+ (0.08 + 0.0402m)
(2,5)	(3,3)	+ (1.51 + 0.0686m)
(3,6)	(4,4)	+ (2.83 + 0.0858m)
(4,7)	(5,6)	- (2.46 + 0.0298m)
(5,8)	(6,7)	- (2.42 + 0.0264m)
(6,9)	(7,8)	- (2.39 + 0.0236m)
(7,10)	(8,9)	- (2.36 + 0.0220m)
(1,5)	(2,3)	- (1.28 - 0.0112m)
(2,6)	(3,4)	+ (0.22 + 0.0476m)
(3,7)	(4,5)	+ (1.58 + 0.0678m)
(4,8)	(5,6)	+ (2.86 + 0.0830m)
(5,9)	(6,7)	+ (4.10 + 0.0954m)
(6,10)	(7,8)	+ (5.30 + 0.1062m)
(1,6)	(2,3)	+ (0.20 + 0.0852m)
(2,7)	(4,4)	- (2.32 - 0.0108m)
(3,8)	(5,5)	- (0.90 - 0.0364m)
(4,9)	(6,6)	+ (0.43 + 0.0552m)
(5,10)	(7,7)	+ (1.69 + 0.0702m)
(1,7)	(3,3)	- (2.47 - 0.0388m)
(2,8)	(4,4)	+ (0.52 + 0.0984m)
(3,9)	(5,6)	- (3.35 - 0.0080m)
(4,10)	(6,7)	- (1.99 - 0.0298m)
(1,8)	(3,3)	- (0.97 - 0.1092m)
(2,9)	(4,5)	- (2.00 - 0.0652m)
(3,10)	(5,6)	+ (0.75 + 0.1074m)
(1,9)	(3,3)	+ (0.52 + 0.1798m)
(2,10)	(5,5)	- (5.77 - 0.0200m)
(1,10)	(3,4)	- (2.11 - 0.1390m)

Table III. 0.95 quantiles for the second largest of m products of gamma variates $\{G_{n_1} \times G_{n_2}\}$.

The entries in A column are the parameters (n_1,n_2) of the gamma variates.
The entries in B column are the parameters of gamma variates whose quantiles may be obtained from Figure 3.
The entries in C column indicate what must be added to the B quantile to obtain the A quantile.

A	B	C
(1,3)	(2,2)	- (1.84 + 0.0332m)
(2,4)	(3,3)	- (1.63 + 0.0222m)
(3,5)	(4,4)	- (1.53 + 0.0172m)
(4,6)	(5,5)	- (1.47 + 0.0142m)
(5,7)	(6,6)	- (1.41 + 0.0126m)
(6,8)	(7,7)	- (1.38 + 0.0114m)
(7,9)	(8,8)	- (1.35 + 0.0102m)
(8,10)	(9,9)	- (1.33 + 0.0094m)
(1,4)	(2,2)	+ (0.47 + 0.0448m)
(2,5)	(3,3)	+ (2.31 + 0.0504m)
(3,6)	(4,5)	- (2.94 + 0.0334m)
(4,7)	(5,6)	- (2.87 + 0.0276m)
(5,8)	(6,7)	- (2.79 + 0.0242m)
(6,9)	(7,8)	- (2.72 + 0.0220m)
(7,10)	(8,9)	- (2.67 + 0.0202m)
(1,5)	(2,3)	- (1.26 - 0.0192m)
(2,6)	(3,4)	+ (0.74 + 0.0504m)
(3,7)	(4,5)	+ (2.44 + 0.0680m)
(4,8)	(5,6)	+ (3.90 + 0.0824m)
(5,9)	(6,7)	+ (5.31 + 0.0936m)
(6,10)	(7,8)	+ (6,68 + 0.1028m)
(1,6)	(2,3)	+ (1.05 + 0.0936m)
(2,7)	(4,4)	- (2.31 - 0.0160m)
(3,8)	(5,5)	- (C.52 - 0.0376m)
(4,9)	(6,6)	+ (1.07 + 0.0566m)
(5,10)	(7,7)	+ (2.55 + 0.0702m)
(1,7)	(3,3)	- (2.24 - 0.0492m)
(2,8)	(4,4)	+ (1.59 + 0.1040m)
(3,9)	(5,6)	- (3.36 - 0.0130m)
(4,10)	(6,7)	- (1.70 - 0.0330m)
(1,8)	(3,3)	+ (0.05 + 0.1224m)
(2,9)	(4,5)	- (1.32 - 0.0716m)
(2,10)	(5,6)	+ (1.99 + 0.1110m)
(1,9)	(3,4)	- (3.13 - 0.0844m)
(3,10)	(5,5)	- (5.75 - 0.0300m)
(1,10)	(3,4)	- (0.83 - 0.1566m)

Table IV. 0.99 quanties for the second largest of m products of gamma variates $\{G_{n_1} \times G_{n_2}\}$.

The entries in A column are the parameters (n_1,n_2) of the gamma variates.
The entries in B column are the parameters of gamma variates whose quantiles may be obtained from Figure 4.
The entries in C column indicate what must be added to the B quantile to obtain the A quantile.

A	B	C
(1,3)	(2,2)	- (1.10 + 0.0326m)
(2,4)	(3,3)	- (1.09 + 0.0228m)
(3,5)	(4,4)	- (1.09 + 0.0182m)
(4,6)	(5,5)	- (1.09 + 0.0154m)
(5,7)	(6,6)	- (1.07 + 0.0138m)
(6,8)	(7,7)	- (1.07 + 0.0124m)
(7,9)	(8,8)	- (1.07 + 0.0114m)
(8,10)	(9,9)	- (1.07 + 0.0104m)
(1,4)	(2,2)	- (0.12 - 0.0330m)
(2,5)	(3,3)	+ (1.03 + 0.0612m)
(3,6)	(4,5)	- (2.17 + 0.0320m)
(4,7)	(5,6)	- (2.16 + 0.0294m)
(5,8)	(6,7)	- (2.15 + 0.0262m)
(6,9)	(7,8)	- (2.14 + 0.0238m)
(7,10)	(8,9)	- (2.13 + 0.0218m)
(1,5)	(2,3)	- (1.22 - 0.0068m)
(2,6)	(3,4)	- (0.06 - 0.0408m)
(3,7)	(4,5)	+ (1.06 + 0.0218m)
(4,8)	(5,6)	+ (2.16 + 0.0776m)
(5,9)	(6,7)	+ (3.25 + 0.0904m)
(6,10)	(7,8)	+ (4.34 + 0.1014m)
(1,6)	(2,3)	- (0.24 - 0.0668m)
(2,7)	(4,4)	- (2.25 - 0.0048m)
(3,8)	(5,5)	- (1.10 - 0.0308m)
(4,9)	(6,6)	+ (0.02 + 0.0500m)
(5,10)	(7,7)	+ (1.12 + 0.0652m)
(1,7)	(2,3)	+ (0.76 + 0.1340m)
(2,8)	(4,4)	- (0.12 - 0.0858m)
(3,9)	(5,6)	+ (3.26 - 0.0026m)
(4,10)	(6,7)	- (2.13 - 0.0244m)
(1,8)	(3,3)	- (1.44 - 0.0880m)
(2,9)	(3,4)	- (2.29 - 0.0530m)
(3,10)	(5,5)	- (0.02 - 0.0962m)
(1,9)	(3,3)	- (0.45 - 0.1510m)
(2,10)	(4,5)	- (0.15 - 0.1330m)
(1,10)	(3,4)	- (2.66 - 0.1106m)

Table V. 0.95 quantiles for the third largest of m products of gamma variates $\{G_{n_1} \times G_{n_2}\}$.

The entries in A column are the parameters (n_1,n_2) of the gamma variates.
The entries in B column are the parameters of gamma variates whose quantiles may be obatined from Figure 5.
The entries in C column indicate what must be added to the B quantile to obtain the A quantile.

A	B	C
(1,3)	(2,2)	- (1.43 + 0.0318m)
(2,4)	(3,3)	- (1.33 + 0.0220m)
(3,5)	(4,4)	- (1.28 + 0.0174m)
(4,6)	(5,5)	- (1.26 + 0.0144m)
(5,7)	(6,6)	- (1.23 + 0.0124m)
(6,8)	(7,7)	- (1.21 + 0.0116m)
(7,9)	(8,8)	- (1.19 + 0.0108m)
(8,10)	(9,9)	- (1.18 + 0.0098m)
(1,4)	(2,2)	+ (0.09 + 0.0372m)
(2,5)	(3,3)	+ (1.55 + 0.0432m)
(3,6)	(4,4)	+ (2.88 + 0.0796m)
(4,7)	(5,5)	+ (4.14 + 0.0920m)
(5,8)	(6,7)	- (2.44 + 0.0244m)
(6,9)	(7,8)	- (2.41 + 0.0220m)
(7,10)	(8,9)	- (2.38 + 0.0204m)
(1,5)	(2,3)	- (1.29 - 0.0120m)
(2,6)	(3,4)	+ (0.25 + 0.0438m)
(3,7)	(4,5)	+ (1.61 + 0.0630m)
(4,8)	(5,6)	+ (2.90 + 0.0774m)
(5,9)	(6,7)	+ (4.16 + 0.0886m)
(6,10)	(7,8)	+ (5.37 + 0.0988m)
(1,6)	(2,3)	+ (0.23 + 0.0878m)
(2,7)	(4,4)	- (2.32 - 0.0094m)
(3,8)	(5,5)	- (0.89 - 0.0338m)
(4,9)	(6,6)	+ (0.45 + 0.0514m)
(5,10)	(7,7)	+ (1.74 + 0.0652m)
(1,7)	(3,3)	- (2.46 - 0.0348m)
(2,8)	(4,4)	+ (0.57 + 0.0908m)
(3,9)	(5,6)	- (3.36 - 0.0074m)
(4,10)	(6,7)	- (1.97 - 0.0274m)
(1,8)	(3,3)	- (0.93 - 0.1004m)
(2,9)	(4,5)	- (1.97 - 0.0600m)
(3,10)	(5,6)	+ (0.80 + 0.0996m)
(1,9)	(3,3)	+ (0.60 + 0.1656m)
(2,10)	(4,5)	+ (0.92 + 0.1402m)
(1,10)	(3,4)	- (2.06 - 0.1278m)

Table VI. 0.99 quantiles for the third largest of m products of gamma variates $\{G_{n_1} \times G_{n_2}\}$.

The entries in A column are the parameters (n_1,n_2) of the gamma variates.
The entries in B column are the parameters of gamma variates whose quantiles may be obtained from Figure 6.
The entries in C column indicate what must be added to the B quantile to obtain the A quantile.

Multivariate Statistical Analysis
R.P. Gupta (ed.)

THE EFFECTS OF ELLIPTICAL DISTRIBUTIONS ON SOME STANDARD PROCEDURES INVOLVING CORRELATION COEFFICIENTS: A REVIEW*

Robb J. Muirhead

Department of Statistics
University of Michigan
Ann Arbor, Michigan
U.S.A. 48104

In this paper we study the effects of elliptical distributions on some procedures involving correlation coefficients. These distributions, whose contours of equal density have the same elliptical shape as the normal, provide attractive alternatives to multivariate normality. We examine the way they effect the distributions of ordinary, multiple and canonical correlation coefficients and standard inference based on them.

1. INTRODUCTION

To a large extent the area of robustness is concerned with the development of statistical procedures which are relatively insensitive to distributional assumptions. It is thus important to understand what happens to procedures which are based on the multivariate normal distribution when the distribution which is being sampled is, in fact, non-normal. This paper is essentially a review of some results in this area, relating to various correlation coefficients.

Many of the commonly used procedures in classical, normal-based multivariate analysis (and elsewhere) are of an asymptotic nature e.g. the usual χ^2 approximation to a likelihood ratio statistic. It is of interest to know what effect non-normal populations have on such standard asymptotic results. When a statistic of interest is a function of a sample covariance matrix its asymptotic distribution can be readily derived from the asymptotic joint normality of the elements of this matrix. For normal populations the covariance matrix in the asymptotic distribution of the sample covariance matrix is particularly simple and is given, for example, in Anderson (1958). When the underlying distribution is non-normal the limiting covariance matrix depends on the fourth order mixed moments of the distribution. Interpreting fourth order moments and understanding how they affect asymptotic distributions is, however, an extremely difficult problem.

It would seem that if a procedure is sensitive to the assumption of multivariate normality then this should be apparent by studying distributions which are "close" to the normal distribution. In this paper we will concentrate on a particular class of non-normal models, namely the class of elliptical distributions. These distributions, whose contours of equal density have the same ellipti-

---

*The preparation of this paper was supported by National Science Foundation Grant MSF 78-18583.

cal shape as the normal, provide attractive and intuitively appealing alternatives to multivariate normality. We will examine the way they affect the distributions of ordinary, multiple and canonical correlation coefficients, and standard inferences based on them.

2. ELLIPTICAL DISTRIBUTIONS

The class of elliptical distributions was first studied in detail by Kelker (1970). Here we review the definition and some properties.

Elliptical distributions can be defined in various ways; we will assume the existence of a density function.

Definition The $m \times 1$ random vector $\underset{\sim}{X}$ has an elliptical distribution with parameters $\underset{\sim}{\mu}(m \times 1)$ and $V(m \times m)$ if its density function is of the form

$$c_m (\det V)^{-\frac{1}{2}} h((\underset{\sim}{x}-\underset{\sim}{\mu})' V^{-1} (\underset{\sim}{x}-\underset{\sim}{\mu}))$$

for some function h, where V is positive definite.

If $\underset{\sim}{X}$ has an elliptical distribution we will write $\underset{\sim}{X} \sim E_m(\underset{\sim}{\mu}, V)$. If $\underset{\sim}{\mu}=\underset{\sim}{0}$, $V=\alpha I_m$ the distribution of $\underset{\sim}{X}$ is called spherical. A more general definition is that $\underset{\sim}{X}$ has a spherical distribution if $\underset{\sim}{X}$ and $H\underset{\sim}{X}$ have the same distribution, for all $m \times m$ orthogonal matrices H.

Familiar examples of elliptical distributions are:

(i) The m-variate normal $N_m(\underset{\sim}{\mu}, \Sigma)$ distribution

$$(2\pi)^{-\frac{1}{2}m} (\det \Sigma)^{-\frac{1}{2}} \exp[-\tfrac{1}{2}(\underset{\sim}{x}-\underset{\sim}{\mu})'\Sigma^{-1}(\underset{\sim}{x}-\underset{\sim}{\mu})]$$

(ii) The m-variate t distribution on n degrees of freedom

$$\frac{\Gamma[\frac{1}{2}(n+m)]}{\Gamma(\frac{1}{2}n)(n\pi)^{\frac{1}{2}m}} (\det V)^{-\frac{1}{2}} (1 + \frac{1}{n} \underset{\sim}{x}' V^{-1} \underset{\sim}{x})^{-\frac{1}{2}(n+m)}$$

(iii) The ε-contaminated normal distribution

$$(1-\varepsilon) N_m(\underset{\sim}{0}, \Sigma) + \varepsilon N_m(\underset{\sim}{0}, \sigma^2 \Sigma)$$

We now list some of the basic properties of elliptical distributions. If $\underset{\sim}{X} \sim E_m(\underset{\sim}{\mu}, v)$ then:

1) The characteristic function of $\underset{\sim}{X}$ is

$$\phi(\underset{\sim}{t}) = E[e^{i\underset{\sim}{t}'\underset{\sim}{X}}] = e^{i\underset{\sim}{t}'\underset{\sim}{\mu}} \,\Psi(\underset{\sim}{t}'V\underset{\sim}{t})$$

for some function Ψ.

2) If they exist, $E(\underset{\sim}{X}) = \underset{\sim}{\mu}$ and $Cov(\underset{\sim}{X}) = \alpha V$ for some constant α. In terms of the characteristic function this constant is $\alpha = -2\Psi'(0)$.

If $\underset{\sim}{X} \sim E_m(\underset{\sim}{\mu}, V)$ and $\underset{\sim}{X}$, $\underset{\sim}{\mu}$ and V are partitioned as

$$\underset{\sim}{X} = \begin{pmatrix} \underset{\sim}{X}_1 \\ \underset{\sim}{X}_2 \end{pmatrix}, \quad \underset{\sim}{\mu} = \begin{pmatrix} \underset{\sim}{\mu}_1 \\ \underset{\sim}{\mu}_2 \end{pmatrix}, \quad V = \begin{bmatrix} V_{11} & V_{12} \\ V_{21} & V_{22} \end{bmatrix}$$

where $\underset{\sim}{X}_1$ and $\underset{\sim}{\mu}_1$ are kx1 and V_{11} is kxk then:

3) $\underset{\sim}{X}_1 \sim E_k(\underset{\sim}{\mu}_1, V_{11})$, and its density function has the same functional form as the density function of $\underset{\sim}{X}$.

4) The conditional distribution of $\underset{\sim}{X}_1$ given $\underset{\sim}{X}_2$ is k-variate elliptical with

$$E(\underset{\sim}{X}_1 | \underset{\sim}{X}_2) = \underset{\sim}{\mu}_1 + V_{12} V_{22}^{-1} (\underset{\sim}{X}_2 - \underset{\sim}{\mu}_2) \quad \text{and}$$

$$Cov(\underset{\sim}{X}_1 | \underset{\sim}{X}_2) = g(\underset{\sim}{X}_2)\,(V_{11} - V_{12}V_{22}^{-1}V_{21})$$

for some function g.

5) If $\underset{\sim}{X} \sim E_m(\underset{\sim}{0}, I_m)$ with density function $c_m h(\underset{\sim}{x}'\underset{\sim}{x})$ and

$$X_1 = r \sin\theta_1 \sin\theta_2 \cdots \sin\theta_{m-2} \sin\theta_{m-1}$$

$$X_2 = r \sin\theta_1 \sin\theta_2 \cdots \sin\theta_{m-2} \cos\theta_{m-1}$$

.
.
.

$$X_{m-1} = r \sin\theta_1 \cos\theta_2$$

$$X_m = r \cos\theta_1$$

$(r>0,\ 0<\theta_i \le \pi \quad i = 1,\ldots,m-2,\ 0<\theta_{m-1} \le 2\pi)$ then $r, \theta_1, \ldots, \theta_{m-1}$ are independent, θ_k has density function proportional to $\sin^{m-1-k}\theta_k$, and $r^2 = \underset{\sim}{X}'\underset{\sim}{X}$ has density function

$$\frac{c_m \pi^{\frac{1}{2}m}}{\Gamma(\frac{1}{2}m)} (r^2)^{\frac{1}{2}m-1}\, h(r^2)$$

Many results which hold for normal distributions hold also for elliptical and spherical distributions. The following property was proved by Kariya and Eaton (1977).

6) Let $\underset{\sim}{X} \sim E_m(\underset{\sim}{0}, I_m)$ with $P(\underset{\sim}{X}=\underset{\sim}{0}) = 0$.

(a) If $W = \dfrac{\underset{\sim}{\alpha}'\underset{\sim}{X}}{||\underset{\sim}{X}||}$ where $\underset{\sim}{\alpha}'\underset{\sim}{\alpha} = 1$ then

$$Y = (m-1)^{\frac{1}{2}} \frac{W}{(1-W^2)^{\frac{1}{2}}} \sim t_{m-1}$$

(b) If B is an mxm symmetric idempotent matrix of rank k then

$$Z = \frac{\underset{\sim}{X}' B \underset{\sim}{X}}{\|\underset{\sim}{X}\|^2} \sim \text{Beta}(\tfrac{1}{2}k, \tfrac{1}{2}(m-k)) .$$

Both (a) and (b) are proved by noting that Y and Z are functions of $\underset{\sim}{X}/\|\underset{\sim}{X}\|$ which is uniformly distributed on the unit sphere in m-space.

There is a growing literature on spherical and elliptical distributions; as well as the papers by Kelker (1970) and Kariya and Eaton (1977) already mentioned, a useful review paper by Devlin, Gnanadesikan and Kettenring (1976) gives many additional references.

3. ASYMPTOTIC NORMALITY OF A SAMPLE COVARIANCE MATRIX

In order to derive asymptotic distributions of statistics which are functions of the elements of a sample covariance matrix we need the asymptotic distribution of this matrix.

Let $\underset{\sim}{X}_1, \underset{\sim}{X}_2, \ldots$ be a sequence of independent and identically distributed mx1 vectors with finite fourth moments and mean $\underset{\sim}{\mu}$ and covariance matrix Σ. Put $n=N-1$ and let $S(n)$ be the sample covariance matrix

$$S(n) = \frac{1}{n} \sum_{i=1}^{N} (\underset{\sim}{X}_i - \overline{\underset{\sim}{X}}_N)(\underset{\sim}{X}_i - \overline{\underset{\sim}{X}}_N)'$$

where $\overline{\underset{\sim}{X}}_N = N^{-1} \sum_{i=1}^{N} \underset{\sim}{X}_i$. For any pxq matrix T, vec(T) denotes the pqx1 vector formed by stacking the columns of T under each other. A straight-forward application of the multivariate central limit theorem shows that, as $n \to \infty$, the asymptotic distribution of $\text{vec}(V(n))$, where

$$V(n) = n^{\frac{1}{2}}[S(n) - \Sigma] ,$$

is m^2-variate normal with mean $\underset{\sim}{0}$ and covariance matrix

$$V = \text{Cov}\,[\text{vec}((\underset{\sim}{X}_1 - \underset{\sim}{\mu})(\underset{\sim}{X}_1 - \underset{\sim}{\mu})')] .$$

The elements of V, expressed in terms of the cumulants of the distribution of $\underset{\sim}{X}_1$, are given by

$$\text{Cov}(u_{ij}(n), u_{k\ell}(n)) = \kappa_{1111}^{ijk\ell} + \kappa_{11}^{ik}\kappa_{11}^{j\ell} + \kappa_{11}^{i\ell}\kappa_{11}^{jk} \tag{1}$$

Here $\kappa^{i\,j\,k\,\ell}_{r_1 r_2 r_3 r_4}$ denote the cumulants of the joint distribution of X_i, X_j, X_k, X_ℓ. If their joint characteristic function is $\phi(t_i, t_j, t_k, t_\ell)$ these are given by

$$\log \phi = \sum_{r_1,\ldots,r_4=0} \kappa^{i\,j\,k\,\ell}_{r_1 r_2 r_3 r_4} \frac{(it_i)^{r_1} (it_j)^{r_2} (it_k)^{r_3} (it_\ell)^{r_4}}{r_1!\, r_2!\, r_3!\, r_4!}$$

In this formula the convention is that if any of the variables are identical the subscripts are amalgamated e.g. $\kappa^{iijk}_{1111} = \kappa^{ijk}_{211}$, $\kappa^{ii}_{11} = \kappa^{i}_{2} = \mathrm{Var}(X_i)$. The covariances (1) have been given by Cook (1951); see also Kendall and Stuart (1969). The skewness γ_1^i and kurtosis γ_2^i of the (marginal) distribution of X_i are

$$\gamma_1^i = \frac{\kappa_3^i}{(\kappa_2^i)^{\frac{3}{2}}}, \qquad \gamma_2^i = \frac{\kappa_4^i}{(\kappa_2^i)^2}$$

In general, it is very difficult to get any feeling for how a particular non-normal model affects standard inference procedures because of the effect of the fourth order cumulants which appear in the asymptotic distributions. If attention is restricted to the class of elliptical distributions, however, things are much simpler because all fourth order cumulants are determined by a simple parameter. If $\underset{\sim}{X} \sim E_m(\underset{\sim}{\mu}, V)$ with characteristic function $\phi(\underset{\sim}{t}) = e^{i\underset{\sim}{\mu}'\underset{\sim}{t}}\psi(\underset{\sim}{t}'V\underset{\sim}{t})$, covariance matrix $\Sigma = -2\psi'(0)V = (\sigma_{ij})$, and finite fourth moments, then:

(a) The marginal distributions of X_j $(j=1,\ldots,m)$ all have zero skewness and the <u>same</u> kurtosis

$$\gamma_2^j = \frac{\kappa_4^j}{(\kappa_2^j)^2} = \frac{3[\psi''(0) - \psi'(0)^2]}{\psi'(0)^2} = 3\kappa \text{ , say } (j=1,\ldots,m)$$

(b) All fourth order cumulants are determined by this <u>kurtosis parameter</u> κ as

$$\kappa^{ijk\ell}_{1111} = \kappa(\sigma_{ij}\sigma_{k\ell} + \sigma_{ik}\sigma_{j\ell} + \sigma_{i\ell}\,\sigma_{jk}) \tag{2}$$

It then follows that if $\underset{\sim}{X}_1, \underset{\sim}{X}_2, \ldots$ is a sequence of independent random vectors with identical elliptical distributions, having covariance matrix $\Sigma = (\sigma_{ij})$ and kurtosis parameter κ then the asymptotic distribution of

$$U(n) = n^{\frac{1}{2}}[S(n) - \Sigma] .$$

is normal with mean 0 and covariances

$$\mathrm{Cov}\,(u_{ij}(n),u_{k\ell}(n)) = \kappa(\sigma_{ij}\sigma_{k\ell} + \sigma_{ik}\sigma_{j\ell} + \sigma_{i\ell}\sigma_{jk}) + \sigma_{ik}\sigma_{j\ell} + \sigma_{i\ell}\sigma_{jk} \;.$$

Note that when the underlying distribution is normal, $\kappa = 0$.

4. THE ORDINARY CORRELATION COEFFICIENT

4.1 The Case of Independence

Consider N pairs of variables (X_i,Y_i) , $i = 1,\ldots,N$ with sample correlation coefficient

$$r = \frac{\sum_{i=1}^{N} (X_i-\bar{X})(Y_i-\bar{Y})}{\left[\sum_{i=1}^{N} (X_i-\bar{X})^2 \sum_{i=1}^{N} (Y_i-\bar{Y})^2\right]^{\frac{1}{2}}}$$

The assumption that is commonly made is that the N pairs are independent $N_2(\underset{\sim}{\mu},\Sigma)$ variables, where

$$\Sigma = \begin{bmatrix} \sigma_1^2 & \rho\sigma_1\sigma_2 \\ \rho\sigma_1\sigma_2 & \sigma_2^2 \end{bmatrix}$$

In this case the X's are independent of the Y's when $\rho = 0$. If, in general, we assume that the X's are independent of the Y's , the normality assumption is not important as long as one set of these variables has a spherical distribution.

Theorem 4.1 (Kariya and Eaton (1977))

If $\underset{\sim}{X} = (X_1,\ldots,X_N)'$ and $\underset{\sim}{Y} = (Y_1,\ldots,Y_N)'$ are independent $(N>2)$, where $\underset{\sim}{X}$ has a spherical distribution with $P(\underset{\sim}{X}=\underset{\sim}{0})=0$ and $\underset{\sim}{Y}$ has any distribution with $P(\underset{\sim}{Y}=k\underset{\sim}{1}$ for some $k)=0$ then

$$(N-2)^{\frac{1}{2}}\,\frac{r}{(1-r^2)^{\frac{1}{2}}} \sim t_{N-2}$$

4.2 Asymptotic Distributions

Let $S(n) = (s_{ij}(n))$ be the 2x2 covariance matrix formed from a sample of size $N = n+1$ from a bivariate elliptical distribution with covariance matrix

$$\Sigma = \begin{bmatrix} 1 & \rho \\ \rho & 1 \end{bmatrix}$$

and kurtosis parameter κ . Put

$$U = \begin{bmatrix} u_{11} & u_{12} \\ u_{21} & u_{22} \end{bmatrix} = n^{\frac{1}{2}} [S(n) - \Sigma] ;$$

the asymptotic distribution of $\underset{\sim}{u} \equiv (u_{11}\ u_{12}\ u_{22})'$ is normal, mean $\underset{\sim}{0}$ and covariance matrix

$$V = \begin{bmatrix} 2+3\kappa & (2+3\kappa)\rho & 2\rho^2+\kappa(1+2\rho^2) \\ (2+3\kappa)\rho & \kappa(1+2\rho^2)+(1+\rho^2) & (2+3\kappa)\rho \\ 2\rho^2+\kappa(1+2\rho^2) & (2+3\kappa)\rho & 2+3\kappa \end{bmatrix} \quad (3)$$

In terms of the elements of U the sample correlation coefficient $r(n)$ can be expanded as

$$\begin{aligned} r(n) &= s_{12}(n)[s_{11}(n)s_{22}(n)]^{-\frac{1}{2}} \\ &= (\rho + n^{-\frac{1}{2}}u_{12})(1 + n^{-\frac{1}{2}}u_{11})^{-\frac{1}{2}} (1 + n^{-\frac{1}{2}}u_{22})^{-\frac{1}{2}} \\ &= \rho + n^{-\frac{1}{2}}(u_{12} - \tfrac{1}{2}\rho u_{11} - \tfrac{1}{2}\rho u_{22}) + 0_p(n^{-1}) \quad . \end{aligned}$$

Hence

$$\begin{aligned} n^{\frac{1}{2}}[r(n)-\rho] &= u_{12} - \tfrac{1}{2}\rho u_{11} - \tfrac{1}{2}\rho u_{22} + 0_p(n^{-\frac{1}{2}}) \\ &= \underset{\sim}{\alpha}'\underset{\sim}{u} + 0_p(n^{-\frac{1}{2}}) \end{aligned}$$

where $\underset{\sim}{\alpha} = (-\tfrac{1}{2}\rho 1 - \tfrac{1}{2}\rho)'$. The asymptotic distribution of $\underset{\sim}{\alpha}'\underset{\sim}{u}$ is normal, mean 0 and variance

$$\underset{\sim}{\alpha}'V\underset{\sim}{\alpha} = (1+\kappa)(1-\rho^2)^2 \quad ,$$

proving the following result.

<u>Theorem 4.2</u>

Let $r(n)$ be the correlation coefficient formed from a sample of size $n+1$ from a bivariate elliptical distribution with correlation coefficient ρ and kurtosis parameter κ . Then, as $n\to\infty$.

$$\frac{n^{\frac{1}{2}} [r(n) - \rho]}{1 - \rho^2} \to N(0,1+\kappa)$$

in distribution.

When the elliptical distribution in Theorem 4.2 is normal the kurtosis parameter κ is zero and the asymptotic distribution is $N(0,1)$. In this situation Fisher (1915) suggested the statistic

$$z = \tanh^{-1} r = \tfrac{1}{2} \ln \frac{1+r}{1-r} ,$$

since this approaches normality much faster than r , with an asymptotic variance which is independent of ρ . In this connection a useful reference is Hotelling (1953). For elliptical distributions, a similar result holds.

Theorem 4.3

Assume that the conditions of Theorem 4.2 hold and put

$$z(n) = \tanh^{-1} r(n) , \; \xi = \tanh^{-1} \rho .$$

Then, as $n\to\infty$,

$$n^{\frac{1}{2}}[z(n) - \xi] \to N(0,1+\kappa) \quad \text{in distribution.}$$

When the elliptical distribution is normal we have $\kappa = 0$ and the asymptotic distribution is $N(0,1)$. In this particular case z is the maximum likelihood estimate of ξ . For general non-normal distributions Gayen (1951) has obtained expressions for the mean, variance, skewness and kurtosis of z . These have been used by Devlin, Gnanadesikan and Kettenring (1976) to study Fisher's z-transformation for some specified elliptical distributions. They state that "the main effect of the elliptically constrained departures from normality appear to be to increase the variability of z" and conclude that the distribution of z can be approximated quite well in many situations, even for small sample sizes, by taking z to be normal with mean ξ and variance

$$\mathrm{Var}(z) = \frac{1}{n-2} + \frac{\kappa}{n+2} \; (n=N-1) .$$

For related work the reader is referred to Kowalski (1972) and Duncan and Layard (1973).

5. THE MULTIPLE CORRELATION COEFFICIENT

5.1 The Case of Independence

Suppose that the mx1 random vector $(Y,\underset{\sim}{X}')'$, where $\underset{\sim}{X}$ is $(m-1)\times 1$, has covariance matrix

$$\Sigma = \begin{bmatrix} \sigma_{11} & \underset{\sim}{\sigma}_{12}' \\ \underset{\sim}{\sigma}_{12} & \Sigma_{22} \end{bmatrix}$$

where Σ_{22} is (m-1)x(m-1). The multiple correlation coefficient between Y and $\underset{\sim}{X}$ is

$$\bar{R} = \left(\frac{\underset{\sim}{\sigma}'_{12}\,\Sigma_{22}^{-1}\,\underset{\sim}{\sigma}_{12}}{\sigma_{11}}\right)^{\frac{1}{2}}$$

Now consider N mx1 vectors

$$\begin{pmatrix} Y_1 \\ \underset{\sim}{X}_1 \end{pmatrix}, \ldots, \begin{pmatrix} Y_N \\ \underset{\sim}{X}_N \end{pmatrix}$$

and form the mxN matrix

$$Z = \begin{bmatrix} Y_1 & \cdots\cdots\cdots & Y_N \\ \underset{\sim}{X}_1 & & \underset{\sim}{X}_N \end{bmatrix} = \begin{bmatrix} \underset{\sim}{Y}' \\ X' \end{bmatrix}$$

where $\underset{\sim}{Y}$ is Nx1 and X is Nx(m-1) . Let A be the usual matrix of sums of squares and products

$$A = Z(I_N - \frac{1}{N}\underset{\sim}{1}\underset{\sim}{1}')Z' = \begin{bmatrix} a_{11} & \underset{\sim}{a}'_{12} \\ \underset{\sim}{a}_{12} & A_{22} \end{bmatrix} \quad (4)$$

where A_{22} is (m-1)x(m-1) and $\underset{\sim}{1} = (1,1,..,1)'$ is Nx1 . The sample multiple correlation coefficient is

$$R = \left(\frac{\underset{\sim}{a}'_{12}\,A_{22}^{-1}\,\underset{\sim}{a}_{12}}{a_{11}}\right)^{\frac{1}{2}} \quad (5)$$

An assumption which is commonly made is that the N columns of Z are independent $N_m(\underset{\sim}{\mu},\Sigma)$ random vectors. In this case the Y's are independent of the $\underset{\sim}{X}$'s when $\bar{R} = 0$. If, in general, we assume that the Y's are independent of the $\underset{\sim}{X}$'s , the normality assumption is not important as long as $\underset{\sim}{Y}$ has a spherical distribution. The following theorem is implicit in Kariya and Eaton (1977) although not explicitly stated.

<u>Theorem 5.1</u>

Let $\underset{\sim}{Y}$ have an N-variate spherical distribution with $P(\underset{\sim}{Y}=\underset{\sim}{0}) = 0$ and let X be an Nx(m-1) random matrix independent of $\underset{\sim}{Y}$ and of rank m-1 with probability one. If R is the sample multiple correlation coefficient given by (5) than $R^2 \sim \text{Beta}(\frac{1}{2}(m-1),\frac{1}{2}(N-m))$, or equivalently,

$$\frac{N-m}{m-1} \cdot \frac{R^2}{1-R^2} \sim F_{n-1,N-m} \ .$$

Proof

Write the matrix A given by (4) as $A=Z(I-M)Z'$ where $M = N^{-1} \underset{\sim}{1}\underset{\sim}{1}'$ and $Z = \begin{bmatrix} \underset{\sim}{Y}' \\ X' \end{bmatrix}$. Then

$$R^2 = \frac{\underset{\sim}{Y}'(I-M)X[X'(I-M)X]^{-1} X'(I-M)\underset{\sim}{Y}}{\underset{\sim}{Y}'(I-M)\underset{\sim}{Y}} \tag{6}$$

Since $I-M$ is idempotent of rank $N-1$ there exists an orthogonal NxN matrix such that

$$H(I-M)H' = \begin{bmatrix} I_{N-1} & \underset{\sim}{0} \\ \underset{\sim}{0}' & 0 \end{bmatrix} .$$

Put $\underset{\sim}{U} = H\underset{\sim}{Y}$ and $\underset{\sim}{V} = HX$ and partition $\underset{\sim}{U}, \underset{\sim}{V}$ as

$$\underset{\sim}{U} = \begin{pmatrix} \underset{\sim}{U}^* \\ U_N \end{pmatrix} , \quad V = \begin{pmatrix} V^* \\ \underset{\sim}{V}_N' \end{pmatrix}$$

where $\underset{\sim}{U}^*$ is $(N-1)$x1 and V^* is $(N-1)$x$(m-1)$. Substituting for $\underset{\sim}{Y}$ and X in (6) then gives

$$R^2 = \frac{\underset{\sim}{U}^{*\prime} V^*(V^{*\prime}V^*)^{-1} V^{*\prime} \underset{\sim}{U}^*}{\underset{\sim}{U}^{*\prime} \underset{\sim}{U}^*}$$

Now $V^*(V^{*\prime}V^*)^{-1}V^{*\prime}$ is idempotent of rank $m-1$ and is independent of $\underset{\sim}{U}^*$ which has an $(N-1)$-variate spherical distribution. Conditioning on V^* , property 6(b) of Section 2 with $B = V^*(V^{*\prime}V^*)^{-1}V^{*\prime}$ then shows that $R^2 \sim \text{Beta}(\tfrac{1}{2}(m-1),\tfrac{1}{2}(N-m))$, completing the proof.

5.2 Asymptotic Distributions

Suppose that the mx1 random vector $(Y \ \underset{\sim}{X}')'$, where $\underset{\sim}{X}$ is $(m-1)$x1 , has an elliptical distribution with covariance matrix

$$\Sigma = \begin{bmatrix} 1 & \underset{\sim}{P}' \\ \underset{\sim}{P} & I_{m-1} \end{bmatrix}$$

where $\underset{\sim}{P} = (\bar{R},0,\ldots,0)'$, and kurtosis parameter κ (We are assuming, without loss of generality, that Σ is in canonical form). Let $S(n) = (s_{ij}(n))$ be the covariance matrix formed from a sample of size $N = n+1$ from this distribution and partition $S(n)$ as

$$S(n) = \begin{bmatrix} s_{11} & \underset{\sim}{s}_{12}' \\ \underset{\sim}{s}_{12} & S_{22} \end{bmatrix} ,$$

so that

$$R^2 = \frac{\underset{\sim}{s}_{12}' S_{22}^{-1} \underset{\sim}{s}_{12}}{s_{11}}$$

Put

$$U = \begin{bmatrix} u_{11} & \underset{\sim}{u}_{12}' \\ \underset{\sim}{u}_{12} & U_{22} \end{bmatrix} , \tag{7}$$

where

$$u_{11} = \frac{n^{\frac{1}{2}}(s_{11}-1)}{1-\bar{R}^2}$$

$$\underset{\sim}{u}_{12} = \left(\frac{n}{1-\bar{R}^2}\right)^{\frac{1}{2}} (I-\underset{\sim}{P}\underset{\sim}{P}')^{-\frac{1}{2}}(\underset{\sim}{s}_{12}-\underset{\sim}{P}) ,$$

and

$$U_{22} = n^{\frac{1}{2}}(I-\underset{\sim}{P}\underset{\sim}{P}')^{-\frac{1}{2}} (S_{22}-I)(I-\underset{\sim}{P}\underset{\sim}{P}')^{-\frac{1}{2}} .$$

In terms of the elements of U , R^2 can be expanded as

$$\begin{aligned} R^2 &= s_{11}^{-1} \underset{\sim}{s}_{12}' S_{22}^{-1} \underset{\sim}{s}_{12} \\ &= \bar{R}^2 + \frac{1}{2n^{\frac{1}{2}}} \bar{R}(1-\bar{R}^2)(u_{12} - \tfrac{1}{2}\bar{R}u_{11} - \tfrac{1}{2}\bar{R}u_{22}) + 0_p(n^{-1}) . \end{aligned} \tag{8}$$

Hence, if $\bar{R} \neq 0$,

$$\begin{aligned} \frac{n^{\frac{1}{2}}(R^2-\bar{R}^2)}{2\bar{R}(1-\bar{R}^2)} &= u_{12} - \tfrac{1}{2}\bar{R}u_{11} - \tfrac{1}{2}\bar{R}u_{22} + 0_p(n^{-\frac{1}{2}}) \\ &= \underset{\sim}{\alpha}'\underset{\sim}{u} + 0_p(n^{-\frac{1}{2}}) \end{aligned}$$

where $\underset{\sim}{\alpha} = (-\tfrac{1}{2}\bar{R},1,-\tfrac{1}{2}\bar{R})'$, $\underset{\sim}{u} = (u_{11},u_{12},u_{22})'$. The asymptotic distribution of $\underset{\sim}{u}$ is normal with mean $\underset{\sim}{0}$ and covariance matrix $(1-\bar{R}^2)^{-2}\ V$, where V is given by (3) , with ρ replaced by $\bar{R}$. Hence the asymptotic distribution of $\underset{\sim}{\alpha}'\underset{\sim}{u}$ is normal, mean 0 and variance $\underset{\sim}{\alpha}'V\underset{\sim}{\alpha} = 1+\kappa$, yielding the following result.

Theorem 5.2

When sampling from an elliptical distribution with multiple correlation coefficient $\bar{R} \neq 0$ and kurtosis parameter κ then, as $n \to \infty$,

$$\frac{n^{\frac{1}{2}}(R^2 - \bar{R}^2)}{2\bar{R}(1-\bar{R}^2)} \to N(0, 1+\kappa) \quad \text{in distribution.}$$

We now turn to the asymptotic distribution of R^2 in the null case when $\bar{R} = 0$. In this situation it is clear that in the expansion (7) for R^2 we need the term of order n^{-1}. Defining the matrix U as in (7), but with $\bar{R} = 0$, we have

$$\begin{aligned} R^2 &= s_{11}^{-1}\, s_{12}'\, S_{22}^{-1}\, s_{12} \\ &= (1 + n^{-\frac{1}{2}}u_{11})^{-1}(n^{-\frac{1}{2}}\underset{\sim}{u}_{12}')(I + n^{-\frac{1}{2}}U_{22})^{-1}(n^{-\frac{1}{2}}\underset{\sim}{u}_{12}) \\ &= \frac{1}{n}[1 - n^{-\frac{1}{2}}u_{11} + O_p(n^{-1})]\ \underset{\sim}{u}_{12}'[I - n^{-\frac{1}{2}}U_{22} + O_p(n^{-1})]\underset{\sim}{u}_{12} \ . \end{aligned}$$

Hence

$$nR^2 = \underset{\sim}{u}_{12}'\, \underset{\sim}{u}_{12} + O_p(n^{-\frac{1}{2}}) \ ,$$

where the asymptotic distribution of $\underset{\sim}{u}_{12} = n^{\frac{1}{2}}\underset{\sim}{s}_{12}$ is $(m-1)$-variate normal with mean $\underset{\sim}{0}$ and covariance matrix $(1+\kappa)I$. This yields:

Theorem 5.3

With the assumptions of Theorem 5.2 but with $\bar{R} = 0$,

$$\frac{nR^2}{1+\kappa} \to \chi^2_{m-1} \quad \text{in distribution.}$$

When the elliptical distribution is normal we have $\kappa = 0$ and then the asymptotic distribution of nR^2 is χ^2_{m-1}. This is a special case of a result due to Fisher (1928) who established that if $n \to \infty$ and $\bar{R} \to 0$ in such a way that $n\bar{R}^2 = \delta$ (fixed) then the asymptotic distribution of nR^2 is $\chi^2_m(\delta)$ i.e. non-central χ^2 on m degrees of freedom and noncentrality δ. A similar result holds also for elliptical distributions.

Theorem 5.4

With the assumptions of Theorem 5.2 but with $n\bar{R}^2 = \delta$ (fixed),

$$\frac{nR^2}{1+\kappa} \to \chi^2_{m-1}(\delta^*) \quad \text{in distribution,}$$

where $\delta^* = \delta/(1+\kappa)$.

It is natural to ask whether Fisher's variance-stabilizing transformation, which works so well in the case of an ordinary correlation coefficient, is useful in the context of multiple correlation. The answer is yes, as long as $\bar{R}>0$.

Theorem 5.5

Assume the assumptions of Theorem 5.2 hold and put

$$z = \tanh^{-1} R \ , \ \xi = \tanh^{-1} \bar{R} \ .$$

Then, as $n\to\infty$, $n^{\frac{1}{2}}(z-\xi) \to N(0,1+\kappa)$ in distribution.

For further results on asymptotic distributions for R the reader is referred to Gajjar (1967). Some of the results presented here appear in Muirhead and Waternaux (1980).

Consider the problem of testing $H_0:\bar{R} = 0$ against $H:\bar{R}>0$ based on a sample of size $N = n+1$ from an elliptical distribution with kurtosis parameter κ . From Theorem 5.3 the asymptotic distribution of $nR^2/(1+\kappa)$ when H_0 is true is χ^2_{m-1} so that an approximate test of size α is to reject H_0 if $nR^2/(1+\kappa) > c_{m-1}(\alpha)$ where $c_{m-1}(\alpha)$ denotes the upper α% point of the χ^2_{m-1} distribution. (If κ is not known it could be replaced by a consistent estimate.) The power function of this test could be calculated approximately using Theorem 5.4 for alternatives $\bar{R}$ which are close to zero and Theorem 5.2 for alternatives $\bar{R}$ further away from zero. Theorems 5.2 or 5.5 could also be used for testing K_0: $\bar{R} = \bar{R}_0(>0)$ against K: $\bar{R} \neq \bar{R}_0$. Putting $\xi_0 = \tanh^{-1}\bar{R}_0$ we know from Theorem 5.5 that when K_0 is true the distribution of $z = \tanh^{-1} R$ is approximately $N(\xi_0,\frac{1+\kappa}{n})$ so that an approximate test of size α is to reject K_0 if

$$\frac{n^{\frac{1}{2}}|z-\xi_0|}{(1+\kappa)^{\frac{1}{2}}} \geq d_\alpha$$

where d_α is the two-tailed α% point of the $N(0,1)$ distribution. It should be remembered that the asymptotic normality of R and z holds only if $\bar{R} \neq 0$ and the normal approximation is not likely to be much good if $\bar{R}$ is close to zero.

6. CANONICAL CORRELATIONS

6.1 Asymptotic Distributions of the Sample Coefficients

Let the $(p+q)\times 1$ random vector $(\underset{\sim}{y}_1' \ \underset{\sim}{y}_2')'$, where $\underset{\sim}{y}_1$ is $p\times 1$ and $\underset{\sim}{y}_2$ is $q\times 1$ $(p\leq q)$, have covariance matrix

$$\Sigma = \begin{bmatrix} \Sigma_{11} & \Sigma_{12} \\ \Sigma_{21} & \Sigma_{22} \end{bmatrix} ,$$

where Σ_{11} is pxp , and let $\rho_1^2,\ldots,\rho_p^2$ $(1>\rho_1 \geq \rho_2 \geq \cdots \geq \rho_p \geq 0)$ be the latent roots of $\Sigma_{11}^{-1}\Sigma_{12}\Sigma_{22}^{-1}\Sigma_{21}$. Then $\rho_1,\ldots,\rho_p$ are the population canonical correlation coefficients. The sample canonical correlation coefficients $r_1,\ldots,r_p$ $(1>r_1>\ldots>r_p>0)$ are the positive square roots of $r_1^2,\ldots,r_p^2$, the latent roots of $S_{11}^{-1}S_{12}S_{22}^{-1}S_{21}$, where S is the sample covariance matrix. Here we review some results of Muirhead and Waternaux (1980) in the area of canonical correlations derived from elliptical samples.

Let S(n) be the covariance matrix formed from a sample of size N = n+1 from an elliptical distribution with covariance matrix (in canonical form)

$$\Sigma = \begin{bmatrix} I_p & P \\ P' & I_q \end{bmatrix} , \text{ where } P = \begin{bmatrix} \rho_1 & & 0 & \vdots \\ & \ddots & & \vdots 0 \\ 0 & & \rho_p & \vdots \end{bmatrix} , \qquad (9)$$

and kurtosis parameter κ . The fourth order cumulants of this distribution are given by (2) and, because Σ has the structure (9) ,

$$\kappa_{22}^{ij} = \kappa(j\neq i, i+p),\ \kappa_{2\,2}^{i\ i+p} = \kappa(1+2\rho_i^2),\ \kappa_{3\,1}^{i\ i+p} = \kappa_{1\,3}^{i\ i+p} = 3\rho_i\kappa ,$$

$$\kappa_{1\,1\,1\,1}^{i\ i+p\ j\ j+p} = \rho_i\rho_j\kappa ,$$

and all other fourth order cumulants are zero.

Suppose that ρ_i is a simple nonzero population coefficient. The corresponding sample coefficient then has the expansion

$$r_i^2 = a_{ii} + \sum_{j\neq i} \frac{a_{ij}\,a_{ji}}{\rho_i^2 - \rho_j^2} + O_p(n^{-\frac{3}{2}})$$

where $A = (a_{ij}) = S_{11}^{-1}S_{12}S_{22}^{-1}S_{21}$. By taking Taylor expansions of the a_{ij} about $S_{11} = I_p$, $S_{12} = P$, $S_{22} = I_q$ an expansion in terms of the elements of S is obtained. Put

$$Z = (z_{ij}) = \begin{bmatrix} Z_{11} & Z_{12} \\ Z_{21} & Z_{22} \end{bmatrix} , \text{ where}$$

$$Z_{11} = n^{\frac{1}{2}}(I-PP')^{-\frac{1}{2}}(S_{11}-I)(I-PP')^{-\frac{1}{2}}$$

$$Z_{12} = n^{\frac{1}{2}}(I-PP')^{-\frac{1}{2}}(S_{12}-P)(I-P'P)^{-\frac{1}{2}} , \text{ and}$$

$$Z_{22} = n^{\frac{1}{2}}(I-P'P)^{-\frac{1}{2}}(S_{22}-I)(I-P'P)^{-\frac{1}{2}} .$$

In terms of these variables we then have

$$\frac{n^{\frac{1}{2}}(r_i^2-\rho_i^2)}{2\rho_i(1-\rho_i^2)} = A_i + O_p(n^{-\frac{1}{2}})$$

where

$$A_i = z_{i,p+i} - \tfrac{1}{2}\rho_i z_{ii} - \tfrac{1}{2}\rho_i z_{p+i,p+i} \quad (i=1,\ldots,p) \ .$$

The A_i's are asymptotically independent $N(0,1+\kappa)$ variables, which yields:

Theorem 6.1

When sampling from an elliptical distribution with distinct non-zero canonical correlation coefficients and kurtosis parameter κ, the asymptotic joint distribution of

$$u_i = \frac{n^{\frac{1}{2}}(r_i^2-\rho_i^2)}{2\rho_i(1-\rho_i^2)} \quad (i=1,\ldots,p)$$

is $N_p(\underset{\sim}{0},(1+\kappa)I_p)$.

When the underlying distribution is normal, $\kappa = 0$ and the u_i's are asymptotically independent $N(0,1)$ variables (Hsu(1941)).

It is interesting to look at the problem of estimating the parameters $\xi_1,\ldots,\xi_p$ defined via the familiar transformation $\xi_i = \tanh^{-1}\rho_i$, where ρ_i is a simple, nonzero coefficient $(i=1,\ldots,p)$. When sampling from a normal population, the maximum likelihood estimate of ξ_i is $\hat{\xi}_i = \tanh^{-1} r_i$, with $E(\hat{\xi}_i) = \xi_i+O(n^{-1})$ and $\mathrm{Var}(\hat{\xi}_i) = \frac{1}{n}+O(n^{-2})$. For general non-normal populations, $E(\hat{\xi}_i) = \xi_i+O(n^{-1})$ while the variance depends on the fourth order cumulants. For elliptical distributions with kurtosis parameter κ , $\mathrm{Var}(\hat{\xi}_i) = (1+\kappa)/n+O(n^{-2})$ so that the $\tanh^{-1}$ transformation stabilizes the variance here to the extent that it does not depend on the population canonical correlation coefficients but only on κ , a measure of the departure from normality.

6.2 Testing the Significance of the Residual Coefficients

Here we consider the problem of testing the null hypothesis H_{p-k} that the smallest p-k population canonical correlation coefficients are zero

$$H_{p-k}: \rho_{k+1} = \ldots = \rho_p = 0 \ .$$

Assuming normality, the likelihood ratio statistic is

$$L_{p-k} = \prod_{i=k+1}^{p} (1-r_i^2)$$

and when H_{p-k} is true,

$$T_{p-k} = -n \log L_{p-k} \to \chi^2_{(p-k)(q-k)}$$

in distribution as $n\to\infty$ (Bartlett (1938)). Assuming that $\rho_1 > .. > \rho_k > 0$, T_{p-k} can be expanded when H_{p-k} is true as

$$T_{p-k} = \sum_{i=k+1}^{p} \sum_{j=k+1}^{q} z^2_{i,p+j} + O_p(n^{-\frac{1}{2}})$$

where $z_{i,p+j} = n^{\frac{1}{2}} s_{i,p+j}$. When the underlying distribution is elliptical with kurtosis parameter κ the asymptotic joint distribution of the $z_{i,p+j}$ in the above sum is $(p-k)(q-k)$-variate normal with mean 0 and covariance matrix $(1+\kappa)I$. This yields:

<u>Theorem 6.2</u>

When sampling from an elliptical distribution with kurtosis parameter κ, then

$$\frac{T_{p-k}}{1+\kappa} \to \chi^2_{(p-k)(q-k)}$$

in distribution when H_{p-k} is true.

Note that if $\hat{\kappa}$ is a consistent estimate of κ when H_{p-k} is true then

$$T^*_{p-k} = T_{p-k}/(1+\hat{\kappa}) \to \chi^2_{(p-k)(q-k)}$$

in distribution.

The results of a Monte Carlo experiment comparing the sampling distributions of T_{p-k} and T^*_{p-k} are given in Muirhead and Waternaux (1980). These results emphatically demonstrate that "the usual test statistic T_{p-k} should be used with extreme care, if at all, for testing the null hypothesis when the multivariate population is non-normal, and in particular is elliptical with longer tails than normal." A better procedure to use the statistic T^*_{p-k} (which amounts to making a simple adjustment to T_{p-k}), since this performs much better than T_{p-k} for long-tailed elliptical distributions. A caveat is probably in order; a test based on T^*_{p-k} may have poorer power properties compared with tests based on some robust estimate of the covariance matrix although if κ is small there probably is not very much difference. It should be remembered that the procedure suggested is an ad-hoc one and in most such procedures there is a trade-off between simplicity and optimality.

A simple method for estimating κ , as well as some non-null distributions of T_{p-k} , may be found in Muirhead and Waternaux (1980).

REFERENCES

[1] Anderson, T.W. (1958) An Introduction to Multivariate Statistical Analysis John Wiley and Sons, New York.

[2] Bartlett, M.S. (1938) Further Aspects of the Theory of Multiple Regression Proc. Camb. Phil. Soc. 34, 33-40.

[3] Cook, M.B. (1951) Bivariate k-statistics and cumulants of their Joint Sampling Distributions Biometrika 38, 179.

[4] Devlin, S.J., Gnanadesikan, R., and Kettenring, J.R. (1976) Some Multivariate Applications of elliptical distributions. In Essays in Probability and Statistics (S. Ideka et al., Eds.), 365-395, Shinko Tsusho Co., Ltd., Tokyo.

[5] Duncan, G.T. and Layard, M.W.J. (1973) A Monte-Carlo study of asymptotically robust tests for correlation coefficients. Biometrika 60, 551-558.

[6] Fisher, R.A. (1915) Frequency distribution of the values of the correlation coefficient in samples from an indefinitely large population. Biometrika 10, 507-521.

[7] Fisher, R.A. (1928) The general sampling distribution of the multiple correlation coefficient. Proc. Roy. Soc. Lond. A 121, 654-673.

[8] Gayen, A.K. (1951) The frequency distribution of the product-moment correlation coefficient in random samples of any size drawn from non-normal universes. Biometrika 38, 219-247.

[9] Hotelling, H. (1953) New light on the correlation coefficient and its transforms. J.R. Statist. Soc. B 15, 193-224.

[10] Hsu, P.L. (1941) On the limiting distribution of the canonical correlations. Biometrika 32, 38-45.

[11] Kariya, T. and Eaton, M.L. (1977) Robust tests for spherical symmetry. Ann. of Statistics 5 206-215.

[12] Kelker, D. (1970) Distribution theory of spherical distributions and a location - scale parameter generalization. Sankhya, A. 32, 419-430.

[13] Kowalski, C.J. (1972) On the effects of non-normality on the distribution of the sample product-moment correlation coefficient. Appl. Statistics 21, 1-12.

[14] Kendall, M.G. and Stuart, A. (1969) The Advanced Theory of Statistics, Vol. I. Hafner, New York.

[15] Muirhead, R.J. and Waternaux, C.M. (1980) Asymptotic distributions in canonical correlation analysis and other multivariate procedures for non-normal populations. Biometrika 67.

Multivariate Statistical Analysis
R.P. Gupta (ed.)

A REVIEW OF STEP-DOWN PROCEDURES FOR MULTIVARIATE ANALYSIS OF VARIANCE*

Govind S. Mudholkar
Department of Statistics
University of Rochester
Rochester, New York
U.S.A.

Perla Subbaiah
Department of Mathematics
Oakland University
Rochester, Michigan
U.S.A.

The step-down procedure for testing the multivariate general linear hypothesis was constructed by J. Roy (1958). Mudholkar and Subbaiah (1975, 1979a) developed the simultaneous confidence intervals associated with this method, and with its variation, the finite intersection tests, proposed by P.R. Krishnaiah (1965a). Some performance properties of J. Roy's procedure are examined by Subbaiah and Mudholkar (1978a). Dempster (1963) discussed the theory and propriety of using the principal variables in step-wise methods. Mudholkar and Subbaiah (1979c) show that the P-values of the tests in the step-down procedure can be combined to obtain asymptotic equivalents of the likelihood ratio test, which is known to be optimal in Bahadur's sense. In this essay these and some other aspects of the step-down methods in MANOVA are reviewed.

1. INTRODUCTION

In most investigations several measurements are obtained on each of a number of individuals or objects, and in general these measurements are of unequal importance to the investigator. However most multivariate methods fail to account for this and treat the variables symmetrically. A notable exception is a class of techniques known as step-down or step-wise procedures, which have several advantages. They are practically simple to implement as they require only the well understood and widely available univariate distributions; they provide a greater resolution and detailed analysis as compared with the usual invariant tests; they can be adapted to deal with the case where the number of response variables is larger than the number of observations; and they can be modified to yield large sample equivalents of the likelihood ratio tests which are known to be optimal. Because of their flexibility and simplicity step-down procedures quickly found their way in a wide variety of applied research; e.g., Sheehan (1968), Wodtke (1967), Donnerstein et al (1972), Smith et al (1972); see also Bock (1975). In practice they are used even when an a priori ordering among the variables is absent.

*Research sponsored in part by the Air Force Office of Scientific Research, Air Force Systems Command, USAF under Grant No. AFOSR-77-3360. The United States Government is authorized to reproduce and distribute reprints for governmental purposes notwithstanding any copyright notation hereon.

Now, suppose that the response variables in an investigation are ordered in a decreasing order of importance or scientific relevance and it is of interest to test certain hypothesis concerning the parameters of the joint distribution of the variables. The step-down or step-wise approach to this hypothesis testing problem consists of first decomposing the hypothesis into component hypotheses involving the parameters of the conditional distributions of the variables conditioned upon the variables earlier in the order. The component hypotheses are then tested sequentially using the conditioning variables as the covariates and the overall hypothesis is rejected if one of these tests is significant.

Even though the step-wise approach is implicit in Rao (1948, 1956), or in the test for additional information discussed in Rao (1973), it was explicitly enunciated and introduced by Rao and Bargmann (1958) in the context of the problem of testing multiple independence in a multivariate normal population using a random sample. Subsequent developments include the following. Bhapkar (1959) discusses the Roy-Bargmann solution in the linear model set up. J. Roy (1958) presents the step-down tests for comparing two covariance matrices and for MANOVA. Dempster (1963) develops a general theory for step-wise procedures based on principal components. Roy and Srivastava (1964) employ step-down techniques in the analysis of hierarchical and p-block multi-response designs. Krishnaiah (1965, 1968, 1969) presents an important variation of the step-down procedure for MANOVA, and considers comparisons of several covariance matrices. Das Gupta (1970) views step-down methods as multiple decision rules. Mietlowski (1974) describes a step-down procedure for testing the homogeneity of variances of p correlated variates with a specific correlation structure.

In this essay we review some aspects of step-wise methods in the framework of the most common multivariate problem, namely, the multivariate analysis of variance. The MANOVA model, the step-down procedure due to J. Roy and two variations of it are given in Section 2. In Section 3 the simultaneous confidence bounds associated with the three procedures are discussed. The performance properties of the step-down tests are summarized in Section 4. The use of principal variables in step-down procedures when the response variables lack a priori ordering or when the number of variables is unduly large is described in Section 5. Finally, in Section 6 we present ways of constructing overall tests by combining the P-values of the component tests. It is observed that many of these modified step-down procedures are asymptotically equivalent to the likelihood ratio test which is known to be optimal in terms of the exact slope.

2. MANOVA MODEL AND THE STEP-DOWN PROCEDURE

The multivariate analysis of variance is probably the most common technique used in the analysis of multivariate data. The models and methods of MANOVA are standard topics in textbooks, handbooks and monographs on multivariate analysis.

The MANOVA problem may be described in many ways: in the regression terminology as in Anderson (1958), in the canonical form discussed in Das Gupta, Anderson, and Mudholkar (1964), in the general linear model framework considered by Mudholkar, Davidson, and Subbaiah (1974b), or in the most common one-way classification set-up as in Rao (1973). We now present the multivariate general linear model and summarize the step-down procedure of J. Roy (1958) and two of its modifications.

2.1 The MANOVA Model: Let $\underset{\sim}{Y}(n \times p)$ denote a matrix whose rows are independently distributed according to a p-variate normal distribution with the same positive definite dispersion matrix $\underset{\sim}{\Sigma}$ and means given by

$$E(\underset{\sim}{Y}) = \underset{\sim}{A}\underset{\sim}{\Theta} \tag{2.1}$$

where $\underset{\sim}{A}(n \times m)$ is a known design matrix of rank $r \leq n - p$, and $\underset{\sim}{\Theta}(m \times p)$ is a matrix of unknown parameters. The MANOVA problem is that of testing the null hypothesis

$$H_0: \underset{\sim}{\Phi} = \underset{\sim}{B}\underset{\sim}{\Theta} = \underset{\sim}{0} \tag{2.2}$$

against the alternative $H_1: \underset{\sim}{\Phi} \neq \underset{\sim}{0}$, where $\underset{\sim}{B}(t \times m)$ is a matrix of full rank. This restriction on the rank of $\underset{\sim}{B}$ will not be assumed in the following sections. The invariant tests of H_0 (Lehmann (1959)) depend on the eigenvalues of $\underset{\sim}{H}\,\underset{\sim}{E}^{-1}$ where $\underset{\sim}{H}$ and $\underset{\sim}{E}$, the S.S.P. matrices "due to hypothesis", and "due to error", respectively, are given by

$$\begin{aligned} \underset{\sim}{H} &= \hat{\underset{\sim}{\Phi}}'\underset{\sim}{W}^{-1}\hat{\underset{\sim}{\Phi}} , \\ \underset{\sim}{E} &= \underset{\sim}{Y}'[\underset{\sim}{I} - \underset{\sim}{A}(\underset{\sim}{A}'\underset{\sim}{A})^{-}\underset{\sim}{A}']\underset{\sim}{Y} , \end{aligned} \tag{2.3}$$

$(\underset{\sim}{A}'\underset{\sim}{A})^{-}$ being a generalized inverse of $(\underset{\sim}{A}'\underset{\sim}{A})$, and $\hat{\underset{\sim}{\Phi}} = \underset{\sim}{B}(\underset{\sim}{A}'\underset{\sim}{A})^{-}\underset{\sim}{A}'\underset{\sim}{Y}$, $\underset{\sim}{W} = \underset{\sim}{B}(\underset{\sim}{A}'\underset{\sim}{A})^{-}\underset{\sim}{B}'$. Roy's largest root $\lambda_{max}(\underset{\sim}{H}\,\underset{\sim}{E}^{-1})$, Hotelling-Lawley's trace $tr(\underset{\sim}{H}\,\underset{\sim}{E}^{-1})$, Wilks' likelihood ratio $|\underset{\sim}{E}|/|\underset{\sim}{H} + \underset{\sim}{E}|$, and Bartlett-Pillai's trace $trE(\underset{\sim}{H} + \underset{\sim}{E})^{-1}$ are among the well known invariant test procedure.

2.2 The Step-down Procedure: Suppose that the p components of the response variables are arranged in a decreasing order with respect to their importance or relevance, and denote $\underset{\sim}{Y} = [\underset{\sim}{y}_1, \ldots, \underset{\sim}{y}_p]$, $\underset{\sim}{\Theta} = [\underset{\sim}{\theta}_1, \ldots, \underset{\sim}{\theta}_p]$, and $\underset{\sim}{Y}_i = [\underset{\sim}{y}_1, \ldots, \underset{\sim}{y}_i], \underset{\sim}{\Theta}_i = [\underset{\sim}{\theta}_1, \ldots, \underset{\sim}{\theta}_i]$, for $i = 1, \ldots, p$. Then the conditional distributions of the elements of $\underset{\sim}{y}_{i+1}$ given $\underset{\sim}{Y}_i$ are univariate normal with common variance $\sigma^2_{i+1} = |\Sigma_{i+1}|/|\Sigma_i|$ and means,

$$E(y_{i+1}|Y_i) = A\eta_{i+1} + Y_i\beta_i \quad , \qquad (2.4)$$

$i = 0,1,\ldots,p-1$, where $\beta_0 = 0$, $\beta_i = \Sigma_i^{-1}(\sigma_{1,i+1},\ldots,\sigma_{i,i+1})'$, $\eta_{i+1} = \theta_{i+1} - \Theta_i\beta_i$, and Σ_i is the first principal minor of order i containing the first i rows and i columns of Σ . It may be noted that the MANOVA hypothesis is true if and only if $BN = 0$, where $N = [\eta_1,\eta_2,\ldots,\eta_p]$. i.e.

$$H_0 \equiv \bigcap_{i=1}^{p} \{H_{0i}: B\eta = 0\} \quad . \qquad (2.5)$$

The step-down procedure consists of the univariate analysis of covariance tests for the hypothesis $H_{01},\ldots,H_{0p}$ conducted sequentially. At every stage of the procedure the responses occuring earlier in the ordering are regarded as the concomitant variables. If $\hat{\eta}_i$ and $s_i^2/(n - r - i + 1)$ are the Gauss-Markoff estimates of η_i and σ_i^2 respectively, and $C_i\,\sigma_i^2$ is the variance covariance matrix of $\beta\hat{\eta}_i$, then the F-statistic for testing H_{0i} is

$$F_i = \frac{(\beta\hat{\eta}_i)'C_i^{-1}(\beta\hat{\eta}_i)/t}{s_i^2/(n - r - i + 1)} \quad . \qquad (2.6)$$

Under H_0 , the p step-down statistics are independently distributed (J. Roy (1958)), and therefore the hypothesis H_0 can be tested at significance level α by distributing the type I error such that $(1 - \alpha) = \prod_{i=1}^{p}(1 - \alpha_i)$, α_i being the level corresponding to H_{0i} . If f_i is the upper $100\alpha_i$ percentage point of the F distribution with d.f. $(t, n - r - i + 1)$, then H_0 is accepted if and only if $F_i \leq f_i$, $i = 1,2,\ldots,p$. We may obtain simultaneous confidence bounds on $a'B\eta_i$,

$$a'B\eta_i \in a'B\hat{\eta}_i \pm \{(a'C_ia)\; s_i^2\; f_i t/(n - r - i + 1)\}^{\frac{1}{2}} \;, \qquad (2.7)$$

$i = 1,2,\ldots,p$, for all $a \in \mathbf{R}^t$ with confidence coefficient $(1 - \alpha)$. It may be noted that these bounds are on functions containing the nuisance parameter Σ .

2.3 Generalized Step-down Procedure: In many an investigation an a priori ordering of the p variables may be absent, but it may be possible only to identify subsets such that the variables within each subset are of comparable importance and the subsets can be ordered. The simplest and most common example of such grouping involves one subset composed of the measurements of primary importance and another subset consisting of readily available, inexpensive "add-on" variables (Jensen (1972)).

Specifically, let k denote the number of subsets, with p_i variables in the ith subset, $i = 1,2,\ldots,k$; $\sum_{i=1}^{k} p_i = p$, and let $\sum_{i=1}^{j} p_i = q_j$. Let $\underset{\sim}{Y} = (\underset{\sim}{Y}_1,\underset{\sim}{Y}_2,\ldots,\underset{\sim}{Y}_k)$ denote the $n \times p$ matrix of observations of order $n \times p$, where $\underset{\sim}{Y}_i$ is the $n \times p_i$ matrix of n observations on the p_i variables in the ith subset. There is a slight overlap of notation regarding the $\underset{\sim}{Y}_i$ and $\underset{\sim}{\textcircled{H}}_i$'s in this section with those in Section 2.2, but it is unlikely to cause ambiguity in the scope of this review. $\underset{\sim}{\Sigma} = (\underset{\sim}{\Sigma}_{ij})$, $\underset{\sim}{\Sigma}_{ij}$ being the $p_i \times p_j$ matrix of covariances of the variables in the ith and j^{th} subsets. The matrix $\underset{\sim}{\textcircled{H}}$ in $E\underset{\sim}{Y} = A\underset{\sim}{\textcircled{H}}$ can be similarly expressed as $\underset{\sim}{\textcircled{H}} = (\underset{\sim}{\textcircled{H}}_1 ,\underset{\sim}{\textcircled{H}}_2,\ldots,\underset{\sim}{\textcircled{H}}_k)$. J. Roy's step-down procedure for testing H_0: $\underset{\sim}{B}\underset{\sim}{\textcircled{H}} = \underset{\sim}{0}$ may be extended by decomposing H_0 into k component hypotheses, and testing them sequentially. Sepcifically, the conditional distributions of the rows of $\underset{\sim}{Y}_i$ $(n \times p_i)$ given $\underset{\sim}{Y}_1,\underset{\sim}{Y}_2,\ldots,\underset{\sim}{Y}_{i-1}$ are independent p_i-variate normal distributions with common dispersion matrix

$$\underset{\sim}{\Sigma}_{ii*} = \underset{\sim}{\Sigma}_{ii} - (\underset{\sim}{\Sigma}_{i1},\underset{\sim}{\Sigma}_{i2},\ldots,\underset{\sim}{\Sigma}_{i,i-1})\, \Sigma_{i-1}^{-1}(\underset{\sim}{\Sigma}_{i1},\underset{\sim}{\Sigma}_{i2},\ldots,\underset{\sim}{\Sigma}_{i,i-1})' \quad ,$$

and means given by

$$E(\underset{\sim}{Y}_i|\underset{\sim}{Y}_1,\ldots,\underset{\sim}{Y}_{i-1}) = \underset{\sim}{A}\ \underset{\sim}{\xi}_i + (\underset{\sim}{Y}_1,\ldots,\underset{\sim}{Y}_{i-1})\underset{\sim}{B}_{i-1} \quad , \tag{2.8}$$

where $\underset{\sim}{\xi}_i = \underset{\sim}{\textcircled{H}}_i - (\underset{\sim}{\textcircled{H}}_1,\underset{\sim}{\textcircled{H}}_2,\ldots\underset{\sim}{\textcircled{H}}_{i-1})\underset{\sim}{B}_{i-1}$, $\underset{\sim}{B}'_{i-1} = (\underset{\sim}{\Sigma}_{i1},\underset{\sim}{\Sigma}_{i2},\ldots,\underset{\sim}{\Sigma}_{i,i-1})\, \underset{\sim}{\Sigma}_{i-1}^{-1}$,

and $\underset{\sim}{\Sigma}_i$ is the first principal minor of order q_i containing the q_i rows and q_i columns of $\underset{\sim}{\Sigma}$. The hypothesis H_0 is true if and only if $H_{0(i)}$: $\underset{\sim}{B}\ \underset{\sim}{\xi}_i = 0$, $i = 1,\ldots,k$. If $\underset{\sim}{T} = \underset{\sim}{H} + \underset{\sim}{E}$, then the "total" and "error" S.S.P. matrices corresponding to the hypothesis $H_{0(i)}$ are given, as in Rao (1973, p.553) by

$$\begin{aligned} \underset{\sim}{T}_{ii*} &= \underset{\sim}{T}_{ii} - (\underset{\sim}{T}_{i1},\underset{\sim}{T}_{i2},\ldots,\underset{\sim}{T}_{i,i-1})\underset{\sim}{T}_{i-1}^{-1}(\underset{\sim}{T}_{i1},\underset{\sim}{T}_{i2},\ldots,\underset{\sim}{T}_{i,i-1})' \\ \underset{\sim}{E}_{ii*} &= \underset{\sim}{E}_{ii} - (\underset{\sim}{E}_{i1},E_{i2},\ldots,E_{i,i-1})E_{i-1}^{-1}(E_{i1},E_{i2},\ldots,E_{i,i-1})' \end{aligned} \tag{2.9}$$

where $\underset{\sim}{T}_i$ and $\underset{\sim}{E}_i$ have the same dimensions as $\underset{\sim}{\Sigma}_i$. Now the k component hypotheses $H_{0(i)}$, $i = 1,2,\ldots,k$ may be tested using any MANOVA statistics, not necessarily the same for each component hypothesis. Then it can be shown that these statistics are independently distributed when H_0 is true. The block structure analogues of J. Roy's step-down procedure based upon these k statistics are the generalized step-down procedures.

If at each stage of this generalized procedure we use the largest root or trace or one of symmetric gauge function tests, then using the results in Mudholkar, Davidson, and Subbaiah (1974b) it is easy to obtain associated sumultaneous confidence bounds. However, as in the confidence bounds given by J. Roy, these will also involve the nuisance parameter $\underset{\sim}{\Sigma}$.

2.4 Finite Intersection Tests: The finite intersection tests for MANOVA introduced by Krishnaiah (1965a, 1965b), differ from the step-down procedure of 2.2 in that the univariate general linear hypothesis arising at each stage of the procedure is treated using tests analogous to Tukey's studentized range test or Dunnett's many-to-one comparison test instead of the variance ratio test. In the union intersection terminology, the finite intersection tests of H_0 are based on the same response-wise finite decomposition of H_0 as in (2.5); but the decomposition, unlike the decomposition in J. Roy's step-down procedure, is finite with respect to contrasts as well. That is,

$$\begin{aligned}\{H_0: \underset{\sim}{B}\underset{\sim}{H} = \underset{\sim}{0}\} &\equiv \bigcap_{i=1}^{p} \{H_{0i}: \underset{\sim}{B}\underset{\sim}{\eta} = \underset{\sim}{0}\} \\ &\equiv \bigcap_{i=1}^{p} \bigcap_{j=1}^{t} \{H_{0ij}: \underset{\sim}{b}_j'\underset{\sim}{\eta}_i = 0\} ,\end{aligned} \tag{2.10}$$

where $\underset{\sim}{b}_1',\ldots,\underset{\sim}{b}_t'$ are the t rows of $\underset{\sim}{B}$. Here we note that B need not be of full rank. The hypothesis H_{0ij} is tested using the statistic

$$F_{ij} = \frac{(\underset{\sim}{b}_j'\hat{\underset{\sim}{\eta}}_i)^2(n - r - i + 1)}{c_{ij}\, s_i^2} \tag{2.11}$$

for $i = 1,\ldots,p$, $j = 1,\ldots,t$, where $\underset{\sim}{b}_j'\hat{\underset{\sim}{\eta}}_i$ is the least squares estimate of $\underset{\sim}{b}_j'\underset{\sim}{\eta}_i$ and $s_i^2/(n - r - i + 1)$ is unbiased estimate of σ_i^2 . The finite intersection procedure consists of accepting H_0 if and only if

$$F_{ij} \leq f_i \ , \ j = 1,\ldots,t \ ; \ i = 1,2,\ldots,p \ . \tag{2.12}$$

The critical constants $f_1,\ldots,f_p$ are obtained such that

$$\Pr(F_{ij} \leq f_i,\ j = 1,\ldots,t;\ i = 1,\ldots,p|H_0)$$
$$= \prod_{i=1}^{p} \Pr(F_{ij} \leq f_i,\ j = 1,\ldots,t|H_0)$$
$$= \prod_{i=1}^{p} (1 - \alpha_i) = 1 - \alpha \ . \tag{2.13}$$

Under H_0, the t statistics $(F_{i1},\ldots,F_{it})$ occuring at the ith stage, $i = 1,\ldots,p$, are distributed according to t-variate F distribution with $(1,\ n - r - i + 1)$ d.f., and the statistics at different stages are independent. Various approximations to the critical constants occuring at different stages are available and are known to be of reasonable accuracy (Armitage and Krishaiah (1978), Cox et al (1979). Alternatively the distribution problem arising here may be dealt with Bonferroni-approximations. The simultaneous confidence bounds on $\underset{\sim}{b}_j'\underset{\sim}{\eta}_i$ associated with the finite intersection tests are

$$\underset{\sim}{b}_j'\underset{\sim}{\eta}_i \in \underset{\sim}{b}_j'\hat{\underset{\sim}{\eta}}_i \pm \{f_i\ c_{ij}\ s_i^2/(n - r - i + 1)\}^{\frac{1}{2}} \ , \tag{2.14}$$

where, again, $\underset{\sim}{\eta}_i$ involve the nuisance parameter $\underset{\sim}{\Sigma}$.

3. THE MANOVA MULTIPLE COMPARISONS USING THE STEP-DOWN PROCEDURES

As noted earlier MANOVA confidence bounds (2.7), or their analogues for the generalized step-down procedure and finite intersection tests, are on parametric functions not only of $\underset{\sim}{\textcircled{H}}$, the parameters of original interest, but also involve the nuisance parameter $\underset{\sim}{\Sigma}$. In view of this Rao (1964, p. 356) observes that "the general drawback of the step-down procedures is that they do not enable us to study the configuration of the mean values of the different populations, which in practical problems is more important than merely establishing differences in the mean values". Mudholkar and Subbaiah (1975, 1979a) derive simultaneous confidence bounds on estimable functions of $\underset{\sim}{\textcircled{H}}$ associated with various step-down methods described in Section 2 and discuss their properties. Some of these results are now summarized. The details for the simultaneous confidence bounds in the most common one-way classification model are briefly discussed at the end of the section.

3.1 Step-down Procedure Confidence Bounds: The step-down tests for the significance of t estimable functions $\underset{\sim}{\Phi} = \underset{\sim}{B}\underset{\sim}{\textcircled{H}}$ defined in (2.2) are given in Section 2.1. In the familiar one-way classification model of MANOVA these functions are the contrasts among the group mean vectors. In MANOVA, simultaneous confidence bounds can be obtained on a class of parametric functions called the extended linear functions of the form $tr(\underset{\sim}{A}\underset{\sim}{\Phi})$ introduced by Mudholkar, Davidson, and Subbaiah (1974a). This class includes the better known bilinear functions $\underset{\sim}{a}'\underset{\sim}{\Phi}\underset{\sim}{d} = tr(\underset{\sim}{d}\underset{\sim}{a}'\underset{\sim}{\Phi})$ as particular cases. Here for simplicity only the bounds on the bilinear functions are briefly described.

Let $F(1 - \alpha_i; t, n - r - i + 1)$ denote the $(1 - \alpha_i)$ percentile of the variance ratio distribution with $(t, n - r - i + 1)$ degrees of freedom, $i = 1,2,\ldots,p$. Let $c_i = F(1 - \alpha_i; t, n - r - i + 1)t/(n - r - i + 1)$, $i = 1,2,\ldots,p$; $c_1^* = c_1, c_i^* = c_i(1 + \sum_{j=1}^{i-1} c_j^*)$, $i = 2,\ldots,p$. Then with simultaneous confidence coefficient at least $100(1-\alpha)\%$, $(1-\alpha) = \prod_{i=1}^{k} (1 - \alpha_i)$, we have

$$\underset{\sim}{a}'\underset{\sim}{\Phi}\underset{\sim}{d} \in \underset{\sim}{a}'\hat{\underset{\sim}{\Phi}}\underset{\sim}{d} \pm (\underset{\sim}{a}'\underset{\sim}{W}\underset{\sim}{a})^{\frac{1}{2}} \sum_{j=1}^{p} |h_j| \sqrt{c_j^*} \tag{3.1}$$

for all $\underset{\sim}{a} \in \mathbb{R}^t$, $\underset{\sim}{d} \in \mathbb{R}^p$, where $\underset{\sim}{h}' = (h_1, h_2, \ldots, h_p) = \underset{\sim}{d}'\underset{\sim}{L}$, $\underset{\sim}{L}$ being the lower triangular matrix such that $\underset{\sim}{L}\,\underset{\sim}{L}' = \underset{\sim}{E}$.

The bounds given by (3.1) are on bilinear functions involving all the response variables. The bounds on the estimable functions of $\underset{\sim}{\phi}_i$ (the ith column of $\underset{\sim}{\Phi}$) corresponding to the ith response variable, obtained by taking unity as the ith and zeroes as the remaining elements in $\underset{\sim}{d}$ at (3.1), are

$$\underset{\sim}{a}'\underset{\sim}{\phi}_i \in \underset{\sim}{a}'\hat{\underset{\sim}{\phi}}_i \pm (\underset{\sim}{a}'\underset{\sim}{W}\underset{\sim}{a})^{\frac{1}{2}} \sum_{k=1}^{i} |\ell_{ik}| \sqrt{c_k^*} , \tag{3.2}$$

for $i = 1,2,\ldots,p$, where $\ell_{ij}'s$ are the elements of the lower triangular matrix $\underset{\sim}{L}$. In terms of the explicit form of $\underset{\sim}{\phi}_i$, i.e., in terms of the elements of original $\underset{\sim}{\Theta}$, we have

$$\underset{\sim}{b}_j'\underset{\sim}{\theta}_i \in \underset{\sim}{b}_j'\hat{\underset{\sim}{\theta}}_i \pm (\underset{\sim}{b}_j'(\underset{\sim}{A}'\underset{\sim}{A})^{-}\underset{\sim}{b}_j)^{\frac{1}{2}} \sum_{k=1}^{i} |\ell_{ik}| \sqrt{c_k^*} , \tag{3.3}$$

for $j = 1,2,\ldots,t$, and $1,2,\ldots,p$. These results are developed in Mudholkar and Subbaiah (1975) and are illustrated using Barnard's data (1935) on skull characters.

3.2 The Bounds with Finite Intersection Tests: The confidence regions based on the finite intersection tests are derived in Mudholkar and Subbaiah (1979a) using reasoning similar to that in Mudholkar and Subbaiah (1975). They illustrate the results in terms of the Iris data of Fisher (1936).

Let F_{ij} denote the statistic, described in (2.11), for testing significance of $\underset{\sim}{b}_j'\underset{\sim}{\eta}_i$, and the critical constant f_i be determined such that

$\Pr(F_{i1} < f_i, \ldots, F_{it} < f_i | H_0) = 1 - \alpha_i$, $i = 1,2,\ldots,p$. Suppose $c_i = f_i/(n - r - i + 1)$, and $c_i^* = c_i(1 + \sum_{j=1}^{i-1} c_j^*)$, for $i = 1,2,\ldots,p$ with $c_1^* = c_1$.

Then the simultaneous confidence intervals

$$\underset{\sim}{b}_j' \underset{\sim}{\textcircled{H}} \underset{\sim}{d} \in \underset{\sim}{b}_j^* \hat{\underset{\sim}{\textcircled{H}}} \underset{\sim}{d} \pm (\underset{\sim}{b}_j'(A'A)^- \underset{\sim}{b}_j)^{\frac{1}{2}} \sum_{i=1}^{p} |h_i| \sqrt{c_i^*} \quad , \tag{3.4}$$

for $j = 1,2,\ldots,t$ and $\underset{\sim}{d} \in \mathbb{R}^p$, $\underset{\sim}{h}' = (h_1, h_2, \ldots, h_p) = \underset{\sim}{d}'\underset{\sim}{L}$, $\underset{\sim}{E} = \underset{\sim}{L}\underset{\sim}{L}'$, $\underset{\sim}{L}$ being the lower triangular matrix such that $\underset{\sim}{E} = \underset{\sim}{L}\underset{\sim}{L}'$, hold with confidence coefficient at least $(1 - \alpha) = \Pi(1 - \alpha_i)$. We note that the rows $\underset{\sim}{b}_1', \underset{\sim}{b}_2', \ldots, \underset{\sim}{b}_t'$ of $\underset{\sim}{B}$ need not be independent vectors.

As a consequence of (3.4) we have

$$\underset{\sim}{b}_j' \underset{\sim}{\theta}_i \in \underset{\sim}{b}_j' \hat{\underset{\sim}{\theta}}_i \pm (\underset{\sim}{b}_j'(\underset{\sim}{A}'\underset{\sim}{A})^- \underset{\sim}{b}_j)^{\frac{1}{2}} \sum_{k=1}^{i} |\ell_{ik}| \sqrt{c_k^*} \ . \tag{3.5}$$

These bounds are similar to those given in (3.3). At each stage of the step-down procedure if the interest is focused upon a finite set of comparisons such as all pair-wise comparisons or comprisons with a control instead of all contrasts, then equation (3.5) yields shorter bounds.

3.3 <u>Multiple Comparisons in the Generalized Step-down Tests</u>: If at each stage of the generalized step-down procedure a multivariate general linear hypothesis is tested using one of the union intersection MANOVA statistics such as the largest root or Hotelling-Lawley's trace, then it is possible to derive the associated simultaneous confidence bounds for all the linear combinations of the estimable functions of $\underset{\sim}{\textcircled{H}}$. Here the confidence bounds, when the largest root statistic is used at each stage, are summarized.

In the notation of Section 2.3, if $Ch_{max}[(\underset{\sim}{T}_{ii^*} - \underset{\sim}{E}_{ii^*})\underset{\sim}{E}_{ii^*}^{-1}] \le c_i$, $i = 1,2,\ldots,k$, then the simultaneous confidence bounds

$$\underset{\sim}{a}'\underset{\sim}{\Phi}\underset{\sim}{d} \in \underset{\sim}{a}'\hat{\underset{\sim}{\Phi}}\underset{\sim}{d} \pm \sum_{i=1}^{k} [\underset{\sim}{h}_i'\underset{\sim}{h}_i \cdot \underset{\sim}{a}'\underset{\sim}{W}\underset{\sim}{a} \cdot c_i^*]^{\frac{1}{2}} \tag{3.6}$$

hold for all $\underset{\sim}{a} \in \mathbb{R}^t$, $\underset{\sim}{d} \in \mathbb{R}^p$ with confidence coefficient at least $(1 - \alpha)$, where $c_1^* = c_1$, $c_i^* = c_i(1 + \sum_{j=1}^{i-1} c_j^*)$, for $i = 2,\ldots,k$, $\underset{\sim}{h}' = \underset{\sim}{d}'\underset{\sim}{L} = (\underset{\sim}{h}_1', \underset{\sim}{h}_2', \ldots, \underset{\sim}{h}_k')$, $\underset{\sim}{L}$ being the lower triangular matrix such that $\underset{\sim}{E} = \underset{\sim}{L}\,\underset{\sim}{L}'$.

These bounds are illustrated in Subbaiah and Mudholkar (1979) using Barnard's data on skull characters by grouping the variables into two subsets.

3.4 <u>One-Way MANOVA</u>: In order to clarify the ideas and the formulae we consider the one-way classification MANOVA model. Suppose that we have a random sample $\underset{\sim}{Y}_{ij}$, $j = 1,2,\ldots,n_i$ of size n_i from a p-variate normal distribution with mean $E(\underset{\sim}{Y}_{ij}) = \underset{\sim}{\mu}_i$ and covariance matrix $\underset{\sim}{\Sigma}$, $i = 1,2,\ldots,k$. The usual null hypothesis in this case is $H_0: \underset{\sim}{\mu}_1 = \ldots = \underset{\sim}{\mu}_k$, and the magnitudes of various contrasts among the vectors $\underset{\sim}{\mu}_i$'s are of interest. In the notation of Section 2.1, $\underset{\sim}{Ⓗ} = (\underset{\sim}{\mu}_1, \underset{\sim}{\mu}_2, \ldots, \underset{\sim}{\mu}_k)'$ with $m = k$, and $(\underset{\sim}{A}'\underset{\sim}{A}) = \text{diag}(n_1, n_2, \ldots n_k)$. Here for $\underset{\sim}{B}$, postulated in (2.2), we may take any $(k-1)xk$ matrix having $(k - 1)$ linear independent contrasts as the rows, e.g., take the k-vector $(-1,0,\ldots,0,+1,0,\ldots,0)$ having -1 in the first, $+1$ in the $(i + 1)^{th}$ and zeroes in other places be the ith row of $\underset{\sim}{B}$, for $i = 1,2,\ldots,k-1$. Thus $\underset{\sim}{\Phi} = \underset{\sim}{B}\underset{\sim}{Ⓗ}$ of order $(k - 1)x\ p$ may be considered for computing the confidence bounds, given in (3.1), associated with the step-down procedure. In case of finite intersection tests $\underset{\sim}{B}$ need not be of full rank; for example, $\underset{\sim}{B}$ may be of order $\binom{k-1}{2}xk$ which gives all pair-wise differences of the k mean vectors. It is thus easy to construct $\underset{\sim}{W} = \underset{\sim}{B}(\underset{\sim}{A}'\underset{\sim}{A})^{-1}\underset{\sim}{B}'$ needed for computation of the bounds in (3.4) and (3.5).

3.5 <u>Analysis of Block Designs</u>: In the case of the Hotelling's problem usually treated using T^2, the step-down procedure and the finite intersection tests are the same. The T^2-statistic has varied applications including its use in the analysis of randomized block designs under nonstandard conditions as given by Graybill (1954). Scheffé (1959) also obtains a test based upon Hotelling's T^2 for a mixed model with interactions for the randomized block experiments. The step-down variations of their analysis, the associated confidence bounds and computational considerations are discussed and illustrated by Mudholkar and Subbaiah (1976). The problem where the response variables can only be grouped into subsets and then ordered is treated in Subbaiah and Mudholkar (1978b). Other block designs can be treated along these lines.

4. PERFORMANCE PROPERTIES OF THE STEP-DOWN TESTS

The studies devoted to the overall performance of the step-down procedure are very few. Hogg (1961) presents a class of nested problems which share the step-wise character of Roy-Bargmann's and J. Roy's set-up. He explains the role of sufficiency in the independence of the step-down statistics under the null hypotheses, but does not consider the performance aspects of the procedures. Das Gupta (1970) uses the results obtained by Anderson (1962,1963) in a multiple decision theoretic formulation of some problems involving nested families of hypotheses to establish a weakly unbiased character of the step-down method as a multiple decision rule.

Subbaiah and Mudholkar (1978a, 1978b) examine the properties of a particular case of the step-down procedure in the Neyman-Pearson framework of hypothesis testing. They consider Hotelling's T^2 problem, a particular case of MANOVA, and identify an ivariance structure in which the step-down statistics are the maximal invariants. The invariance reduction clarifies the noncentrality parameters in the power functions of the step-down procedures and permits power comparisons of various step-down procedures and Hotelling's T^2 test.

Specifically, the Hotelling's problem in canonical form pertains to testing $H_0: \underset{\sim}{\mu} = \underset{\sim}{0}$ vs. $H_1: \underset{\sim}{\mu} \neq \underset{\sim}{0}$ for a p-dimensional normal distribution $N(\underset{\sim}{\mu},\underset{\sim}{\Sigma})$ using $\underset{\sim}{Y}_1, \underset{\sim}{Y}_2, \ldots, \underset{\sim}{Y}_n$, a sample of size n ; or equivalently using the sufficient statistics $(\bar{\underset{\sim}{Y}}, \underset{\sim}{S})$, $\bar{\underset{\sim}{Y}}$ being the sample mean and $\underset{\sim}{S} = \sum_{i=1}^{n} (\underset{\sim}{Y}_i - \bar{\underset{\sim}{Y}}) \ (\underset{\sim}{Y}_i - \underset{\sim}{Y})'$. Then Hotelling's T^2 is given by $T^2 = n(n - 1)\bar{\underset{\sim}{Y}}'\underset{\sim}{S}^{-1}\bar{\underset{\sim}{Y}}$, and the step-down statistics F_i given in equation (2.6) reduce to

$$F_i = (n - i)(T_i^2 - T_{i-1}^2)/[(n - 1) + T_{i-1}^2] \quad , \qquad (4.1)$$

where $T_0^2 = 0$, and T_i^2 denotes Hotellings T^2 statistic for testing $(\mu_1,\ldots,\mu_i)=\underset{\sim}{0}'$ based on the first i variates $i = 1,2,\ldots,p$. The following is a brief summary of the power properties of the step-down methods.

(i) If the component tests are unbiased, then the step-down procedure is unbiased. If the component tests are consistent (i.e., the power increases to one as $n \to \infty$), then the step-down procedure is consistent for any fixed alternative hypothesis.

(ii) The step-down statistics F_i in (4.1) are maximal invariants under the group of transformations $\bar{\underset{\sim}{Y}}^* = \underset{\sim}{L}\ \bar{\underset{\sim}{Y}}$, $\underset{\sim}{S}^* = \underset{\sim}{L}\ \underset{\sim}{S}\ \underset{\sim}{L}'$, where $\underset{\sim}{L}$ is a non-singular lower triangular matrix. The power functions of the invariant tests involve only $\eta_1^2, \eta_2^2, \ldots, \eta_p^2$, where $\underset{\sim}{\eta} = (\eta_1,\eta_2,\ldots,\eta_p)' = \underset{\sim}{B}^{-1}\underset{\sim}{\mu}$, $\underset{\sim}{B}$ being the lower triangular matrix such that $\underset{\sim}{\Sigma} = \underset{\sim}{B}\ \underset{\sim}{B}'$.

(iii) The power function of the step-down procedure for testing the hypothesis $\underset{\sim}{\mu} = \underset{\sim}{0}$ is an increasing function of η_i^2 where $\eta_1^2,\ldots,\eta_{i-1}^2$ are fixed and $\eta_{i+1}^2 = \ldots = \eta_p^2 = 0$, $i = 1,\ldots,p$.

(iv) A Monte Carlo study of the power functions of the T^2 and step-down test when $p = 2$ indicates the following: (a) when $\alpha_1 = \alpha_2$, step-down procedure is slightly superior to the T^2 test along the coordinate axes, i.e., $\eta_1 = 0$, or $\eta_2 = 0$. On the other hand, the T^2 test dominates the step-down procedure along the equiangular line $\eta_1 = \eta_2$. (b) The power of the step-down procedure for fixed (η_1,η_2) is an increasing function of α_1 if $\eta_1^2 > \eta_2^2$, and a decreasing function of α_1 if $\eta_1^2 < \eta_2^2$. (c) When $\eta_1^2 \neq \eta_2^2$, a selection of (α_1, α_2) is possible such that the step-down procedure at (η_1^2, η_2^2) is more powerful than Hotelling's T^2 test; but when $\eta_1^2 = \eta_2^2$, such a selection of (α_1, α_2) does not seem possible. (d) The precision of the inferences on the means of the variables is a decreasing function of their order in the step-down procedure.

5. STEP-DOWN TESTS ON PRINCIPAL VARIABLES

If there is a prior order among the response variables, then the step-down methods are clearly appropriate. However, they can be usefully implemented on the principal components or principal variables even when there is no such ordering readily available. Among obvious reasons for using step-wise methods even in the absence of an ordering are their flexibility, simplicity, and the detail and resolution of inference available from the better understood independent univariate tests constituting them. The detail of inference is also available from other union intersection methods such as the largest root criterion. But in view of the results in the following section, the step-down tests are expected to be more effective in detecting general departures from the null hypothesis. The use of the principal variables in step-down tests and related theory is developed by Dempster (1963). His proposal of using certain linear combinations of the response variables in step-down tests not only provides a meaningful alternative to the absent prior order among the variables but makes the procedure applicable when p , the number of variables is larger than n , the number of observations, and when the normal distribution underlying the model is singular.

Now suppose that the covariance matrix Σ in the MANOVA model of Section 2.1 is possibly singular with rank $(\Sigma) = f \leq p$, and denote $s = \min(f,n)$. Let H and E be the S.S.P. matrices given by (2.3) and $T = H + E$. Let $C = C(T)$ be an $s \times p$ matrix depending on T such that $CTC' = I_s$. Then using the distribution theory developed by James (1954) Dempster shows that under H_0 , $C H C'$ and $C E C'$ are both distributed independently of T according to a distribution which is free of both $C(T)$ and Σ . Moreover, the step-down statistics F_i given in (2.6), obtained using $C H C'$ and $C E C'$ in place of H and E (i.e., obtained by replacing the $n \times p$ matrix Y of observations by $n \times s$ matrix $X = YC'$) are independently distributed as the variance-ratio statistics. Note that $X = YC'$ transforms the p-dimensional independent observations Y_i' into independently distributed s-dimensional $X_i' = Y_i'C$, $i = 1,2,...,n$. The components of X_i are known as the principal variables and the above variance ratios are the step-down statistics associated with

them. If Σ is nonsingular, $n > p$, and $\underset{\sim}{C}$ is such that $\underset{\sim\sim\sim}{CTC'} = \underset{\sim}{I}_p$, then
the coordinates of $\underset{\sim}{X}_i$ are the principal components of $\underset{\sim}{Y}_i$ with the coefficients obtained form $\underset{\sim}{T}$. The interesting distributional result is that the null
distributions of the step-down statistics are unaffected even though the
coefficients of the linear combinations are data based.

In practice, it is unnecessary to reconstruct data in order to obtain the
step-down statistics associated with the principal variables. If $\underset{\sim\sim\sim}{CEC'} = \underset{\sim}{E}*$,
and $\underset{\sim\sim\sim}{CTC'} = \underset{\sim}{C}(\underset{\sim}{H} - \underset{\sim}{E})\underset{\sim}{C}' = \underset{\sim}{T}*$, the Gram-Schmidt reduction of $\underset{\sim}{E}*$ and $\underset{\sim}{T}*$ give
diagonal matrices

$$\underset{\sim1}{L}\underset{\sim}{E}*\underset{\sim1}{L}' = \underset{\sim1}{D}_1 \quad , \quad \text{and} \quad \underset{\sim2}{L}\underset{\sim}{T}*\underset{\sim2}{L}' = \underset{\sim2}{D}_2 \quad , \tag{5.1}$$

where $\underset{\sim}{L}_1$ and $\underset{\sim}{L}_2$ are lower triangular matrices with unit diagonal elements.
In terms of the elements $d_{ii}^{(1)}$ and $d_{ii}^{(2)}$ of these diagonal matrices, the step-
down statistics may be expressed as

$$F_i = \frac{1 - R_i}{R_i} \cdot \frac{n - r - i + 1}{t} \quad , \tag{5.2}$$

where $R_i = d_{ii}^{(1)}/d_{ii}^{(2)}$, $i = 1,2,\ldots,s$.

It may be noted that the elegant distribution theory associated with the
step-down statistics of the principal variables in the null case breaks down if
the null hypothesis is false. Consequently the multiple comparison procedures
with their use is unavailable.

6. MODIFIED STEP-DOWN TESTS

The step-down procedures as discussed so far consist of several tests of
significance based upon the step-down statistics and when possible the post-hoc
analysis using associated confidence bounds. Their use requires decisions
regarding the order in which the variables are used for the tests, and regarding
the distribution of α the type I error probability as α_i , $(1-\alpha) = \Pi(1-\alpha_i)$,
the significance levels associated with the component tests. In this section
we discuss some modifications of the step-down procedures in which the P-values
of the component tests are combined to yield single test statistics for the
overall null hypothesis. In modified procedures at each stage the P-value of the
component test is examined instead of comparing the test statistic with a
critical constant; thus avoiding the problem of specifying α_i . Some of the
resulting procedures have maximal exact Bahadur slopes and are asymptotically

equivalent to the MANOVA likelihood ratio statistic.

Let P_i be the null probability of the variance ratio statistic F_i at the ith stage given by (2.3), exceeding its observed value, $i = 1,2,\ldots,p$. It is easy to see that under H_0 the P-values P_i, $i = 1,2,\ldots,p$ are independent uniform (0,1) random variables, and may be combined using some combination statistic. The classical problem of combining independent tests is excellently reviewed in Oosterhoff (1969), Liptak (1958), or in Mudholkar and George (1979). The well known combination statistics include $\psi_T = \min(-2 \log P_i)$ due to Tippett, $\psi_F = -2 \Sigma \log P_i$ due to Fisher, $\psi_N = \Sigma\Phi^{-1}(1 - P_i)$ due to Liptak, (Φ^{-1} is the inverse of the c.d.f. of the standard normal distribution) and $\psi_L = \Sigma \log[P_i/(1 - P_i)]$ introduced by George (1977); (also see Mudholkar and George (1979), George and Mudholkar (1980)). These combination statistics have simple null distributions: ψ_T is distributed as the smallest order statistic of p independent uniform (0,1) random variables; ψ_F is a χ^2_{2p}-variable, ψ_N is $N(0,p)$ and ψ_L, distributed as a convolution of p logistic random variables, is excellently approximated by a scaled student t variable $a \cdot t_\nu$, where $a = \Pi\{p(5p + 2)/[3(5p + 4)]\}^{\frac{1}{2}}$ and $\nu = 5p + 4$.

Oosterhoff (1969) in his monograph describes and uses a number of asymptotic approaches to studying the combination tests. However, an effective asymptotic scheme, which is not discussed in the monograph, is Bahadur's (e.g., 1971) method of comparing the exact slopes. This method, unlike other asymptotic methods yields measures which describe the operating characteristics of the major combination tests over broad sets of alternatives, and successfully narrows the class of the contenders substantially. If the null hypothesis is false, then for any fixed alternative the P-value P of any reasonable test converges to zero (exponentially) as the sample size n tends to infinity. The exact slope $C(\underset{\sim}{\theta})$ at an alternative $\underset{\sim}{\theta}$, which measures the rate of the decline of P with respect to n , is the a.s. limit of $-(2/n) \log P$, provided that it exists. The greater the exact slope better is the test; and a test with maximal exact slope is called Bahadur optimal or B-optimal.

In Mudholkar and Subbaiah (1979b) it is shown that the P-values of the step-down tests for the Hotelling's T^2 problem combined using Fisher's combination method yields an overall test which has the same exact slope as the T^2-test, and that it is B-optimal. The same optimality holds if the logit method is used for combining the P-values (George and Mudholkar (1980)). Table 1 gives the exact power of T^2-test for $p = 2$, $\alpha = .01$, and $n = 20$. Tables 2,3 give the corresponding empirical power functions for Fisher's and Tippett's combinations of the step-down P-values, estimated using 3000 samples. Tables 4 and 5 give the comparisons of the power functions of various combinations for $p = 3$ and 4 along certain directions. It may be noted that the Tippett's combination of the P-values is almost equivalent to the step-down procedure. From the empirical studies it may be concluded that the Fisher's combination of the P-values is close to T^2-test even in small samples.

TABLE 1

EXACT POWER FUNCTION OF HOTELLING'S T^2 TEST PROCEDURE

$\alpha = .01,\ p = 2,\ n = 20$

η_2 \ η_1	0.0	0.1	0.2	0.5	0.8	1.0	1.2	1.5
1.5	.998	.998	.998	.999	1.00	1.00	1.00	1.00
1.2	.962	.963	.967	.984	.996	.999	1.00	1.00
1.0	.846	.851	.863	.929	.981	.995	.999	1.00
0.8	.605	.614	.640	.790	.935	.981	.996	1.00
0.5	.195	.206	.237	.466	.790	.929	.984	.999
0.2	.027	.032	.050	.237	.640	.863	.967	.998
0.1	.014	.018	.032	.206	.614	.851	.963	.998
0.0	.010	.014	.027	.195	.605	.846	.962	.998

TABLE 2

POWER FUNCTION OF FISHER'S COMBINATION OF THE STEP-DOWN TESTS ESTIMATED FROM THE MONTE CARLO EXPERIMENT

$\alpha = .01,\ p = 2,\ n = 20$

η_2 \ η_1	0.0	0.1	0.2	0.5	0.8	1.0	1.2	1.5
1.5	.997	.998	.998	.999	1.00	1.00	1.00	1.00
1.2	.958	.950	.963	.990	.997	1.00	1.00	1.00
1.0	.815	.824	.837	.928	.983	.996	1.00	1.00
0.8	.577	.582	.623	.794	.949	.981	.998	1.00
0.5	.77	.194	.221	.496	.800	.928	.983	.999
0.2	.023	.030	.049	.241	.655	.850	.963	.998
0.1	.014	.023	.034	.198	.593	.939	.955	.996
0.0	.011	.013	.028	.185	.594	.844	.962	.997

TABLE 3

POWER FUNCTION OF TIPPETT'S COMBINATION OF THE STEP-DOWN TESTS ESTIMATED FROM THE MONTE CARLO EXPERIMENT

$\alpha = .01,\ p = 2,\ n = 20$

η_2 \ η_1	0.0	0.1	0.2	0.5	0.8	1.0	1.2	1.5
1.5	.998	.997	.998	.995	.997	.998	1.00	1.00
1.2	.965	.954	.956	.954	.962	.990	.998	.999
1.0	.963	.860	.839	.835	.916	.965	.995	.999
0.8	.614	.614	.615	.627	.818	.933	.989	1.00
0.5	.203	.209	.201	.349	.691	.893	.977	1.00
0.2	.025	.027	.041	.226	.676	.876	.978	.999
0.1	.017	.019	.030	.204	.644	.883	.973	.998
0.0	.011	.010	.030	.208	.656	.885	.981	.999

TABLE 4

POWER FUNCTIONS OF T^2-TEST AND
MODIFIED STEP-DOWN TESTS
$\alpha = .01$, $p = 3$, $n = 20$

Configuration	Tests	η 0.0	0.1	0.2	0.4	0.6	0.8	1.0
$(\eta, \eta, \eta)'$	Fisher	.013	.017	.066	.366	.818	.987	.999
	Tippett	.010	.016	.049	.198	.518	.826	.972
	T^2	.012	.018	.062	.340	.791	.981	.999
	T^2-exact	.010	.019	.058	.344	.795	.980	.999
$(\eta, 0, 0)'$	Fisher	.010	.012	.025	.085	.243	.498	.753
	Tippett	.009	.009	.025	.092	.298	.600	.846
	T^2	.011	.012	.024	.080	.237	.498	.758
	T^2-exact	.010	.013	.022	.080	.235	.491	.753

TABLE 5

POWER FUNCTIONS OF T^2-TEST AND
MODIFIED STEP-DOWN TESTS
$\alpha = .01$, $p = 4$, $n = 20$

Configuration	Tests	η 0.0	0.1	0.2	0.4	0.6	0.8	1.0
$(\eta, \eta, \eta)'$	Fisher	.012	.027	.066	.458	.886	.996	1.00
	Tippett	.012	.018	.043	.195	.491	.808	.966
	T^2	.010	.027	.065	.422	.854	.993	1.00
	T^2-exact	.010	.019	.063	.400	.860	.992	1.00
$(\eta, 0, 0, 0)'$	Fisher	.015	.017	.030	.072	.196	.400	.634
	Tippett	.012	.012	.021	.081	.262	.536	.816
	T^2	.015	.014	.026	.060	.184	.389	.636
	T^2-exact	.010	.012	.019	.063	.182	.400	.659

Recently Mudholkar and Subbaiah (1979c) have investigated the B-optimality of the combinations of P-values. Among other results they have shown that the P-values of J. Roy's step-down tests, or of the analogous tests based on the principal variables proposed by Dempster, combined using the logit or Fisher's methods yield asymptotic B-equivalents of the MANOVA likelihood ratio test, i.e., they are B-optimal.

REFERENCES

[1] Anderson, T.W. (1958). An Introduction to Multivariate Statistical Analysis, John Wiley & Sons, Inc., New York.

[2] Anderson, T.W. (1962). The choice of the degrees of a polynomial regression and a multiple decision problem. Ann. Math. Statist. 33, 255-265.

[3] Anderson, T.W. (1963). Determination of the order of dependence in normally distributed time series. Time Series Analysis (M. Rosenblatt, ed.), John Wiley and Sons, New York.

[4] Armitage, J.V., and Krishnaiah, P.R. (1978). Approximate percentage points for applications of finite intersection tests for multiple comparisons of means. Technical Report Aerospace Research Laboratory, Wright Patterson Air Force Base, Ohio.

[5] Bahadur, R.R., Some Limit Theorem in Statistics, Regional Conference Series in Applied Mathematics, No. 4, SIAM, Philadelphia, 1971.

[6] Barnard, M.M. (1935). The secular variations of skull characters in four series of Egyptian skulls, Annals of Eugencies, 6, 352-371.

[7] Bhapkar, V.P. (1959). A note on multiple independence under multivariate normal linear models, Ann. Math. Statist., 30, 1248-1251.

[8] Bock, R.D. (1975). Multivariate Statisitcal Methods in Behavioral Research, McGraw-Hill, Inc., New York.

[9] Cox, C.M., Krishnaiah, P.R.Lee, J.C., Reising, J., and Schuurmann, F.J. (1978). A study of finite intersection tests for multiple comparisons of means, Department of Mathematics Technical Report, University of Pittsburgh.

[10] Das Gupta, S. (1970). Step-down multiple decision rules, Essays in Probability and Statistics, (Bose et al., ed.) University of North Carolina Press.

[11] Das Gupta, S., Anderson, T.W., and Mudholkar, G.S. (1964). Montonicity of the power functions of some tests of the multivariate linear hypothesis. Ann. Math. Statist., 35, 200-205.

[12] Dempster, A.P. (1963). Multivariate theory for general step-wise methods, Ann. Math. Statist., 36, 873-883.

[13] Donnerstein, E., Donnerstein, M., Simon, S., and Ditrichs, R. (1972), Variables in interracial agression: anonymity, expected realiation, and a riot, J. of Personality and Social Psychology, 22, 236-245.

[14] Fisher, R.A. (1936). The use of multiple measurements in taxonomic problems. Ann. Eugen., 7, 179-188.

[15] George, E.O. (1977) Combining Independent One-sided Statistical Tests - Some Theory and Applications. Unpublished Ph.D. Dissertation, University of Rochester, Rochester, New York.

[16] George, E.O. and Mudholkar, G.S. (1980). The logit method for combining independent tests. J. Amer. Statist. Assoc., to appear.

[17] Hogg, R.V. (1961). On the resolution of statistical hypotheses, J. Amer. Statist. Assoc., 56, 978-989.

[18] Jensen,D.R. (1972). Some simultaneous multivariate procedures using Hotelling's T^2 statistics, Biometrics, 28, 39-53.

[19] Krishnaiah, P.R. (1965a). Multiple comparison tests in multiresponse experiments, Sankhya Ser. A, 27, 65-72.

[20] Krishnaiah, P.R. (1965b). On the simultaneous ANOVA and MANOVA tests. Ann. Inst. Statist. Math., 17, 33-35.

[21] Krishnaiah, P.R. (1968). Simultaneous test for the equality of convariance matrices against certain alternatives. Ann. Math. Statist., 39, 1303-1309.

[22] Krishnaiah, P.R. (1969). Simultaneous test procedures under general MANOVA models. In"Multivariate Analysis-II" (P.R. Krishnaid, ed.) Academic Press, New York.

[23] Lehmann, E.L. (1959). Testing Statistical Hypotheses, John Wiley and Sons, Inc., New York.

[24] Liptak, T., (1958). On the combination of independent tests. Magyar Tud. Akad. Mat. Kutato Int. Kozl., 3, 171-197.

[25] Mietlowski, W.L. (1974). On the comparison of the dispersions of correlated vectors. Unpublished Ph.D. Dissertation, University of Rochester, Rochester, New York.

Multivariate Statistical Analysis
R.P. Gupta (ed.)

ON CONSTRAINED LEAST SQUARES ESTIMATION

P. Scobey and D.G. Kabe

Department of Mathematics
St. Mary's University
Halifax, N. S., Canada

Given the univariate normal least squares linear regression model $y = x\beta+e$, $E(e)=0$, $E(ee')=A\sigma^2$, full rank A known and σ^2 unknown, subject to equality (inequality) linear restrictions $F\beta=w$, F and w known, a quadratic programming approach to obtain the least squares estimate of β is presented. Three stage least squares theory under inequality constraints is developed. Bayesian point estimate of β, given the restrictions $F\beta\geq w$, is derived. An application to finite sampling theory is mentioned.

1. INTRODUCTION

Let $y = X\beta+e$, where y is an N component column observed data vector, X is $N\times q$ and of rank $q<N$, β $q\times N$, $E(e) = 0$, $E(ee') = A\sigma^2$, be the usual univariate normal linear regression model. Further, let β be subjected to $g<q$ linear inequality restrictions $F\beta\geq w$, where $g\times q$ F of rank g is a known matrix and w is a known vector. Then the problem of obtaining the least squares estimate of β has been considered in the literature by several authors. This estimate of β is a solution to a quadratic programming problem. Judge and Takayama [5] assert that this estimate has a mixed type of sampling distribution which is partly continuous and partly discrete. Lovell and Prescott [8] show that the standard t test for testing the significance of the regression coefficients can be misleading. Zellner [10], and Judge and Takayama [5] claim that the moments of this estimate are difficult to derive. Armstrong and Frome [1] apply branch and bound algorithm to derive the best estimate of β. Liew [7] assumes that only a part of $F\beta = (F_1'F_2')'\beta$ say $F_2\beta = w_2$ is binding and omits the part $F_1\beta>w_1$, and reduces the case to the equality constrained case. He extends the procedure to the three stage least squares case. David [2] follows Liew's [7] procedures and extends it to the Bayesian case.

For a less than full rank linear regression model the inequality restricted case does not appear to be treated in the literature at all. Further, even for the full rank case the distribution theory and the hypothesis testing theory is practically unexplored.

We consider the normal case (full rank case) and obtain the least squares estimate of β by using a classical quadratic programming approach due to Kabe

* This research is supported by a National Research Council Grant - A-4018.

[6]. This approach is based on quadratic forms, subjected to linear restrictions, minimization theory which is stated in the next section. Section 3 develops the inequality constrained least squares theory including the three stage least squares theory. Section 4 considers an application to finite population sampling theory and Bayesian point estimation is derived in section 5. The advantage of the present exposition over those already available in the literature is that the present exposition explicitly points out the exact parallelism that exists between the inequality constrained case and the equality constrained case.

Sometimes the same symbol denotes different quantities, however, its meaning is made explicit in the context.

2. QUADRATIC PROGRAMMING TECHNIQUE

Let x be an N component vector, μ an N component vector, Δ an $N\times N$ positive semidefinite symmetric matrix and Δ^+ its Moore-Penruse inverse, D a given $q\times N$ matrix, range space of D' is contained in the range space of Δ. Then Kabe [6] shows that

$$\text{Min } (x-\mu)'\,\Delta(x-\mu)\ ,\ \text{subject to } Dx = v\ , \tag{1}$$

is given by

$$(v-D\mu)'(D\Delta^+D')^+(v-D\mu)\ , \tag{2}$$

and occurs at

$$x = \Delta^+D'(D\Delta^+D')^+\,v + (I-\Delta^+D'(D\Delta^+D')^+D)\mu \tag{3}$$

In case (1) is modified to

$$\text{Min } (x-\mu)'\Delta\,(x-\mu)\ ,\ \text{subject to } Dx = v\ ,\ x \geq 0\ , \tag{4}$$

then the minimum occurs at

$$x = \Delta^+D'(D\Delta^+D')^+v + (I-\Delta^+D'(D\Delta^+D')\mu+t\ , \tag{5}$$

where $Dt = 0$. If the range space of D' is not contained in the range space of Δ , then (4) is still minimized at (5), where now Δ^+ in (5) is to be replaced by $(\Delta+\alpha D''D)^+$, where α is so chosen that the range space of D' is contained in the range space of $(\Delta+\alpha D'D)$.

Note that (5) is a general solution to the linear matrix equation $Dx = v$. Actually, (5) is a one to one linear transformation from the N components of x

to q components of v, provided rank of D is q, and $(N-q)$ components of t as $Dt = 0$. We may prove this linear transformation as follows. We assume $\Delta = I$ and $\mu = 0$, and find an $(N-q)\times N$ semiorthogonal matrix A orthogonal to D such that $A'A = (I-D^+D)$. Now we set

$$\begin{pmatrix} D \\ A \end{pmatrix} x = \begin{pmatrix} v \\ w \end{pmatrix} , \tag{6}$$

where w is arbitrary and write x as

$$x = \begin{pmatrix} D \\ A \end{pmatrix}' \left[\begin{pmatrix} D \\ A \end{pmatrix} \begin{pmatrix} D \\ A \end{pmatrix}' \right]^{-1} \begin{pmatrix} v \\ w \end{pmatrix} \tag{7}$$

$$= D^+v+A^+w = D^+v+t, \; Dt = 0 \; .$$

In case $\Delta \neq I$, $\mu \neq 0$, then (7) modifies to (5).

If Δ is nonsingular, then μ and D always belong to the range space of Δ. When Δ is singular μ or D or both may not belong to the range space of Δ. When μ does not belong to the range space of Δ, whether D belongs to the range space of Δ or not, we must restrict the solution x to the range space of Δ, i.e., x must satisfy $Wx = 0$, where W is the $(N-p)\times N$ matrix of the $(N-p)$ latent vectors of Δ corresponding to the $(N-p)$ zero roots of Δ, where we assume the rank of Δ to be $p \geq q$, q being the rank of D. Although (5) theoretically is a representation of an optimal solution x to (4), from computational point of view it is difficult to apply in practice and sometimes it is as tedious to apply as the several other algorithms developed for solving quadratic programming problems. Our algorithm requires good skill and problem solving capacity to determine t in practice. For further details of this algorithm refer to Kabe [6].

We have the following optimality criterion for our algorithm. Suppose we have determined a suitable t, say t_0, such that $Dt_0 = 0$. For this value of t_0 let the optimal solution (5) be represented by x_0. If x_0 is optimal, then the derivative of $(x_0+\theta t_1-\mu)'\Delta\,(x_0+\theta t_1-\mu)$ with respect to θ vanishes at $\theta = 0$, where t_1 is any solution of $Dt_1 = 0$.

We illustrate our algorithm by two numerical examples.

Example 1. Min $a'x+\frac{1}{2}x'\,\Delta x$, subject to, (8)

$$Dx = v, \; x \geq 0 \; , \text{ where} \tag{9}$$

$$a' = [2,1,1], \Delta = \begin{bmatrix} 1 & 2 & -1 \\ 2 & 4 & -2 \\ -1 & -2 & 1 \end{bmatrix}, \tag{10}$$

$$Dx = d'x = [1,2,1]x = 4 \quad , \quad x \geq 0 \quad . \tag{11}$$

Now the rank of Δ is unity and

$$36\ \Delta^{+} = \begin{bmatrix} 1 & 2 & -1 \\ 2 & 4 & -2 \\ -1 & -2 & 1 \end{bmatrix} \tag{12}$$

The latent vector corresponding to the zero root is proportional to

$$w' = [3,-4,-5] \quad , \tag{13}$$

and thus d does not belong to the range space of Δ and a does not belong to the range space of Δ. We first assume $t = 0$ in (5) and set

$$x = \Delta^{+}d(d'\Delta^{+}d)^{-1}4 \quad , \tag{14}$$

as a solution to $d'x = 4$. This solution is

$$x = [1,2,-1]' \quad , \tag{15}$$

and indeed is an optimal solution to (8) without the restrictions $x \geq 0$. Now we must determine a t such that

$$x = [1,2,-1]' + t \geq 0 \quad t_1 + 2t_2 + t_3 = 0 \quad . \tag{16}$$

Since $w'x = 0$ yields $w't = 0$, our t also satisfies

$$3t_1 - 4t_2 - 5t_3 = 0 \quad , \tag{17}$$

and a solution to (16) and (17) is

$$t_1 = 3k \ , \ t_2 = -4k \ , \ t_3 = 5k \quad . \tag{18}$$

It follows that

$$x = [1+3k \ , \ 2-4k, \ -1+5k]' \tag{19}$$

is an optimal solution for some k which yields $x \geq 0$.
Now for determining the three components of x we have only two equations $d'x = 4$, $w'x = 0$ and hence one component of x must be arbitrarily chosen. We choose $x_2 = 0$, i.e., $k = \frac{1}{2}$, and find that

$$x = \left[\frac{5}{2} , 0, \frac{3}{2}\right]' \tag{20}$$

is a possible optimal solution. Now by setting $x = (\frac{5}{2} + 3\theta, -4\theta, \frac{3}{2} + 5\theta)'$, where $t_1 = (3, -4, 5)$ is a solution of $d't_1 = 0$, we find that the derivative of

$$a'x + \tfrac{1}{2}x'\Delta x \tag{21}$$

with respect to θ vanishes to $\theta = 0$ and hence (20) is an optimal solution to (8) .

Example 2. Gue and Thomas ([4], p.162, example 9) wish to

$$\text{Min } x'\Delta x = 3x_1^2 + 2x_1 x_2 + 3x_2^2 \quad \text{, subject to,} \tag{22}$$

$$x_1 + 3x_2 \geq 1 \text{ , } 2x_1 + x_2 \leq 5 \quad , \quad x_1,x_2 = 0 \text{ or } 1 \text{ .} \tag{23}$$

Since this is a convex quadratic programming problem in two variables the solution lies on the boundary if the solution is integral. However, there is no integral solution on the boundary and hence slack variables have to be added to apply our technique. Thus we write (22) as

$$\text{Min } x'Gx, \; G = \begin{bmatrix} \Delta & 0 \\ 0 & 0 \end{bmatrix} \quad \text{, subject to,} \tag{24}$$

$$\begin{bmatrix} 1 & 3 & 1 & 0 \\ 2 & 1 & 0 & 1 \end{bmatrix} x = \begin{bmatrix} 1 \\ 5 \end{bmatrix} \quad , \quad x_1,x_2 = 0 \text{ or } 1. \tag{25}$$

Now (24) and (25) satisfy the conditions of (4) and (5) and in this case (5) yields

$$\begin{aligned} x &= G^+D'(DG^+D')^+(1,5)' + t \\ &= \frac{1}{5}[14, -3, 0, 0] + t \quad , \end{aligned} \tag{26}$$

where $Dt = 0$ yields

$$t_1 + 3t_2 + t_3 = 0\ ,\ 2t_1 + t_2 + t_4 = 0\ . \tag{27}$$

Obviously we must choose $t_2 = 3/5$, and we may take $t_3 = 0$, which yields $t_1 = -9/5$, $t_4 = 3$. Thus

$$x = [1,\ 0,\ 0,\ 3]', \tag{28}$$

is a possible optimal solution. However, since x_1 and x_2 are restricted to be nonnegative integers the optimality criteria given by us is not applicable here. Gue and Thomas [4] show that (28) is the optimal solution.

3. LEAST SQUARES ESTIMATION

For the linear regression model

$$y = X\beta + e\ , \tag{29}$$

we have that

$$(y-X\beta)'A^{-1}(y-X\beta)$$

$$= (y-X\hat{\beta})'A^{-1}(y-X\hat{\beta})+(\beta-\hat{\beta})'T^{-1}(\beta-\hat{\beta})\ , \tag{30}$$

$$\hat{\beta} = TX'A^{-1}y\ ,\quad T = (x'A^{-1}X)^{-1} \tag{31}$$

If β is restricted by g linear inequalities

$$F\beta \geq w\ , \tag{32}$$

then by adding a vector of slack variables θ , we write (32) as

$$(F \quad -I)(\beta' \quad \theta') = w\ . \tag{33}$$

Now consider the minimum value problem

$$\text{Min } (\beta-\hat{\beta})'\ T^{-1}(\beta-\hat{\beta})\ , \tag{34}$$

subject to (33). On setting

$$\beta-\hat{\beta} = z\ ,\quad x = (z',\theta')'\ , \tag{35}$$

we find that (34) reduces to

$$\text{Min } x' \begin{pmatrix} T^{-1} & 0 \\ 0 & 0 \end{pmatrix} x, \text{ subject to,} \tag{36}$$

$$(F \quad -I)x = w-F\hat{\beta}\ . \tag{37}$$

However now (36) and (37) are in the frame work of (1) and (3) or (1) and (5) and hence a solution to (36) and (37) may be written as

$$\begin{bmatrix} \bar{\beta} \\ \theta \end{bmatrix} = \begin{bmatrix} \hat{\beta}+TF'(FTF')^{-1}(w-F\hat{\beta})+t_1 \\ t_2 \end{bmatrix} , \tag{38}$$

where t_1 is a solution to

$$Ft_1 = t_2 \ , \ t_2 \geq 0 \ . \tag{39}$$

Thus $\bar{\beta}$ is a representation of an optimal solution to

$$\underset{\beta}{\text{Min}} \ (y-x\beta)'A^{-1}(y-x\beta) \ , \text{ subject to } \ F\beta \geq w \ . \tag{40}$$

We note that

$$\beta_0 = \hat{\beta}+TF'(F\ T\ F')^{-1}(w-F\hat{\beta}) \quad , \tag{41}$$

is the optimal solution to (40) when all the inequalities hold as equalities. Note the parallelism between (38) and (41) for the estimation of β .

To derive Liew's [7] three stage least squares theory, we write (34) and (38) as

$$\text{Min} \ (\beta-\hat{\beta})'T^{-1}(\beta-\hat{\beta}) \ , \text{ subject to } \ F\beta \geq w \ , \tag{42}$$

occurs at

$$\bar{\beta} = \hat{\beta}+T\ F'\lambda+t_1 \quad , \tag{43}$$

$$\lambda = (F\ T\ F')^{-1}(w-F\hat{\beta}) \ . \tag{44}$$

The dual of (42) is

$$\text{Max} \ w'\lambda+(\bar{\beta}-\hat{\beta})'T^{-1}(\bar{\beta}-\hat{\beta}) \quad , \tag{45}$$

subject to

$$F'\lambda+x'A^{-1}y = T^{-1}\bar{\beta}, \ \lambda \geq 0 \ , \tag{46}$$

where $\bar{\beta}$ is any solution to (42). Liew [7] solves the dual first and then the primal. We have solved the primal first and the dual is automatically solved.

Liew [7] considers a system of p structural simultaneous equations and writes the reduced form of this system as

$$y = Z\ \delta+e \quad , \tag{47}$$

where y is $pt\times 1$, Z is $pt\times q$, δ is $q\times 1$, $E(e) = 0$, and e has a certain $pt\times pt$ positive definite unknown covariance matrix. However, an estimate of this covariance matrix is available and to use this estimate effectively Liew [7] further reduces (47) to

$$\bar{X}'y = \bar{X}'Z\,\delta + \bar{X}'e \quad , \tag{48}$$

where $\bar{X} = (I \otimes X)$ is $tp\times tr$, $A \otimes B = (Ab_{ij})$,

I is $t\times t$, X is $p\times r$. The three stage least squares problem now is

$$\text{Min } (\bar{X}'y-\bar{X}'Z)'(S \otimes (X'X))(\bar{X}'y-\bar{X}'Z) \quad , \tag{49}$$

subject to

$$F\,\delta \geq w \quad . \tag{50}$$

Obviously, from (38), the solution for δ is

$$\bar{\delta} = \hat{\delta} + (Z'\bar{X}(S^{-1} \otimes (X'X)^{-1})\bar{X}'Z)^{-1}F'\lambda + t_1 \quad , \tag{51}$$

$$Ft_1 = t_2 \ , \ t_2 \geq 0 \quad , \tag{52}$$

$$\bar{\delta} = [Z'\bar{X}(S^{-1} \otimes (X'X)^{-1})\bar{X}'Z]^{-1} [Z'\bar{X}(S^{-1} \otimes (X'X)^{-1})\bar{X}'y] \quad , \tag{53}$$

$$\lambda = \{F[Z'\bar{X}(S^{-1} \otimes (X'X)^{-1})\bar{X}'Z]^{-1}F'\}^{-1}(w-F\,\hat{\delta}) \ . \tag{54}$$

The two stage least squares theory given by Liew [7] follows by replacing $t\times t$ S in (49), (50), (51), (52), (53) by an identity matrix of the same dimensions.

4. FINITE SAMPLING APPLICATION

Let a finite population consist of N elements $y' = (y_1,\ldots,y_N)$ and let a simple random sample of size n be drawn from this universe. Then $J_N'y = \bar{Y}$ is the population mean corresponding to the sample mean $\bar{y}$, where J_N denotes an N component column vector of unities.

The auxilliary variable vector is $x' = (x_1,\ldots,x_q)$. Tht total matrix of the qN auxilliary variables measured from their respective means is denoted by the $N\times q$ matrix X, i.e.,

$$X = \begin{bmatrix} x_{11} - \bar{X}_1 & . & . & . & x_{q1} - \bar{X}_q \\ . & . & . & . & . \\ x_{1N} - \bar{X}_1 & . & . & . & x_{qN} - \bar{X}_q \end{bmatrix} , \tag{55}$$

where $\bar{X}_i = \sum_j x_{ij}/N$, $j = 1,\ldots,N$. Corresponding to $\bar{X}_1$ there is a simple random sampling mean of x denoted by $\bar{x}_i$, $i = 1,\ldots,q$. A regression estimate of the population mean is then given by

$$\bar{\mu}_1 = \bar{y} - \sum_{i=1}^{q} \hat{\beta}_i (\bar{x}_i - \bar{X}_i) , \tag{56}$$

where $\hat{\beta} = (\hat{\beta}_1,\ldots,\hat{\beta}_q)'$ minimizes $(Y-X\beta)'(Y-X\beta)$,
i.e., $\hat{\beta} = (X'X)^{-1}X'Y$, $Y = y - J_N\bar{Y}$.

Now Des Raj [3] constructs q similar estimates of the population mean, namely

$$\bar{\alpha}_i = \bar{y} - k_i(\bar{x}_i - \bar{X}_i) , \ i = 1,\ldots,q , \tag{57}$$

where k_i are known suitable constants such that the sign of k_i is the same as the sign of $\hat{\beta}_i$. On multiplying (57) by θ_i and adding the results we find that

$$\bar{\mu}_2 = \bar{y} - \sum_{i=1}^{q} k_i\theta_i(\bar{x}_i - \bar{X}_i) , \tag{58}$$

is an unbiased estimator of the population mean $\bar{Y}$ provided taht $J_q'\theta = 1$, $\theta' = (\theta_1,\ldots,\theta_q)$. It follows that

$$\bar{\mu}_2 = \bar{y} - \sum_{i=1}^{q} \bar{\beta}_i (\bar{x}_i - \bar{X}_i) , \tag{59}$$

is an unbiased estimator of the population mean where $\bar{\beta} = (\bar{\beta}_1,\ldots,\bar{\beta}_q)'$ is a solution to

$$\text{Min } (Y - X\beta)'(Y-X\beta) , \text{ subject to,} \tag{60}$$

$$m\beta' - 1, \ \beta \geq 0 , \ m' = (\frac{1}{k_1},\ldots,\frac{1}{k_q}) . \tag{61}$$

From (5) a solution to (60) is

$$\bar{\beta} = \hat{\beta} + A^{-1}m(1-m'\hat{\beta})/m'A^{-1}m+t , \tag{62}$$

where $m't = 0$, $X'X = A$. The minimum value of (60) is

$$(Y-X\hat{\beta})'(Y-X\hat{\beta}) + (1-m'\hat{\beta})^2/m'Am + t'At \quad . \tag{63}$$

Obviously

$$V(\bar{\mu}_2) - V(\bar{\mu}_1) = \frac{N-n}{N-1}\ \frac{1}{n}\ \left[(1-m'\hat{\beta})^2/m'Am + t'At\right] /N \quad , \tag{64}$$

as

$$V(\bar{\mu}_1) = \frac{N-n}{N-1}\ \frac{1}{n}\ \left[(Y-X\hat{\beta})'(Y-X\hat{\beta})/N\right] \quad . \tag{65}$$

5. BAYESIAN ESTIMATION

If we assume a normal set up and a suitable prior, then the posterior density of β is a multivariate t density. Now a result given by Maritz ([9], p.4, equation 1.3.3) shows that the required empirical Bayes best point estimate of β , given the restrictions $F\beta \geq w$, is the conditional mean of the posterior density of β , given $F\beta \geq w$.

If the posterior density is

$$f(\beta)= K(1+(\beta-\beta_0)'A\ (\beta-\beta_0))^{-\frac{1}{2}q} \tag{66}$$

where K denotes constant terms and β_0 is a function of our y values, then the conditional mean of β is

$$\bar{\beta} = \beta_0 + A^{-1}F'(FA^{-1}F')^{-1}(w-F\beta_0)+t_1 \quad , \tag{67}$$

$$Ft_1 = t_2 \ , \ t_2 \geq 0 \quad . \tag{68}$$

The result (67) is obtained by

$$\text{Min}\ (\beta-\beta_0)'A(\beta-\beta_0) \quad , \text{ subject to } \ F\beta \geq w \tag{69}$$

REFERENCES

[1] Armstrong, R.D. and Frome, E.L. (1976). A branch and bound solution of a restricted least squares problem. Technometrics 18, 447-450.

[2] Davis, W.W. (1978). Bayesian analysis of the linar model subject to linear inequality constraints. J. Amer. Statist. Asso. 73, 573-579.

[3] Des, Raj (1965. On a method of using multiauxilliary information in sample surveys. J. Amer. Statist. Asso. 60, 270-277.

[4] Gue, R.L. and Thomas, M.E. (1968). Mathematical Methods in Operations Research. New York, Macmillan.

[5] Judge, G.G. and Takayama, T. (1966). Inequality restrictions in regression analysis. J. Amer. Statist. Asso. 61, 166-188.

[6] Kabe, D.G. (1973). A note on a quadratic programming problem. Industrial Math. 23, 61-66.

[7] Liew, Chong Kiew (1976). A two stage least squares estimation with inequality restrictions on parameters. Review Econ. and Statist. 58, 234-238.

[8] Lovell, M.C. and Prescott, E. (1970). Multiple regression with inequality constraints, pretesting bias, hypothesis testing and efficiency. J. Amer. Statist. Asso. 65, 913-925.

[9] Maritz, J.S. (1970). Empirical Bayes Methods. London, England, Methuen and Company.

[10] Zellner, A. (1961). Linear regression with inequality constraints on the coefficients. Rotterdam: International Center for Management Sciences Report No. 6109.

Multivariate Statistical Analysis
R.P. Gupta (ed.)

INTERVAL ESTIMATION IN MODELS WITH DISTRIBUTED LAGS

K.R. Shah
Department of Statistics
University of Waterloo
Waterloo, Ontario, Canada

and

A. El-Shaarawi
Canada Centre for Inland Waters
Burlington, Ontario
Canada

The problem of the interval estimation of the lag parameters in models with geometric lags is considered. Likelihood intervals and exact confidence intervals are obtained and are compared with the approximate intervals based on the large sample theory.

1. INTRODUCTION

In this paper we study the problem of interval estimation of the lag parameter in the univariate and multivariate models with geometric lags. In section two we derive the likelihood intervals and the exact confidence intervals for the lag parameter. The confidence intervals are compared with the intervals obtained on the basis of the large sample theory. An example studied here shows that the results based on the large sample theory can be misleading. Methods of section 2 are extended to the multivariate case in section 3.

2. UNIVARIATE MODELS

We shall consider the following model:

$$y_t = \sum_{i=0}^{\infty} \beta\gamma^i x_{t-i} + \underset{\sim}{\theta}'\underset{\sim}{z}_t + \varepsilon_t \ ; \ t = 1,2,\ldots,T, \tag{2.1}$$

where x_t and $\underset{\sim}{z}_t$ are the values of the $\ell + 1$ explanatory variables, β, $\underset{\sim}{\theta}$ and γ are the unknown parameters and the ε_t's are the error variables assumed to be independent $N(0,\sigma^2)$.

We may re-write (2.1) as

$$\begin{aligned} y_t &= \gamma^t \sum_{i=t}^{\infty} \beta\gamma^{i-t} x_{t-i} + \sum_{i=0}^{t-1} \beta\gamma^i x_{t-i} + \underset{\sim}{\theta}'\underset{\sim}{z}_t + \varepsilon_t \\ &= \gamma^t \eta + \sum_{i=0}^{t-1} \beta\gamma^i x_{t-i} + \underset{\sim}{\theta}'\underset{\sim}{z}_t + \varepsilon_t \end{aligned} \tag{2.2}$$

where $\eta = \beta \sum_{i=0}^{\infty} \gamma^i x_{-i}$

We note that (2.2) is a non-linear model and the parameters can be estimated using the non-linear least squares methods. Approximate confidence intervals for any sub-set of parameters can be obtained by the usual large sample methods (Dhrymes, 1971). In this paper we give methods of exact inference which would be valid for any sample size.

For a given value of γ , (2.2) is the usual linear model. Let $\hat{\eta}(\gamma)$, $\hat{\beta}(\gamma)$ and $\hat{\sigma}^2(\gamma)$ be the maximum likelihood (m.l.) estimates of the parameters involved in this model. Let $u_t(\gamma) = (y_t - \hat{y}_t(\gamma))/\hat{\sigma}(\gamma)$ where $\hat{y}_t(\gamma)$ is the estimated value of $E(y_t)$. The m.l. estimates are minimal sufficient statistics and the $u_t(\gamma)$'s are ancillary statistics. The joint density function of the $u_t(\gamma)$'s can be used to obtain the marginal likelihood for γ . An expression for this marginal likelihood as given in El-Shaarawi (1977) is

$$L(\gamma,\underset{\sim}{y}) = [\hat{\sigma}(\gamma)]^{-T+\ell+3} \tag{2.3}$$

We can use $L(\gamma,\underset{\sim}{y})$ to obtain a likelihood interval for γ . Of course, this interval does not have a frequency interpretation. Alternatively, one may look for confidence interval for γ . To obtain such an interval we consider the model

$$y_t = \gamma^t\eta + \sum_{i=0}^{t-1} \beta\gamma^i x_{t-i} + \underset{\sim}{\theta}'\underset{\sim}{z}_t + \delta w_t(\gamma) + \varepsilon_t , \tag{2.4}$$

where $w_t(\gamma) = t\gamma^{t-1}\hat{\eta}(\gamma) + \sum_{i=0}^{t-1} i\gamma^{i-1}x_{t-i}\hat{\beta}(\gamma)$. We note that $\hat{\eta}(\gamma)$ and $\hat{\beta}(\gamma)$ are as defined earlier and hence are functions of y_t . If we proceed with model (2.4) in a <u>formal</u> manner we can compute an F Statistic to test the hypothesis $\delta = 0$. It can be shown that for any fixed γ , this statistic $F(\gamma)$ follows an F distribution with 1 and $T - \ell - 3$ degrees of freedom (d.f.). Further, if $\hat{\gamma}$ is the least squares estimate of γ , $F(\hat{\gamma}) = 0$ (El-Shaarawi and Shah (1979)). The set of values of γ for which $F(\gamma)$ is large may be regarded as unreasonable and hence the set of values of γ for which $F(\gamma)$ is less than the appropriate table value consitutes a confidence region for γ .

Both the above methods i.e. the likelihood and the confidence interval methods can be modified to obtain joint confidence regions for the parameters β and γ .

We shall now apply these methods to a model considered by Dhrymes (1971) which is a special case of the model (2.2). We consider

$$y_t = \gamma^t\eta + \sum_{i=0}^{t-1} \beta\gamma^i x_{t-i} + \varepsilon_t . \tag{2.5}$$

We simulated some samples for $\eta = 2$, $\beta = 1$, $x_t = t$, $T = 20$ and $\gamma = .5$. Figures 2.1, 2.2 and 2.3 show the marginal likelihood for γ divided by its maximum value, i.e. relative marginal likelihood $R(\gamma,\underset{\sim}{y})$ and the values of $F(\gamma)$ for three such samples. They all exhibit lack of symmetry which is particularly pronounced in Fig. 2.2. These figures also show the confidence intervals for γ .

We note that the intervals are reasonable for the first and the third samples whereas the interval based on the second sample is too wide. For the third sample, the exact 95% confidence interval is 0.47 to 0.64 whereas the large sample method as described by Dhrymes (1971, p. 106) gives 0.55 to 0.61. To examine this further, we took 500 such samples. The histogram of the values of $\hat{\gamma}$ for these 500 samples is given in Fig. 2.4. We note that the distribution of $\hat{\gamma}$ is not symmetrical. The following gives the results of three sets of 500 samples.

T	γ	Average of $\hat{\gamma}$	s.d. of $\hat{\gamma}$	Exact interval lower upper (average values)	Large sample interval lower upper (average values)
20	.5	.480	.1040	.2034 .6352	.3614 .5951
20	.7	.700	.0282	.6257 .7552	.6606 .7388
50	.7	.700	.0105	.6712 .7214	.6894 .7103

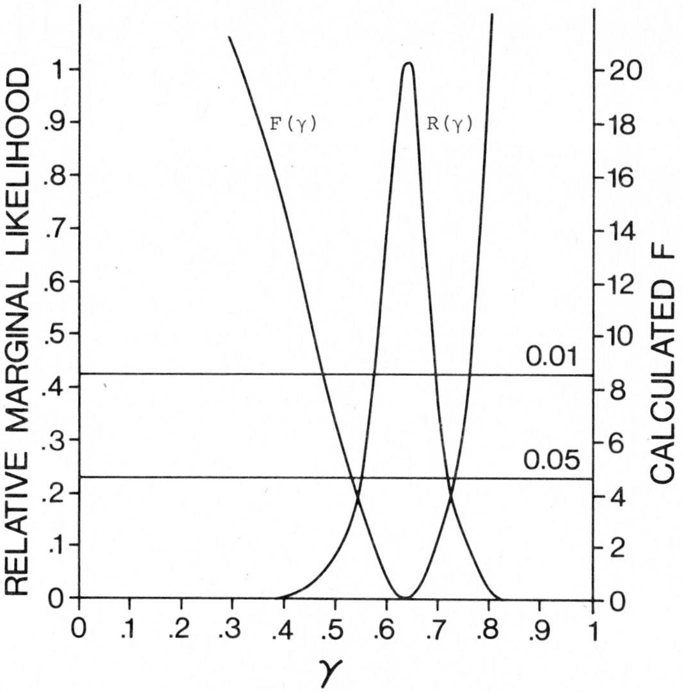

FIGURE 2.1

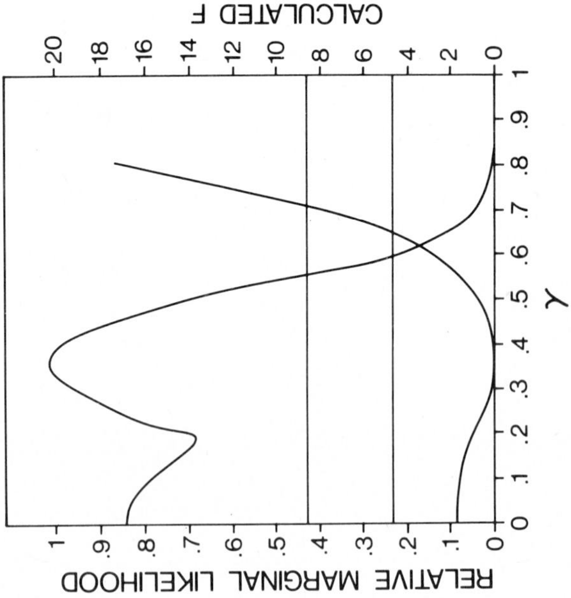

FIGURE 2.2

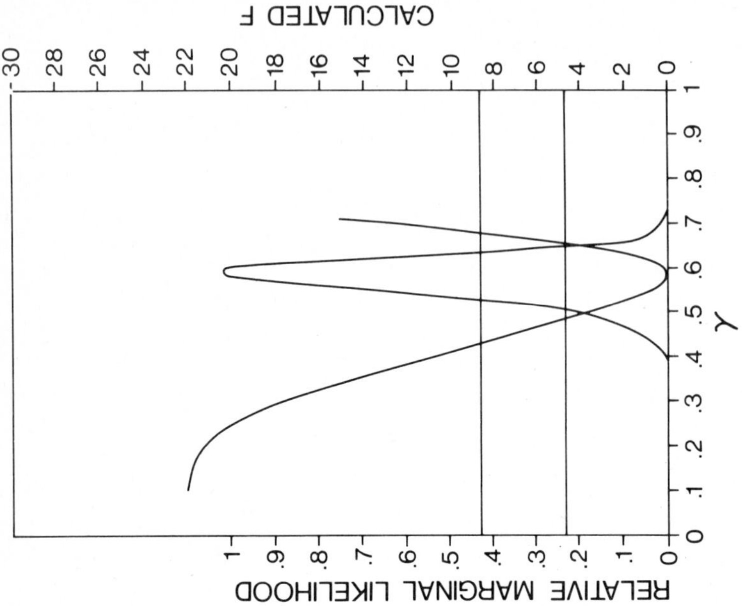

FIGURE 2.3

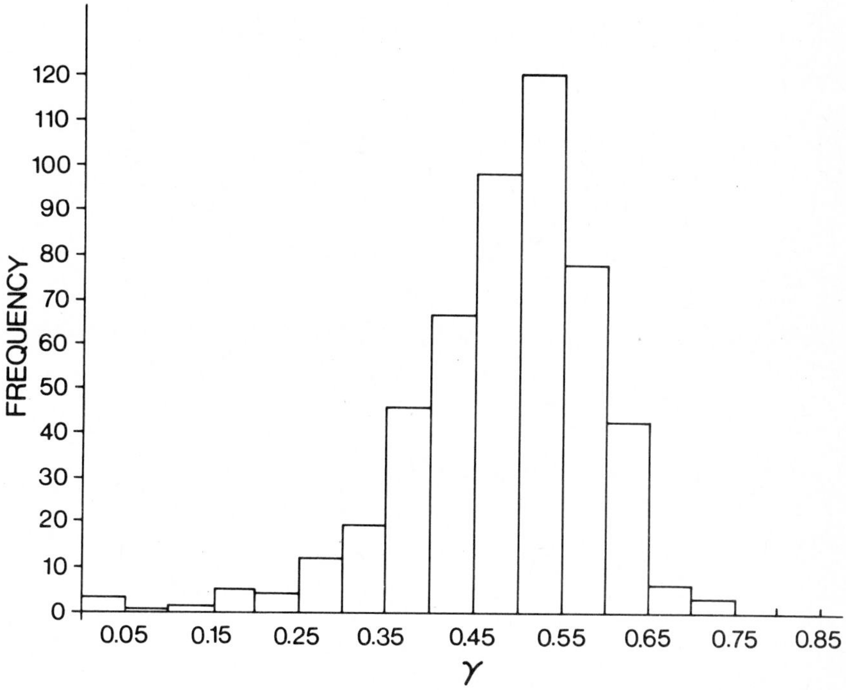

FIGURE 2.4

It is seen that the interval based on the large sample theory is narrower than the one based on the statistic $F(\gamma)$. It was verified that the intervals based on $F(\gamma)$ did include the true parameter value with an approximately correct relative frequency and consequently the intervals based on the large sample theory did not cover the true value with an appropriate relative frequency.

3. MULTIVARIATE MODELS

Here, we consider the model

$$y_{tj} = \sum_{i=0}^{\infty} \beta_j \gamma_j^i x_{t-i} + \underset{\sim}{\theta}_j' \underset{\sim}{z}_t + \varepsilon_{tj} \; ; \qquad (3.1)$$

$$j = 1,2,\ldots,p \; ;$$
$$t = 1,2,\ldots,T \; .$$

where x_t and $\underset{\sim}{z}_t$ are the values of the $\ell + 1$ explanatory variables, β_j, γ_j, $\underset{\sim}{\theta}_j$ are unknown parameters and ε_{tj}'s are the error variables assumed to follow a multivariate normal distribution. We assume $E(\varepsilon_{tj}) = 0$, $\mathrm{Cov}\,(\varepsilon_{tj}, \varepsilon_{t'j'}) = 0$ for $t \neq t'$ and $\mathrm{Cov}\,(\varepsilon_{tj}, \varepsilon_{tj'}) = \sigma_{jj'}$.

We may re-write (3.1) as

$$y_{tj} = \gamma_j^t \sum_{i=t}^{\infty} \beta_j \gamma_j^{i-t} x_{t-i} + \sum_{i=0}^{t-1} \beta_j \gamma_j^i x_{t-i} + \underset{\sim}{\theta}_j' \underset{\sim}{z}_t + \varepsilon_{tj}$$

$$= \gamma_j^t \eta_j + \sum_{i=0}^{t-1} \beta_j \gamma_j^i x_{t-i} + \underset{\sim}{\theta}_j' \underset{\sim}{z}_t + \varepsilon_{tj} \; , \qquad (3.2)$$

where $\eta_j = \beta_j \sum_{i=0}^{\infty} \gamma_j^i x_{-i}$.

For given values of $\gamma = (\gamma_1,\ldots,\gamma_p)'$, (3.2) is the usual multivariate analysis of variance (MANOVA) model. Let $\mu_j' = (\eta_j, \beta_j, \underset{\sim}{\theta}_j')$. The m.l. estimates of μ_j and of $\Sigma = ((\sigma_{ij}))$ given by $\hat{\mu}_j(\gamma)$ and $\hat{\Sigma}(\gamma)$ are minimal sufficient statistics. Let $y_t' = (y_{t1},\ldots,y_{tp})$ and let G be a lower triangular matrix such that $GG' = \hat{\Sigma}(\gamma)$. We note that the random variables $\underset{\sim}{\mu}_t = G^{-1}(\underset{\sim}{y}_t - \hat{\underset{\sim}{y}}_t)$; $t = 1,2,\ldots,T$, have a joint distribution free from the parameters μ_j and Σ and hence these may be used as ancillary statistics. Using the method of Saw (1970) and Kalbfleisch and Sprott (1970), the marginal likelihood for γ turns out to be proportional to

$$\prod_{i=1}^{p} g_{ii}(\gamma)^{-\,(T - (\ell+2) - i)} \qquad (3.3)$$

Likelihood regions for γ may be obtained from (3.3). To obtain confidence region for γ we note that we may compute $F_i(\gamma_i)$ as in section 2 to compute appropriate confidence interval for γ_i . To obtain a joint confidence region we note that $\{\gamma_1,\ldots,\gamma_p\colon F_i(\gamma_i) \leq \lambda$ for $i = 1,2,\ldots p\}$ is a conservative $100(1-\gamma)\%$ confidence region for γ where λ is determined from the relation

$$\sum_{i=1}^{p} P[F_i(\gamma_i) \geq \lambda] = \alpha \; .$$

REFERENCES

[1] Dhrymes, P.J. (1971). Distributed lags, problems of estimation and formulation. Holden-Day, Inc. San Francisco.

[2] El-Shaarawi, A. (1977). Marginal likelihood solution to some problems connected with regression analysis. J.R. Statist. Soc. B, 39, 343-348.

[3] El-Shaarawi, A. and Shah, K.R. (1979) Interval estimation in non-linear models. Submitted for publication.

[4] Kalbfleisch, J.D. and Sprott, D.A. (1970). Applications of likelihood methods to models involving large numbers of parameters (with discussion). J.R. statist soc. B, 29, 357-372.

[5] Saw, J.G. (1970). The multivariate linear hypothesis with non-normal errors and a classical setting for the structure of inference in a special case. Biometrika, 57, 531-535.

Multivariate Statistical Analysis
R.P. Gupta (ed.)

ESTIMATIONS OF REGRESSION CURVES WHEN THE CONDITIONAL DENSITY OF THE PREDICTOR VARIABLE IS IN SCALE EXPONENTIAL FAMILY

R. S. Singh

Department of Mathematics and Statistics
University of Guelph
Guelph, Ontario
Canada

With X denoting the predictor variable and Y denoting the response variable, estimations of the regression curve of Y on X are considered in situations where the conditional density of X given Y is in scale exponential families. Two types of estimators are proposed. Constructions of these estimators do not require any observations on Y but on X alone. Both types of estimators are shown to be strongly consistent. Speeds with which these estimated curves approach to the actual curve as the sample size on X increases are noted.

INTRODUCTION AND PRELIMINARIES

Let (X,Y) be a bivariate random variable, X the independent or the predictor variable and Y , the dependent or the response variable. The regression curve of Y on X is given by

$$r(X) = E(Y \mid X) \quad . \tag{1.0}$$

If it is known that the regression curve r(X) is a polynomial, logarithmic, exponential or likewise known function of X , then possibly r can be estimated by the methods of maximum likelihood principle or least squares theory, provided observations on (X,Y) are available. The question is how to estimate the regression curve r if nothing is known about the functional form of r or if no observations on Y is possible. The object of this paper is to give a partial answer to this question. More precisely, using independent observations on X alone this paper proposes strongly consistent estimators of r and gives speed of consistencies.

Estimation of regression curve arises in almost every practical situations wherever prediction of a variable Y is desired at a given level of another variable X . For example, in medicine, biology, epidemiology, economics, business or in an area of engineering, Y could be the effect variable and X could be the cause variable.

The problem of estimation of regression curve r when nothing is known about the form of r or when no observations on Y are at hand, is not entirely new. Kale (1962), Nadaraya (1964, 1965) and most recently Singh and Tracy (1977) have considered such problems. Whereas in the first three of these papers, the conditional density of X given Y = y is normal with mean y and variance 1, and the unconditional distribution function, say H, of Y possesses a density, Singh and Tracy considered a more general situation, namely where the density of X given $Y = y$ is of the form $C(y)\, u(x)\, e^{yx}$ and H need not possess a density.

In this paper we will consider the following model. We will assume that the conditional density of X given $Y = y$ is of the form

$$p_y(x) = g(x)\, b(y)\, e^{-x/y} \tag{1.1}$$

if $y > 0$ and 0 otherwise, where the function g appearing in (1.1) does not involve y, and is completely <u>known</u>, and $b(y)$ is $\left(\int g(x) \exp(-x/y)\, dx\right)^{-1}$. Thus for a given value of y, $p_y(x)$ is completely <u>known</u>. However, the marginal distribution function, say H, of Y is completely <u>arbitrary</u> and <u>unknown</u>. Consequently, $p(x) = \int p_y(x)\, dH(y)$, the marginal p.d.f. of X, and

$$f(x) = \int b(y)\, e^{-x/y}\, dH(y) \tag{1.2}$$

both are <u>unknown</u>, and so is the regression curve $r(x)$ of Y on X which takes the form

$$r(x) = \frac{\int y\, b(y)\, e^{-x/y}\, dH(y)}{f(x)} \tag{1.3}$$

in our above set up.

Notice that the well known gamma density defined for an $\tau > 0$ as

$$\frac{x^{\tau-1}\, y^{-\tau}\, e^{-x/y}}{\Gamma(\tau)}$$

for x and y positive, and 0 elsewhere, is of the form (1.1).

The object of this paper is to estimate the regression curve r when neither the form of r is known nor is available an observation on Y. More explicitly, using a random sample $\underset{\sim}{X}_n = (X_1, \ldots, X_n)$ on X alone, we will exhibit two types of estimators $\hat{r}$ and r^* of r, prove strong consistency of both estimators and give speeds there.

Our methods of estimation of r stem from our discoveries that $r(x)$ can be represented as

$$r(x) = \frac{\int_x^\infty f(t)\, dt}{f(x)} \tag{1.4}$$

or as

$$r(x) = \frac{E[I(X \geq x)/g(X)]}{f(x)} \tag{1.5}$$

where $I(A)$ is the indicator function of the set A. Thus we have reduced the problem of estimation of r to the estimation of f alone with (1.4), and to the estimation of f and $E[I(X \geq \cdot)/g(x)]$ with (1.5). In the next section we will propose two types of estimators $\hat{r}$ and r^* corresponding to (1.4) and (1.5) respectively, prove strong consistency of $\hat{r}$ and r^* and give there the speeds of consistency.

PROPOSED ESTIMATORS OF THE REGRESSION CURVE AND THEIR CONSISTENCIES

Let $s > 0$ be a positive integer (to be chosen appropriately to suit the situations described in Theorem 2.1 or 2.2 below). Let K_s be a real valued continuous function of bounded variation and vanishing off an interval (a,b), finite or infinite, such that $\int y^k K_s(y)\,dy = 1$ or 0 according as $j = 0$ or $j = 1,2,\ldots,s-1$.

Notice that the standard normal density is an example of K_s for $s = 1$ and 2. For $a = 0$ and $b = 1$ and for $s = 1,2,\ldots,5$, K_s can be taken, respectively, as

$$K_1(t) = I(0 < t < 1)$$
$$K_2(t) = (4 - 6t)I(0 < t < 1)$$
$$K_3(t) = 3(3 - 12t + 10t^2)\, I(0 < t < 1)$$
$$K_4(t) = 8(2 - 15t + 30t^2 - 17.5t^3)\, I(0 < t < 1)$$
$$K_5(t) = 25(1 - 12t + 42t^2 - 56t^3 + 25.2t^4)\, I(0 < t < 1).$$

From these functions, we can also construct examples of functions K_s for $s = 1,2,\ldots,5$ and $a = -1$ and $b = 1$.

Let $0 < \varepsilon = \varepsilon_n$ be a function of n such that ε decreases to zero as n increases to infinity. We hereinafter abbreviate K_s to K . Let

$$\hat{f}(x) = (n\varepsilon)^{-1} \sum_{j=1}^{n} \{K(\frac{X_j - x}{\varepsilon})/g(X_j)\} \tag{2.1}$$

Then by the arguments used in Singh (1977a), $\hat{f}$ is mean square as well as strongly consistent estimator of f . We will <u>indicate</u> the proof of the following lemma which we will need later. Numbers $c_0, c_1, \ldots$ below are absolute constants.

<u>Lemma 2.1</u>. <u>For an integer</u> $s > 0$ <u>and for a number</u> A , <u>let</u>

$$\sup_{x>A} \varepsilon^{-1} \int_{x+a\varepsilon}^{x+b\varepsilon} \left| \int y^{-s} p_y(t)\, dH(y) \right| dt < \infty , \tag{2.2}$$

<u>and with</u> $v(x)$ <u>denoting the total variation of</u> $1/g$ <u>over</u> $(x + a\varepsilon, x + b\varepsilon)$, <u>let</u>

$$\sup_{x>A} v(x) < \infty . \tag{2.3}$$

<u>Then taking</u>

$$\varepsilon = c_0(n^{-1} \log \log n)^{1/(2s+2)}$$

<u>in</u> (2.1), <u>we have</u>

$$\sup_{x>A} |\hat{f}(x) - f(x)| = O(n^{-1} \log \log n)^{s/(2s+2)}$$

<u>with probability one</u>.

<u>Proof</u>. The arguments used in the proof of Theorem 3.1 of Singh (1977a) lead to

$$\sup_{x>A} |E\hat{f}(x) - f(x)| \le c_1 \varepsilon^s \{\sup_{x>A} \varepsilon^{-1} \int_{x+a\varepsilon}^{x+b\varepsilon} |f^{(s)}(t)|dt\} = O(\varepsilon^s) \tag{2.4}$$

where the equality follows from our hypotheses (2.2) and (2.3) since $g(t)\, f^{(s)}(t) = \int (-y)^{-s} p_y(t)\, dH(y))$ is justified due to Theorem 2.9 of Lehmann (1959). Also, if we redefine $Z_x(\cdot)$ in the proof of Theorem 3.2 of (1977a) by $K\{(\cdot - x)/\varepsilon\}/g(\cdot)$, then the arguments there in conjunction with (2.3) here give

$$\sup_{x>A} |\hat{f}(x) - E\hat{f}(x)| = O(\varepsilon^{-2} n^{-1} \log\log n)^{1/2} \tag{2.5}$$

with probability one. Now (2.4) and (2.5) give the desired conclusion. ■

Having noted that $\hat{f}$ are strongly consistent estimators of f, our _first_ type of _proposed_ estimators of the regression curve r are, in view of (1.4),

$$\hat{r}(x) \begin{cases} \dfrac{\int_x^{x+\varepsilon^{-\gamma}} \hat{f}(t)\, dt}{\hat{f}(x)} & \text{if } \hat{f}(x) \neq 0 \\ \\ \text{arbitrary} & \text{if } \hat{f}(x) = 0 \end{cases} \tag{2.6}$$

where $\gamma > 0$.

Now notice that $\hat{h}(x) = n^{-1} \sum_{j=1}^{n} \{I(X_j \geq x)/g(X_j)\}$ is an _unbiased_ estimator of $h(x) = E(I(X \geq x)/g(X))$. The proof of the following lemma follows from the law of iterated logarithms for the empirical distributions and the arguments in the proof of the preceding lemma.

Lemma 2.2. For a number B, _let_

$$1/g \text{ be of bounded variation over } (B,\infty). \tag{2.7}$$

Then

$$\sup_{x>B} |\hat{h}(x) - h(x)| = O(n^{-1} \log\log n)^{1/2}$$

with probability one.

In view of Lemmas 2.1 and 2.2, and the representation (1.5), our _second_ type of _proposed_ estimators of the regression curve r are,

$$r^*(x) = \begin{cases} \dfrac{\hat{h}(x)}{\hat{f}(x)} & \text{if } \hat{f}(x) \neq 0 \\ \\ \text{arbitrary} & \text{if } \hat{f}(x) = 0 \end{cases} \tag{2.8}$$

Now we will show that $\hat{r}$ and r^* are strongly consistent estimators of r. We further give the speeds of these consistencies.

Theorem 2.1. Let s, A, ε _and_ $\hat{f}$ _be as given in Lemma_ 2.1. _Further, for an integer_ $\alpha > 0$, _let_

$$\sup_{x>A} \int y^{\alpha+1}\; p_y(x)\; dH(y) < \infty \tag{2.9}$$

and $\hat{r}$ *be as in* (2.6) *with* $\gamma = s/(\alpha+1)$. *Then for every subset* S *of* (A,∞) *with* $\inf_{x\in S} f(x) > 0$

$$\sup_{x\in S} |\hat{r}(x) - r(x)| = O(n^{-1}\log\log n)^{w} \tag{2.10}$$

with probability one, where $w = s\alpha(\alpha+1)^{-1}(2s+2)^{-1}$.

Proof. From the definition of f we can write

$$\int_{x+\varepsilon^{-\gamma}}^{\infty} f(t)dt = \int b(y)[\int_{x+\varepsilon^{-\gamma}}^{\infty} e^{-t/y}\, dt]dH(y)$$

$$= \int y\; b(y)e^{-x/y}(\exp(-\varepsilon^{-\gamma}/y)dH(y) \quad .$$

Therefore, since for every positive t and nonnegative integer α , $e^{-t} \leq \alpha! t^{-\alpha}$, we have

$$\int_{x+\varepsilon^{-\gamma}}^{\infty} f(t)dt \leq \frac{\alpha!\varepsilon^{\gamma\alpha}}{g(x)} \int y^{\alpha+1}\; p_y(x)dH(y) \quad .$$

Hence in view of (2.3) and (2.9) ,

$$\sup_{x>A} \int_{x+\varepsilon^{-\gamma}}^{\infty} f(t)dt = O(\varepsilon^{\gamma\alpha}) \tag{2.11}$$

Thus (2.4) and (2.11) give

$$\sup_{x>A} |\int_{x}^{x+\varepsilon^{-\gamma}} \hat{f}(t)dt - \int_{x}^{\infty} f(t)dt|$$

$$\leq \sup_{x>A} \int_{x}^{x+\varepsilon^{-\gamma}} |\hat{f}(t)-f(t)|dt$$

$$+ \sup_{x>A} \int_{x+\varepsilon^{-\gamma}}^{\infty} f(t)dt$$

$$= O(\varepsilon^{s-\gamma}) + O(\varepsilon^{\gamma\alpha}) = O(\varepsilon^{s\alpha/(\alpha+1)})$$

with probability one since $\gamma = s/(\alpha+1)$. Thus we conclude that with probability one

$$\int_{x}^{x+\varepsilon^{-\gamma}} \hat{f}(t)dt = \int_{x}^{\infty} f(t)dt + O(\varepsilon^{s\alpha/(\alpha+1)})$$

and

$$\hat{f}(x) = f(x) + O(\varepsilon^{s})$$

both uniformly in $x > A$. Thus, since $S \subset (A,\infty)$ and $\inf_{x \in S} f(x) > 0$, we have with probability one

$$\hat{r}(x) = \frac{\int_x^{x+\varepsilon^{-\gamma}} \hat{f}(t)}{\hat{f}(x)} = \frac{\int_x^{\infty} f(t)\,dt + 0(\varepsilon^{s\alpha/(\alpha+1)})}{f(x) + 0(\varepsilon^{s})}$$

$$= \frac{\int_x^{\infty} f(t)dt}{f(x)} + 0(^{s\alpha/(\alpha+1)})$$

uniformly in $x \in S$. The proof of the theorem is now complete since $\varepsilon = 0(n^{-1} \log\log n)^{s/(2s+2)}$. ∎

Theorem 2.2 below proves strong consistency of r^* as estimators of r and gives speeds of consistency. The proof of the theorem follows by the arguments identical to those given in the latter part of the proof of Theorem 2.1.

Theorem 2.2. Let s, A, ε and $\hat{f}$ be as given in Lemma 2.1 and r^* be defined as in (2.8). Further, let (2.7) with B there replaced by A here hold. Then for every subset S of (A,∞) with $\inf_{x \in S} f(x) > 0$,

$$\sup_{x \in S} |r^*(x) - r(x)| = 0(n^{-1} \log\log n)^{1/2}$$

with probability one.

REMARKS AND EXAMPLES

Notice that if the response variable Y is a bounded random variable then (2.2) with $a = 0$ and (2.9) hold for every value of s and α and for every positive A such that p, the marginal density of the predictor variable, is bounded on (A,∞). In view of (2.3) p is bounded on (A,∞) if f is so.

In the well known example of gamma density mentioned in Section 1, $g(x) = x^{\tau-1}/\Gamma(\tau-1)$ for $x > 0$ and 0 elsewhere. In the exponential density $y^{-1} \exp(-x/y)\, I(x > 0)$, where $y > 0$; $g(x) = I(x > 0)$, and in another case with $p_y(x) = [\sum_{i=0}^{\infty} (i+1) I(i \le x < i+1)]\,[(1 - \exp(-1/y))/y]$ for y positive, $g(x)$ is $\sum_{i=0}^{\infty} I(i \le x < i+1)$. In each of these examples, (2.3) with $a \ge 0$, and (2.7) hold for all positive numbers A and B.

Our methods of estimation of r suggest that in various situations stiatistics $\hat{r}$ and r^* based on a random sample $\underset{\sim}{X}_n$ on X can be exhibited such that for any subset S of the real line, $\hat{r}$ and r^* approach almost surely the actual curve r uniformly on S with the speed arbitrarily close to $0(n^{-1/2})$.

Notice that compared to $\hat{r}$, r^* are much easier to construct and approach the actual regression curve r with much faster speed. From these two points of view, estimators r^* seem to be preferable to estimators $\hat{r}$.

Acknowledgement

This research was supported by the National Research Council of Canada under research grant no. A4631. Part of this research was carried out while the author was visiting the Banaras Hindu University during March - April 1978.

REFERENCES

[1] Kale, B. (1962). A note on a problem in estimation. *Biometrika*, 49, 553-556.

[2] Lehmann, E.L. (1959). *Testing Statistical Hypotheses.* New York: Wiley.

[3] Nadaraya, E.A. (1964). Estimation of a convolution component. *Soobščenija Akad. Nauk. Gruzin. USSR*, xxxiv, 1, 19-24.

[4] Nadaraya, E.A.(1965).On nonparametric estimates of density function and regression curves. *Theor. Prob. Appl.*, 10, 186-190.

[5] Singh, R.S. (1977a). Improvement on some known non-parametric uniformly consistent estimators of derivatives of a denisty. *Ann. Statist.*, 5, 394-399.

[6] Singh, R.S. (1977b). Applications of estimators of a density and its derivatives to certain statistical problems. *J. Roy. Statist. Soc.*, 39, No. 3, 357-363.

[7] Singh, R.S. and Tracy, D.S. (1977). Strongly consistent estimators of kth order regression curves and rates of convergence. *Z. Wahrscheinlichkeitstheorie und verw. Gebiete*, 40, 339-348.

Multivariate Statistical Analysis
R.P. Gupta (ed.)
© North-Holland Publishing Company, 1980

A TECHNIQUE FOR SELECTION OF VARIABLES IN MULTIPLE REGRESSION

Umed Singh
Dept. of Statistics
School of Business
Temple University
Philadelphia, PA.,
U.S.A.

and

Vrinda Tayal
Department of Math. & Statistics
Haryana Agricultural University
Hissar, INDIA

This paper is concerned with decreasing the computational effort using the C_p-statistic of Mallows as the basic criteria for comparing two regressions. A procedure is developed which will indicate 'good' regression with a minimum of computation. Proposed method turns out considerably efficient over existing computational techniques. An example is included to illustrate the essential points and the gain due to the proposed method.

INTRODUCTION

At occasions applied statisticians who use regression analysis have more independent variables from which they wish to choose for inclusion in their final equation. There are many good reasons for attempting a reduction in the number of variables for final equation. Aside from the principle of parsimony there is no point in including variables that have little effect on the dependent variables. Further, it is often difficult and expensive to gather information on many variables. We usually need more than one computer run on any given data to arrive at a final prediction equation and computing time goes up sharply with the increased number of independent variables. The more number of independent variables has a strong bearing on the accuracy of computations and reliability of results.

A number of problems arise because of the non-orthogonality of regression data. If the design matrix 'X' has orthogonal columns, the effects of individual variables are clear and the problems of estimation and subset selection are elementary. Unfortunately, with undesigned experiments the columns of X are rarely orthogonal and there may exist near dependencies which, in turn, cause high correlations between variables or sets of variables. In such cases, it is generally difficult to assess the effects of individual factors as their relations with other factors are often complex. In order to determine the chief causes of something, we need to select the best variables. The problem of determining the 'best' subset of variables has long been of interest to applied statisticians and primarily because of the current availability of high speed computers, this problem has received considerable attention in the recent statistical literature.

The problem of selecting a subset of independent or predictor variables is usually described in an idealized setting. That is, it is assumed that (a) the analyst has data on a large number of potential variables and (b) the analyst has 'good data'. The assumption of 'good data' includes the usual linear model assumptions such as homogeneity of variance, etc. A serious problem which is included under this heading is that of multicollinearity among the independent variables. The consequences of near degeneracy of the matrix of independent variables have been described by a number of authors. For example, see the text by Johnston (1972) or the paper by Mason et al. (1975). In practice, the lack of these assumptions may make a detailed subset selection analysis a meaningless exercise.

The problem of etermining an appropriate equation based on a subset of the original set of variables contains three basic ingredients, namely (i) the computational technique used to provide the information for the analyst, (ii) the criteria used to analyse the variables and select a subset, if that is appropriate, and (iii) the estimation of the coefficients in the final equation. In this paper we shall devote ourselves with decreasing the computational effort using C_p-statistic of Mallows as the basic criterion for comparing two regressions. The problem of determining appropriate equation has been reviewed in length by Hocking (1976).

To provide a basis for the discussion, section 2 contains a review of the consequences of incorrect model specification which provides a theoretical motivation for variable deletion. A number of computational techniques are reviewed and a new technique is described in section 3. A technique is developed which will indicate 'good' regressions with a minimum of computation. In section 4 a numerical example is included to illustrate the essential points.

NOTATION AND BASIC CONCEPTS

2.1 Notation and Assumptions:

It is assumed that there are $n > (t+1)$ observations on a t-vector of input variables, $x' = (x_1,\ldots,x_t)$, and a scalar response, y , such that the j-th response, $j = 1,2,\ldots,n,$ is determined by

$$y_j = \beta_0 + \sum_{i=1}^{t} \beta_i x_{ij} + \varepsilon_j \quad \ldots \qquad (2.1)$$

The errors, ε_j , are assumed identically and independently distributed, usually normal, with mean zero and unknown variance σ^2 .

The model (2.1) is frequently expressed in matrix notation as

$$Y = X\beta + \varepsilon \quad \ldots \qquad (2.2)$$

Here Y is the n-vector of observed responses, X is the design matrix of dimension $n \times (t+1)$, assumed to have rank $(t+1)$, and β is the $(t+1)$ - vector of unknown regression coefficients.

In the variable selection problem, let r denote the number of terms which are deleted from model (2.1), that is, the number of coefficients which are set to zero. The number of terms which are retained in the final equation will be denoted by $p = t + 1 - r$. Statistics associated with a p-term model will be subscripted by p while those associated with the full model will not be subscripted.

2.2 Consequences of Incorrect Model Specification

The section provides a brief review of the consequences of incorrectly specifying the model either in terms of retaining extraneous variables or deleting relevant variables. Consequences are summarized in three results.

Result 1: In the classical linear regression model, omission of a variable specified by the truth introduces bias in all the least squares estimates and decreases the variance of all the least squares estimates.

Result 2: In the classical linear regression model, discarding an independent variable whose parameter value is small (in magnitude) than the theoretical standard deviation of its estimate (from given data) will decrease the mean square error of all the least squares estimates.

Result 3: In the classical regression model, inclusion of an irrelevant variable does not introduce bias in the least squares estimates, and increases the variance and mean square error of all the least squares estimates.

The properties of the least squares estimates have been described by several authors with recent results given by Walls and Weeks (1969), Rao (1971) and Hocking (1974).

COMPUTATION TECHNIQUES

This section contains a discussion of a number of computational procedures which will provide information on some or all of the subset combinations. Attention is focussed primarily on least squares fitting. Mantel (1970) has pointed out that many procedures are now available for selecting variables in multiple regression analyses. The most important are: -

1. Forward Selection (or Step-Up).
2. Backward Elimination (or Step-Down).
3. Convensional Stepwise. In this procedure, due to Efroymson (1960), one has two F-levels, say F_{IN} and F_{OUT}. At each step, a variable is discarded if this would increase the residual sum of squares by not more than F_{OUT} times the current residual mean square. If there is a

choice of variable to be discarded, then the one given the smallest increase in the residual sum of squares is chosen. If no variable can be discarded in this way, then a variable is introduced if this would reduce the residual sum of square by at least F_{IN} times the residual mean square after introducing the new variable. Again, if there is a choice of variable to be introduced, then the one giving the largest decrease in the residual sum of squares is chosen.

4. Forward Selection followed by Conventional Stepwise. This in effect starts the stepwise procedure when all linearly independent variables are in the equation.
5. Optimum Regression, as described by Garside (1965), Hocking and Leslie (1967), Beale, Kendall and Mann (1967), La Motte and Hocking (1970) or Beale (1970). This involves finding the equation minimizing the residual sum of squares for any given number of non-zero regression coefficients. Garside describes an efficient method of enumerating all 2^t equations obtainable by selecting any subset from the t independent variables, assuming that they are not linearly dependent. The other papers describe methods of finding a guaranteed optimum solution to this problem without explicitly enumerating all combinations.

Many of these procedures are discussed in Chapter 6 of Draper and Smith (1966). They can be compared both in terms of computational effect and in terms of the desirability of the end product. Many writers, such as Mantel (1970), plead that there is an inherent virtue in Method 2, i.e. backward elimination. We propose and justify here a method which retains the merits of backward elimination and results in decrease of the computational effort.

3.1 Description of the Method

Any departure from the pairwise orthogonality of regressor variables means the existence of multicollinearity and the severity of multicollinearity increases as $|X'X|$ moves closer to zero. If the columns of X are so colinear as to be close to the case of 'exact linear combinations' but not quite the case, it can be appreciated that $|X'X|$ can be close to zero and

$$\mathrm{Var}(\hat{\underset{\sim}{\beta}}) = (X'X)^{-1}\,\sigma^2 \text{ can 'blow up' since } (X'X)^{-1}$$
$$\text{is defined as } (X'X)^{-1} = |X'X|^{-1}\,(\mathrm{adj}\ X'X)' .$$

This is a serious case of multicollinearity. For the consequences of severe multicollinearity and rules for its detection one may see Tayal (1979). The presence of severe multicollinearity violates the assumption of 'good data' and only biased estimation procedures are the solution and variable selection procedures listed earlier will not be adopted. It is, therefore, assumed that severe multicollinearity is not present in the data.

Consider a simple case of one regressor variable. Let the model be given by

$$Y = \beta_0 + \beta_1 X_1 + \varepsilon \quad \ldots \tag{3.1}$$

Rewriting the model (3.1)

$$Y = \alpha + \beta_1(X_1-\bar{X}_1) + \varepsilon \quad \ldots \tag{3.2}$$

The least squares estimates of the parameters in the model (3.1) are correlated unless $\bar{X}$, is zero. However, the least squares estimates of the parameters in the model (3.2) are uncorrelated. This implies that centering of the regressor variable has eliminated non-orthogonality. Centering removes the non-essential ill-conditioning, thus reducing the variance inflation in the coefficient estimates. In a linear model centering removes the correlation between the constant term and all linear terms. In addition, in a quadratic model centering reduces, and in certain situations completely removes, the correlation between the linear and quadratic terms. In case of several regressor variables centering of the regressor variables will reduce the extent of non-orthogonality among column vectors of X. The variance inflation factor for each term in the model measures the collective impact of simple correlations on the variance of the coefficient of that term. So a procedure for selecting variable which utilizes the above fact is bound to be superior over those procedures which ignore it. The method proposed here does make full advantage of this observation.

Assuming that there are only p real regressor variables. With p regressors the model is given by

$$Y_j = \beta_0 + \sum_{i=1}^{p} \beta_i x_{ij} + \varepsilon_j \text{ , } j = 1,2,\ldots,n \quad \ldots \tag{3.3}$$

If we fit the model (3.3) the estimation space is of dimension 'p+1' and the error space is of dimension 'n-p-1'. These two spaces are orthogonal to each other. Now suppose we introduce 'r' extraneous variables in the model (3.3) then it is given by the model (2.1). Here the estimation space is of dimension (t+1) and the error space is of dimension 'n-t-1'. The 't-p' extra dimensions of estimation space are derived from the first error space of dimension (n-p-1) and is orthogonal to the first estimation space of dimension (p+1). In other words we say that the 'r' regressor variables have been fitted through errors. Since these r regressor variables have been fitted through errors the effect of centering these variables is not likely to effect their corresponding variance inflation factor. Hence, the comparison of the variation inflation factor before centering and after centering will indicate which regressor variables have prediction power. In some situations it may happen that some regressor variables are already almost orthogonal to the remaining other regressors in the model and centering of these variable will make no effect, as a consequence the comparison of their variation inflation factors before and after centering are going to be almost the same except little change in the values due to rounding off errors and this will be taken care by the criterion of absolute t values for the estimated regression coefficients. The basic steps in the proposed procedure are as follows:-

1. Fit the model

   $$Y_j = \beta_o + \sum_{i=1}^{t} \beta_i X_{ij} + \varepsilon_j$$

   for regressor variables in original form and calculate the estimated standard errors of the parameters.

2. Fit the model

   $$Y_j = \beta_o' + \sum_{i=1}^{t} \beta_i(X_{ij}-\bar{X}_i) + \varepsilon_j$$

   and calculate the estimated standard errors of the parameters.

3. Compare the two standard errors of the parameters for the two regression equations and consider those regressor variables variables for dropping whose estimated standard errors are not reduced on centering of the regressor variables and whose $|t|$ values are less than one.

4. Suppose the number of regressor variables retained in the model after step 3 is 'p'. Proceed now with the usual step down procedure or having decided on 'p' variables, calculate C_p , statistics for all possible subsets having (p-1) regressor variables and choose the subset of variables having the smallest C_p statistic. Calculate C_p statistics for all possible subsets having (p-2) regressor variables (choosing from the p-1 variables) and choose the subset with the smallest C_p value and continue the process.

5. From these selected subsets having p , p-1 , p-2, ..., 1 regressor variables, choose the one with the smallest C_p and the corresponding regression equation is the finally selected equation.

ILLUSTRATION

The experimental data considered here is reported in Tayal (1979). The objective of the experiment is to study the soil test values and yield of wheat crop and to evolve a basis for making specific fertilizer recommendations. Variables studies in the experiment are described below:

Description of Variables

Y	Yield of wheat
X_1	Applied Nitrogen (N)
X_2	Applied Phosphorus (P)
X_3	Applied Potash
X_4	Nutrient uptake of N
X_5	Nutrient uptake of P
X_6	Nutrient uptake of K
X_7	Soil test value of N
X_8	Soil test value of P
X_9	Soil test value if K

Observations from seventy eight experimental units are available for analysis. Correlation matrix is reported in Table 1. Thumb rule for detection of multicollinearity reveals that it can be treated to be tolerable. The experimental data are analysed for the selection of variables in linear regression by three different methods and results for each method are presented here.

Table 1

Table 1. Correlation matrix

Y	1.0000									
X_1	0.7920	1.0000								
X_2	0.3920	0.0571	1.0000							
X_3	-0.2670	-0.0000	-0.0328	1.0000						
X_4	0.8980	0.7835	0.2768	-0.0176	1.0000					
X_5	0.8971	0.6228	0.4475	0.0542	0.7812	1.0000				
X_6	0.8852	0.6759	0.4280	-0.1086	0.8150	0.8875	1.0000			
X_7	-0.1363	-0.0159	0.0229	-0.0334	-0.1745	-0.2417	-0.0758	1.0000		
X_8	-0.0482	0.0182	0.0064	0.0000	0.1886	0.0106	-0.0099	0.1097	1.0000	
X_9	0.1482	0.0312	0.0189	-0.0092	0.1874	0.1697	0.1197	-0.3280	-0.2905	1.0000

4.1 Selection of Variables by Backward Elimination

The backward elimination procedure yields the final prediction equation as:

$$\hat{Y} = 3.42 + \underset{(0.008)}{0.43\, X_1} + \underset{(0.009)}{0.29\, X_2} + \underset{(0.032)}{0.15\, X_4} + \underset{(0.293)}{2.34\, X_5}$$

Figures in parenthesis indicate estimated standard errors and the value of R^2, the coefficient of determination, turns out to be 0.921. The sequence in which the variables are dropped in stages is X_8, X_9, X_3, X_7 and X_6. The significance level for Partial F-test is taken at 0.05. Note that to arrive at the final prediction equation the number of fitted equations happens to be (1+9+8+7+6+5) = 36.

4.2 Selection of Variables by Forward Selection

The forward selection procedure is terminated when variables X_4, X_5, X_1 and X_2 are in the regression equation. Interesting the two procedures Step-Down and Step-Up yield the same final prediction equation. The number of regression equations fitted for forward selection procedure is 35.

In order to examine the performance of other selection criteria namely, residual mean squares (RMS), squared multiple correlation coefficient (R_p^2), adjusted R^2 ($\bar{R}_p^2$) and total squared error (C_p) defined as $C_p = \frac{RSS_p}{\hat{\sigma}^2} + 2\,p - n$, where $\hat{\sigma}^2$ is an estimate of error variance from full model and RSS is the residual sum of squares, computations are carried out for all possible regressions. For a given p the fitted equation yielding the best value of these selection criteria is considered. The results are reported in Table 2.

Table 2

P	RMS_p	R_p^2	$\bar{R}_p^2$	C_p
2	24.62	0.8048	0.8022	115.35
3	12.32	0.9036	0.9010	21.48
4	10.13	0.9218	0.9186	5.87
5	9.49	0.9277	0.9238	2.09
6	9.44	0.9291	0.9242	2.76
7	9.50	0.9296	0.9236	4.29
8	9.62	0.9286	0.9227	6.13
9	9.74	0.9299	0.9218	8.00
10	9.88	0.9299	9.9206	8.00

A close examination reveals that all the selection criteria yield p = 5 as the best implying that the final prediction equation should contain only 4 independent variables and the resulting prediction equation turns out to be the same as reported earlier. Graphical representations of C_p, R_p^2, $\bar{R}_p^2$ and RMS_p are given in Figures I and II. In graphical representation we take 'p' as the number of independent variables and not the number of parameters in the regression equation. Examination of C_p, R_p^2, $\bar{R}_p^2$ and RMS_p plots against p in Fig. I and II indicates that the number of variables to enter the final equation should be taken to be four.

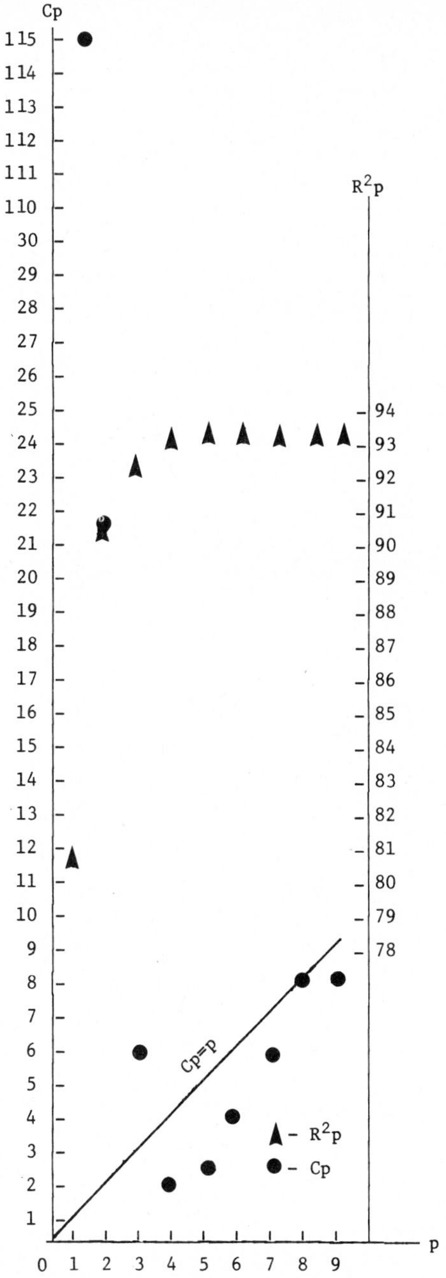

Fig. I. Cp and R^2p (10^2) Plots for Example I

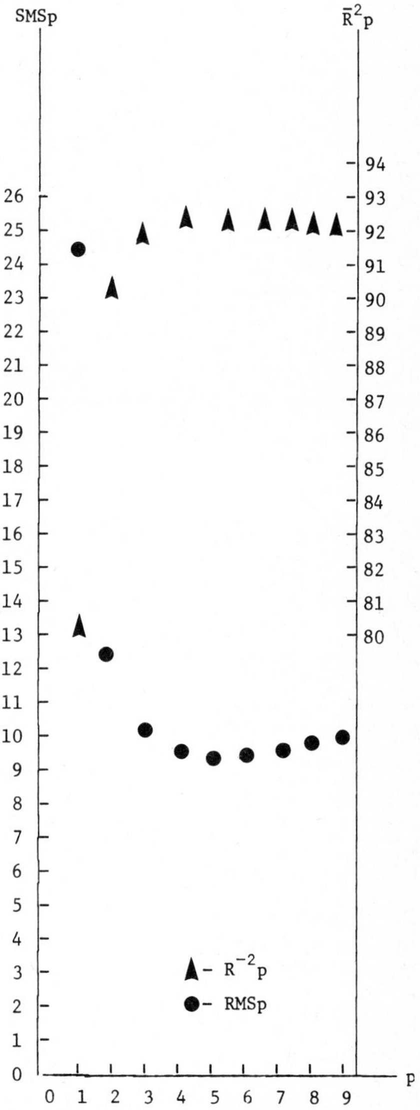

Fig. II. RMSp and $R^{-2}p$ (10^2) Plots for Example I

4.3 Selection of Variables by Proposed Procedure

The computations for the proposed procedure are given in Table 3.

Table 3

Parameter Estimates	$SE(b_i)$ for Non-centered data	$SE(b_i)$ for Centered data	$\|t\|$ values
b_1	0.0979	0.0082	4.05
b_2	0.1245	0.0093	1.98
b_3	0.1208	0.0129	0.40
b_4	0.0358	0.0345	4.09
b_5	0.4210	0.3613	5.11
b_6	0.1900	0.1009	0.96
b_7	0.0132	0.0142	0.76
b_8	0.0792	0.0856	0.76
b_9	0.0042	0.0048	0.07

Examination of the absolute 't' values of estimated regression coefficients shows that $|t|$ values of regression coefficients b_3, b_6, b_7, b_8, and b_9 are less than one. Comparison of the estimated standard errors of two sets of estimated regression coefficients for non-centered and centered data show that regression coefficients whose standard errors are reduced on centering of data are b_1, b_2, b_3, b_5 and b_6. Combining these two criteria of $|t|$ values and reduced standard errors, we see that variables X_7, X_8 and X_9 have no prediction powers and should be dropped simultaneously in search for good regression equation with minimum number of variables. With the remaining variables Step-down procedure is followed in the selection of variables. Obviously the final prediction equation turns out to be the same.

The proposed procedure results in saving of 24 regression equation fittings and arrives at the same result as obtainable through other selection procedures. This is a tremendous decrease in computational efforts and is much faster over other selection procedures. The method described here for elimination of variables should not be judged alone on the example presented here. However, the method has been applied to several other regression problems where elimination of variables was suggested and in every case the performance has been comparable to the one discussed here. At this time we can only say that out limited experience has been encouraging.

REFERENCES

[1] Beale, E.M.L. (1970). Selecting an Optimum Subset. Integer and Nonlinear Programming. Ed. J. Abadie, North Holland Publishing Co. Amsterdam.

[2] Beale, E.M.L., Kendall, M.G., and Mann, D.W. (1967). The Discarding of Variables in Multivariate Analysis. Biometrika 54, 357-366.

[3] Draper, N.R., and Smith, H. (1966). Applied Regression Analysis. John Wiley Sons, Inc., New York.

[4] Efroymson, M.A. (1960). Multiple Regression Analysis. Mathematical Methods for Digital Computers. Ed. A. Ralston and H.S. Wilf. John Wiley Sons, Inc., New York.

[5] Garside, M.J. (1965). The Best Sub-set in Multiple Regression Analysis. Applied Statistics 14, 196-200.

[6] Hocking, R.R. (1974). Misspecification in Regression. The American Statistician 28, 39-40.

[7] Hocking, R.R. (1976). Analysis and Selection of Variables. Biometrics 32, 1-49.

[8] Hocking, R.R., and Leslie, R.B. (1967). Selection of the Best Subset in Regression Analysis. Technometrics 9, 531-540.

[9] Johnston, J. (1972). Econometrics Methods. (2nd edition). McGraw Hill, New York.

[10] La Motte, L.R., and Hocking, R.R. (1970). Computational Efficiency in the Selection of Regression Variables. Technometrics 12, 83-93.

[11] Mallows, C.L. (1973). Some Comments on C_p . Technometrics 15, 661-75.

[12] Mantel, N. (1970). Why Step-down Procedures in Variable Selection. Technometrics 12, 6 1-25.

[13] Mason, R.L., Webster, J.T., and Gunst, R.F. (1975). Sources of Multicollinearity in Regression Analysis. Communication in Statistics 4, 277-92.

[14] Rao, P. (1971). Some Notes on Misspecification in Multiple Regression. The American Statistician 25, 27-9.

[15] Tayal, V. (1979). Comparison of Methods of Selection of Variables in Multiple Regression. Unpublished M.Sc. thesis. Haryana Agricultural University. Hissar, India.

[16] Walls, R.E. and Weeks, D.L. (1969). A note on the Variance of a Predicted Response in Regression. The American Statistician 23, 24-6.

Multivariate Statistical Analysis
R.P. Gupta (ed.)
© North-Holland Publishing Company, 1980

ASYMPTOTIC DISTRIBUTION OF LATENT ROOTS AND APPLICATIONS

M.S. Srivastava
Department of Mathematics
University of Toronto
Toronto, Ontario

E.M. Carter
Department of Mathematics
University of Guelph
Guelph, Ontario

In this paper the asymptotic distribution of the normalized roots of (i) the non-central Wishart distribution, (ii) the non-central multivariate F-distribution (MANOVA) and (iii) the non-central distribution of the canonical correlations are derived up to $O(n^{-3/2})$, and their applications in several multivariate tests are discussed.

1. INTRODUCTION

The asymptotic distribution of the latent roots of 'non-central' matrices plays an important role in many multivariate testing problems. For example, in multivariate analysis of variance even the distribution of the test statistic for testing the dimensionality of the 'mean' matrix, depends on the roots of the mean matrix. And to compare the power of several available test statistics in many situations, the asymptotic distribution of the latent roots of 'non-central' matrices are needed. There are several ways of deriving the asymptotic distribution of the latent roots (or normalized roots). The first term can be obtained by applying Hsu's Theorem (Hsu 1948). To get terms of order $O(n^{-1})$, one can apply perturbation methods. See, for example, Fujikoshi (1977). However, the perturbation method does not work when the function of the roots is asymmetric, such as the distribution of the maximum characteristic roots. To get terms of higher order, as well as to obtain results of asymmetric functions, the methods of Khatri and Srivastava (1977) and Srivastava and Carter (1979) are more suitable. Using these results, we give in this paper the asymptotic distribution of the normalized roots of (i) the non-central Wishart distribution, (ii) the non-central multivariate F-distribution (MANOVA) and (iii) the canonical correlation up to the order $O(n^{-3/2})$. Several applications of these results are also mentioned.

2. DISTRIBUTION OF THE NORMALIZED ROOTS OF A NON-CENTRAL WISHART MATRIX.

2.1. Introduction. Let $X \sim N_{p,n}(M,I_p,I_n)$. That is, let $E(X) = M$ and let the n columns of X be independently and normally distributed with covariance matrix I_p . Then the distribution of $S = n^{-1}X X'$ has a non-central Wishart distribution with non-centrality matrix $M M'$. The distribution of the characteristic roots of S has been derived by Carter and Srivastava (1979). We now give some applications of this distribution.

Example 2.1. Tests for Non-additivity. Consider the following two-way classification model with one observation per cell:

$$y_{ij} = \mu + \tau_i + \beta_j + \gamma_{ij} + \varepsilon_{ij} \ , \ i=1,\ldots,t \ , \ j = 1,\ldots,b \ , \tag{2.1}$$

where τ_i is the 'treatment' effects, β_j the 'block' effects and γ_{ij} are the 'interaction' effects. Without loss of generality, assume $b \le t$. Under the usual assumptions that ε_{ij} are independent $N(0,\sigma^2)$ and $\Sigma_i\tau_i = 0$, $\Sigma_j\beta_j = 0$, $\Sigma_i\gamma_{ij} = 0 = \Sigma_j\gamma_{ij}$, we iwsh to test that $\gamma_{ij} = 0$. That is, that there is no interaction. Letting $\Gamma = (\gamma_{ij})$, we can write

$$\Gamma = \sum_{i=1}^{p} a_i \, \underline{u}_i \, \underline{v}_i' \ , \quad p = b-1 \ , \tag{2.2}$$

where a_i's are the non-zero characteristic roots of $\Gamma\Gamma'$ or $\Gamma'\Gamma$, $\underline{u}_1,\ldots,\underline{u}_p$ are the eigenvectors corresponding to the roots $a_1 \ge \ldots \ge a_p \ge 0$ of $\Gamma\Gamma'$ and $\underline{v}_1,\ldots,\underline{v}_p$ are the eigenvectors corresponding to the roots $a_1 \ge \ldots \ge a_p \ge 0$ of $\Gamma'\Gamma$. (Note that $a_b = 0$.) In the case where

$$\Gamma = a_1 \, \underline{u}_1 \, \underline{v}_1' \tag{2.3}$$

that is, there is only one non-zero root, Johnson and Graybill (1972) and Corsten and Van Eijnsbergen (1972), showed that the likelihood ratio test was given by

$$\Lambda_1 = \frac{\ell_1}{\ell_1 + \ldots + \ell_p} \tag{2.4}$$

where $\ell_1 > \ell_2 > \ldots > \ell_p$ are the roots of $n^{-1}ZZ'$ and

$$Z = (z_{ij}) \ , \ z_{ij} = y_{ij} - \bar{y}_{i\cdot} - \bar{y}_{\cdot j} + \bar{y}_{\cdot\cdot} \ . \tag{2.5}$$

It can be shown that ZZ' has a non-central Wishart distribution $W_p(\sigma^2 I,n,\Omega)$, where $n = t-1$, and the non-zero roots of Ω are the same as the non-zero roots of $\Gamma\Gamma'$.

In the above it is not known, however, how many of the a_i's (except a_b) are zero. Thus, if

$$\Gamma = a_1 \, \underline{u}_1 \, \underline{v}_1' + a_2 \, \underline{u}_2 \, \underline{v}_2' \tag{2.6}$$

and we wish to test the hypothesis that $a_1 = a_2 = 0$, the likelihood ratio test is given by

$$\Lambda_2 = \frac{\ell_1 + \ell_2}{\ell_1 + \ldots + \ell_p} \quad . \tag{2.7}$$

Similarly, if

$$\Gamma = a_1 \underline{u}_1 \underline{v}_1' + \ldots + a_q \underline{u}_q \underline{v}_q' \, , \; q < p \, , \tag{2.8}$$

and we wish to test the hypothesis that $a_1 = a_2 = \ldots = a_q = 0$, the likelihood ratio test is given by

$$\Lambda_q = \frac{\ell_1 + \ldots + \ell_q}{\ell_1 + \ldots + \ell_p} \quad . \tag{2.9}$$

It raises a natural question which of these statistics should be used for testing 'no interaction'. Obviously, it will never be known how many of the a_i's are different from zero. In the absence of such a knowledge Λ_1 becomes a natural contender. However, its performance should be investigated when several of the a_i's are different from zero. Such an investigation can be carried out by obtaining its non-null distribution. This requires the asymptotic distribution of the normalized roots of the non-central Wishart distribution. The asymptotic distribution of Λ_1 along with tabulated values of its power are given at the end of this section. We find that its power is less than satisfactory when several a_i's are different from zero except when a_1 is much larger than other a_i's . Obviously, when several of the a_i's are different from zero, some other test should be used. We thought of using the test statistic

$$L = (\Pi_{i=1}^p \ell_i)^{1/p} / (\Sigma_{i=1}^p \ell_i / p) \; . \tag{2.10}$$

However, this test turned out to have low power. For this reason, the tabulated values of its power are not reported except that we give its asymptotic distribution at the end of this section. Currently, another alternative test is under investigation. This test is given by

$$Q = \Sigma_{i=1}^p (u_i - \bar{u})^2 \, , \text{ where } \; u_i = \ln \ell_i \; \text{ and } \; \bar{u} = p^{-1} \Sigma u_i \; . \tag{2.11}$$

2.2. ASYMPTOTIC DISTRIBUTION OF THE ROOTS OF A NON-CENTRAL WISHART MATRIX

Let $X \sim N_{p.n}(M, I_p \otimes I_n)$. Then the distribution of $S = n^{-1}XX'$ is given by

$$\Gamma_p^{-1}(\tfrac{1}{2}n)(\tfrac{1}{2}n)^{\frac{1}{2}np}|S|^{\frac{1}{2}(n-p-1)} \quad \mathrm{etr}[-\tfrac{1}{2}nS - \tfrac{1}{2}MM']\,{}_0F_1(\tfrac{1}{2}n : \tfrac{1}{4}SMM') .$$

Let $\theta_1 > \ldots > \theta_p$ be the roots of S. By invariance the distribution of $\theta_1,\ldots,\theta_p$ depends only on the roots of MM'. Hence we let $MM' = nA$, where $A = \mathrm{diag}(a_1,\ldots,a_p)$, $a_1 > \ldots > a_q > a_{q+1} = \ldots = a_p = 0$. The asymptotic distribution of $\theta_1,\ldots,\theta_p$ has been derived in detail in Carter and Srivastava (1979). For completeness we give here the distribution of the transformed roots

$$x_i = n^{1/2}(\theta_i - (1+a_i)) , \quad i = 1,\ldots,q ,$$

$$x_j = n^{1/2}(\theta_j - 1) , \quad j = q+1,\ldots,p .$$

That is, the distribution of $x_1,\ldots,x_q$ is given by

$$(2\pi)^{-\frac{1}{2}q} \prod_{i=1}^{q} \sigma_i^{-1}\exp[-x_i^2/(2\sigma_i^2)]$$

$$\prod_{q\le i<j\le p} (x_i - x_j) \prod_{j=q+1}^{p} \exp[-\tfrac{1}{4}x_j^2]$$

$$\pi^{\frac{1}{2}(p-q)^2}\,\Gamma_{p-q}^{-1}(\tfrac{1}{2}(p-q))(2\pi)^{-\frac{1}{2}(p-q)(p-q+1)}\,2^{-\frac{1}{2}(p-q)}$$

$$\{1 + n^{-1/2}T_1 + n^{-1}T_2 + O(n^{-3/2})\} ,$$

where

$$\sigma_i^2 = 2(1+2a_i) ,$$

$$T_1 = \sum_{i=1}^{q} (1+2a_i)^{-1}[\tfrac{1}{6}(1+2a_i)^{-2}(1+3a_i)x_i^3 - (1+2a_i)^{-1}(1+3a_i)x_i$$

$$- \tfrac{1}{2}(p-1)x_i] + \tfrac{1}{2} \sum_{i\le i<j\le q} (a_i - a_j)^{-1}(x_i - x_j)$$

$$+ \tfrac{1}{2}\sum_{i=1}^{q}\sum_{j=q+1}^{p} a_i^{-1}(x_i - x_j) - \tfrac{1}{2}(p+1)\sum_{j=q+1}^{p} x_j + \tfrac{1}{6}\sum_{j=q+1}^{p} x_j^3 ,$$

and

$$T_2 = \sum_{i=1}^{q} [\tfrac{1}{2}(1+2a_i)^{-2}(1+4a_i)(\tfrac{1}{4}x_i^4(1+2a_i)^{-2} - 3x_i^2(1+2a_i)^{-1} + 3)$$

$$+ \tfrac{1}{9}(1+3a_i)^2(1+2a_i)^{-3}(\tfrac{1}{8}x_i^6(1+2a_i)^{-3} - \tfrac{15}{4}x_i^4(1+2a_i)^{-2}$$

$$+ \frac{45}{2} x_i^2(1+2a_i)^{-1}-15) - \frac{1}{3}(p-1)(1+3a_i)(1+2a_i)^{-2}(\frac{1}{4} x_i^4(1+2a_i)^{-2}$$

$$- 3x_i^2(1+2a_i)^{-1}+3) + \frac{1}{4}(p-1)^2(1+2a_i)^{-1}(\frac{1}{2} x_i^2(1+2a_i)^{-1}-1)$$

$$- \frac{1}{2}(p-1)(1+2a_i)^{-1}(\frac{1}{2} x_i^2(1+2a_i)^{-1}-1)] - \frac{1}{4} \sum_{i=1}^{q} \sum_{j=q+1}^{p} a_i^{-2}(x_i-x_j)^2$$

$$+ \frac{1}{8} [\sum_{i=1}^{q} \sum_{j=q+1}^{p} a_i^{-1}(x_i-x_j)]^2 - \frac{1}{4} \sum_{1\le i<j\le q} (x_i-x_j)^2(a_i-a_j)^{-2}$$

$$+ \frac{1}{8} [\sum_{1\le i<j\le q} (a_i-a_j)^{-1}(x_i-x_j)]^2 + [\sum_{k=1}^{q} \frac{1}{6} x_k^3(1+2a_k)^{-3}(1+3a_k)$$

$$- \frac{1}{2} (p-1)(1+2a_k)^{-1}x_k - (1+2a_k)^{-2}(1+3a_k)][\sum_{\substack{i=1\\ i\neq k}}^{q} (\frac{1}{6} x_i^3(1+2a_i)^{-3}(1+3a_i)$$

$$- \frac{1}{2} (p-1)(1+2a_i)^{-1}x_i - (1+2a_i)^{-2}(1+3a_i)) + \frac{1}{2} \sum_{1\le i<j\le q} (a_i-a_j)^{-1}(x_i-x_j)$$

$$+ \frac{1}{2} \sum_{i=1}^{q} \sum_{j=q+1}^{p} a_i^{-1}(x_i-x_j) - \frac{1}{2} (p-1) \sum_{j=q+1}^{p} x_j + \frac{1}{6} \sum_{j=q+1}^{p} x_j^3]$$

$$+ \frac{1}{4} \sum_{1\le i<j\le q} \sum_{k=1}^{q} \sum_{\ell=q+1}^{p} (x_i-x_j)(x_k-x_\ell)(a_i-a_j)^{-1}a_k^{-1}$$

$$+ \sum_{1\le i<j\le q} (a_i+a_j)(a_i-a_j)^{-1} + \frac{1}{2} (p-q) \sum_{i=1}^{q} (1+a_i) - \frac{1}{6} (p-q)$$

$$- \frac{1}{24} p(p-1)(2p+5) + \frac{1}{24} q(q-1)(2q+5) + \frac{1}{4} q(p-1)(3p+1)$$

$$+ \frac{1}{2} (\frac{1}{6} \sum_{i=q+1}^{p} x_i^3 - \frac{1}{2} (p+1) \sum_{i=q+1}^{p} x_i)^2 + \frac{1}{4} (p+1) \sum_{i=q+1}^{p} x_i^2 - \frac{1}{8} \sum_{i=q+1}^{p} x_i^4$$

$$+ \frac{1}{2} [\frac{1}{6} \sum_{j=q+1}^{p} x_j^3 - \frac{p+1}{2} \sum_{j=q+1}^{p} x_j][\sum_{1\le i<j\le q} (a_i-a_j)^{-1}(x_i-x_j)$$

$$+ \sum_{i=1}^{q} \sum_{j=q+1}^{p} a_i^{-1}(x_i-x_j)] .$$

2.3. Asymptotic distribution of Λ_1 . When only one root $a_1 \neq 0$, the asymptotic distribution of Λ_1 is given by Carter and Srivastava (1979). In this subsection, we consider the case when $a_i \neq 0$, $i = 1,2,\ldots,q$, $q < p$. Following as in Carter and Srivastava (1979), the asymptotic distribution of Λ_1 is given by

$$\mu = (1+a_1)(p + \sum_{i=1}^{q} a_i)^{-1} ,$$

$$\sigma_i^2 = 2(1+2a_i) , \quad i = 1,2,\ldots,q ,$$

$$\mu_1 = [(1+a_1)^{-1} - \frac{1}{2}\gamma]\sigma_1^2 ,$$

$$\gamma = 2(p + \sum_{i=1}^{q} a_i)^{-1} ,$$

$$\sigma^2 = \mu_1^2 \sigma_1^{-2} + \frac{1}{2}(p-q)\gamma^2 + \frac{1}{4}\gamma^2 \sum_{i=2}^{q} \sigma_i^2 ,$$

$$\sigma\alpha_1 = -\frac{1}{2}(p+1)\gamma - \frac{1}{2}(p-1)\mu_1(1+2a_1)^{-1} + \frac{1}{2}(p-q)a_1^{-1}(\mu_1+\gamma)$$

$$+ \frac{1}{2}\sum_{i=2}^{q}(\mu_1 + \frac{1}{2}\gamma\sigma_i^2)(a_1-a_i)^{-1} - \frac{1}{2}(1+a_1)^{-2}\sigma_1^2$$

$$+ \frac{1}{4}\gamma^2[\sum_{i=1}^{q}\sigma_i^2 + (p-q)(p-q+1)] ,$$

and

$$\sigma^3\alpha_3 = -\frac{1}{2}(1+a_1)^{-2}\mu_1^2 + \frac{1}{2}(p + \sum_{i=1}^{q} a_i)^{-2}[\mu_1 - (p-q)\gamma - \frac{1}{2}\gamma\sum_{i=2}^{q}\sigma_i^2]^2$$

$$+ \frac{1}{6}(1+3a_1)(1+2a_1)^{-3}\mu_1^3 - \frac{1}{6}\gamma^3\sum_{i=2}^{q}(1+3a_i) .$$

Table 1. Power of the likelihood ratio test for non-additivity at $\alpha = .05$

(p,n)			(4,15)	(6,15)	(3,10)	(5,10)
a_1	a_2	a_3	Power	Power	Power	Power
1.0	0.0	0.0	.1571	.1712	.1003	.1564
1.5	0.0	0.0	.1982	.2193	.1015	.1047
2.0	0.0	0.0	.2853	.2876	.1309	.1777
1.5	1.0	0.0	.1940	.2139	.1566	.1876
2.0	1.0	0.0	.2087	.2526	.1294	.1824
3.0	1.0	0.0	.3328	.4061	.1405	.2223
2.0	1.5	1.0	.1037	.1534	.0765	.1225
3.0	2.0	1.0	.1155	.2024	.0571	.1196
4.0	2.0	1.0	.1677	.3031	.0583	.1400

Table 2. Power of the likelihood ratio test for non-additivity at $\alpha = .01$

(p,n)			(4,15)	(6,15)	(3,10)	(5,10)
a_1	a_2	a_3	Power	Power	Power	Power
1.0	0.0	0.0	.0923	.1034	.0585	.1008
1.5	0.0	0.0	.1070	.1388	.0461	.1047
2.0	0.0	0.0	.1390	.1810	.0454	.1111
1.5	1.0	0.0	.0420	.1175	.1895	.1076
2.0	1.0	0.0	.0487	.1471	.0711	.1064
3.0	1.0	0.0	.0791	.2451	.0662	.1267
2.0	1.5	1.0	.0080	.0746	.0384	.0623
3.0	2.0	1.0	.0062	.1033	.0262	.0593
4.0	2.0	1.0	.0062	.1638	.0254	.0754

In the above tables, the percentage points were taken from Krishnaiah and Schuurmann (1974).

2.4. Asymptotic distribution of L. The asymptotic non-null distribution of L is given by

$$P(n^{1/2}(\ln L - \ln \mu)/\sigma \le \xi) =$$

$$\Phi(\xi) - n^{-1/2}\psi(\xi)[\frac{1}{6}\sum_{i=1}^{q}\frac{(1+3a_i)\mu_i^3}{(1+2a_i)^3}(\xi^2-1) + \frac{(p-q)}{6}\gamma^3(\xi^2-1)$$

$$+ \frac{1}{2}\sum_{i=1}^{q}(p-q)(\mu_i-\gamma)a_i^{-1} - \frac{(p-1)}{2}\sum_{i=1}^{q}\mu_i(1+2a_i)^{-1}$$

$$+ \frac{1}{2}\sum_{1\le i<j\le q}\frac{\mu_i-\mu_j}{a_i-a_j} - \frac{1}{2}(p-q)q\gamma] + O(n^{-1}) , \quad \text{where}$$

$$\mu = \frac{p^p \prod_{i=1}^{q}(1+a_i)}{(p + \sum_1^q a_i)^p} , \quad \sigma_i^2 = 2(1+2a_i) , \quad i = 1,\ldots,q ,$$

$$\mu_i = [(1+a_i)^{-1} - p(p + \Sigma a_i)^{-1}]\sigma_i^2 ,$$

$$\gamma = 2(\sum_{i=1}^{q} a_i)(p + \sum_{i=1}^{q} a_i)^{-1} , \quad \sigma^2 = \sum_{i=1}^{q}\mu_i^2/\sigma_i^2 + \frac{1}{2}(p-q)\gamma^2 .$$

3. ASYMPTOTIC DISTRIBUTION OF THE NORMALIZED ROOTS OF THE NON-CENTRAL MULTIVARIATE F-MATRIX

3.1. Introduction. Let $\underline{x}_1,\ldots,\underline{x}_{N_1}$, $\underline{y}_1,\ldots,\underline{y}_{N_2}$ be independently distributed where $\underline{x}_i \sim N_p(\mu,\Sigma)$, $i = 1,2,\ldots,N_1$ and $\underline{y}_j \sim N_p(\underline{\nu},\Sigma)$, $j = 1,2,\ldots,N_2$.

Let $\underline{e}_p = (1,1,\ldots,1)'$ denote a p-vector of ones; whenever the order of the vector is apparent from the context, we shall drop the subscript p from $\underline{e}$. Kraft, Olkin and van Eeden (1972) considered the following testing problems:

(a) $H_1 : \underline{\mu} = c\,\underline{\nu}$ vs $A_1 \neq H_1$, $-\infty < c < \infty$,

(b) $H_2 : \underline{\mu} = c\,\underline{\nu} + d\,\underline{e}$ vs $A_2 \neq H_2$, $-\infty < c,d < \infty$.

They used the asymptotic theory of $-2 \ln \lambda$ to obtain the null and non-null distributions of the likelihood ratio statistics λ, which under some regularity conditions have chi-square distributions with appropriate degrees of freedom.

The testing problem (a) has been considered several times in the statistical literature, see, for example Anderson (1951). This is a problem on testing the rank of the non-centrality matrix. The asymptotic null and non-null distributions of the likelihood ratio statistic can be obtained from the results of Fujikoshi (1977) up to the order $O(n^{-1})$ and up to the order $O(n^{-3/2})$ from the results of this section. However, the distribution of (b) is not available. In this paper, we give an alternative method of deriving the likelihood ratio test statistic and show that this problem reduces to that of testing the dimensionality of the non-centrality matrix. The distribution can be obtained from the results of this paper.

If c were known, the problem reduces to that of testing

$$H_2 : \underline{\mu} - c\,\underline{\nu} = d\,\underline{e} \quad \text{vs} \quad A_2 \neq H_2 .$$

Let B : $(p-1)\times p$ matrix of rank $(p-1)$ such that $B\,\underline{e} = \underline{0}$. Then given c, H_2 is equivalent to (see Srivastava and Khatri (1979), Chapter IV)

$$B(\underline{\mu} - c\,\underline{\nu}) = \underline{0} ,$$

and the likelihood ratio test rejects H_2 if

$$(N_1^{-1} + c^2 N_2^{-1})^{-1}(\bar{\underline{x}} - c\,\bar{\underline{y}})'B'(BSB')^{-1}B(\bar{\underline{x}} - c\,\bar{\underline{y}}) \geq k_1$$

where

$$\bar{\underline{x}} = N_1^{-1}\sum_{\alpha=1}^{N_1} \underline{x}_\alpha , \quad \bar{\underline{y}} = N_2^{-1}\sum_{\alpha=1}^{N_2} \underline{y}_\alpha ,$$

$$S = \sum_{\alpha=1}^{N_1} (\underline{x}_\alpha - \bar{\underline{x}})(\underline{x}_\alpha - \bar{\underline{x}})' + \sum_{\alpha=1}^{N_2} (\underline{y}_\alpha - \bar{\underline{y}})(\underline{y}_\alpha - \bar{\underline{y}})' ,$$

and k_1 depends on the size of the test. That is, H_2 is rejected if

$$\underline{a}'\begin{pmatrix}\bar{\underline{x}}'\\ \bar{\underline{y}}'\end{pmatrix}B'(BSB')^{-1}B(\bar{\underline{x}}\,,\bar{\underline{y}})\underline{a} > k_1 \,,$$

where

$$\underline{a}' = (a_1,a_2) \quad , \quad a_1 = (N_1^{-1} + c^2N_2^{-1})^{-1/2} \quad , \quad a_2 = c/(N_1^{-1} + c^2N_2^{-1})^{1/2} \,.$$

Since the vector $\underline{a}$ is unknown, the hypothesis H_1 is rejected for large values of

$$\min_{\underline{a}} \underline{a}'\begin{pmatrix}\bar{\underline{x}}'\\ \bar{\underline{y}}'\end{pmatrix}B'(BSB')^{-1}B(\bar{\underline{x}}\,,\bar{\underline{y}})\underline{a} \geq k_\alpha \,.$$

Let $\ell_1 > \ell_2$ be the roots of

$$\begin{pmatrix}\bar{\underline{x}}'\\ \bar{\underline{y}}'\end{pmatrix}B'(BSB')^{-1}B(\bar{\underline{x}}\,,\bar{\underline{y}}) \,.$$

Then, the hypothesis H is rejected if $\ell_2 \geq k_\alpha$.

By using the results fo Srivastava and Khatri (1979), Corollary 1.9.2, p. 19, it can be shown that the above statistic is equivalent to the one obtained by Kraft, Olkin and van Eeden (1972). Thus the problem is that of testing the 'dimensionality' of the non-central matrix.

The above problem can be generalized to that of testing

$$H_2 \;:\; \underline{\mu} = c\,\underline{\nu} + \Gamma\,\underline{d}$$

where $\Gamma : p\times r$ of rank r and is known (see Srivastava and Khatri (1979), Chapter IV).

3.2. Asymptotic distribution of the roots of a non-central F-matrix. Let $X \sim N_{p,n_1}(M,\Sigma\otimes I_{n_1})$ and let $S \sim W_p(\Sigma,n_2)$. Then the distribution of $\theta = \mathrm{diag}(\theta_1,\ldots,\theta_p)$, $\theta_1 > \ldots > \theta_p$, the characteristic roots of $XX'(XX'+S)^{-1}$ is given from James (1964) by

$$\Gamma_p^{-1}(\tfrac{1}{2}\,n_2)\Gamma_p^{-1}(\tfrac{1}{2}\,n_1)(\Gamma_p(\tfrac{1}{2}(n_1+n_2))\pi^{\frac{1}{2}p^2}\;\Gamma_p^{-1}(\tfrac{1}{2}\,p)\prod_{1\leq i<j\leq p}(\theta_i-\theta_j)$$

$$\prod_{i=1}^{p}\{\theta_i^{\frac{1}{2}(n_1-p-1)}(1-\theta_i)^{\frac{1}{2}(n_2-p-1)}\}\exp - \tfrac{1}{2}\,\mathrm{tr}\,MM'$$

$${}_1F_1(\tfrac{1}{2}(n_1+n_2)\;;\;\tfrac{1}{2}\,n_1\;:\;\tfrac{1}{2}\,MM'\,,\theta)\,.$$

As the sum of squares matrix due to treatments in MANOVA is based on group means, and by invariance, it can be assumed that MM' is of the form n_2A, with $A = \mathrm{diag}(a_1 > \dots > a_q > a_{q+1} = \dots = a_p = 0)$. Using the expression in Theorem 4 of Srivastava and Carter (1979) for the ${}_1F_1$ hypergeometric function we obtain the asymptotic distribution of θ as

$$\Gamma_p^{-1}(\tfrac{1}{2}p)\Gamma_p^{-1}(\tfrac{1}{2}n_2)\Gamma_p^{-1}(\tfrac{1}{2}n_1)\Gamma_p(\tfrac{1}{2}(n_1+n_2))$$

$$\Gamma_q^{-1}(\tfrac{1}{2}(n_1+n_2))\Gamma_q(\tfrac{1}{2}n_1)\Gamma_q(\tfrac{1}{2}p)$$

$$(\tfrac{1}{2}n)^{\frac{1}{4}(q+1)+\frac{1}{2}n_2(q-p)}\ \pi^{-\frac{1}{4}q(q-2n_1+3)+\frac{1}{2}p^2}$$

$$\prod_{1\le i<j\le q}\{(\theta_i-\theta_j)^{1/2}(a_i-a_j)^{-1/2}\}\prod_{i=1}^{q}\prod_{j=q+1}^{p}\{(\theta_i-\theta_j)^{1/2}\}$$

$$\prod_{q<i<j\le p}\{(\theta_i-\theta_j)\}\prod_{i=1}^{q}\{a_i^{-\frac{1}{2}(n_1+n_2-q+1)}\theta_i^{-\frac{1}{4}(n_1-p)}\gamma_i^{\frac{1}{2}(n_1+n_2)-\frac{1}{4}(n_1+p-2)}$$

$$(8\xi_i)^{1/2}\exp[\tfrac{1}{2}n_2\gamma_i a_i^{-1}+n_2\theta_i^{1/2}\gamma_i^{1/2}-\tfrac{1}{2}n_2a_i]\}$$

$$\prod_{i=1}^{p}\{\theta_i^{\frac{1}{2}(n_1-p-1)}(1-\theta_i)^{\frac{1}{2}(n_2-p-1)}\}$$

$$\{1+n_2^{-1}[\sum_{i=1}^{q}d_i+\frac{1}{2}\sum_{1\le i<j\le q}(d_{ij}+c_{ij})$$

$$+\sum_{i=1}^{q}\sum_{j=q+1}^{p}c_{ij}-\frac{1}{8}(n_1-p)(n_1+p-2q-2)\sum_{i=1}^{q}(\theta_i\gamma_i)^{1/2}]+O(n^{-2})\},$$

where

$$\delta_i^2 = 1+4(a_i\theta_i)^{-1}+4n_1n_2^{-1}(a_i\theta_i)^{-1},$$

$$\gamma_i^{1/2} = \frac{1}{2}(1+\delta_i)(a_i\theta_i)^{1/2},\quad \xi_i^{-1} = a_i\theta_i\delta_i(1+\delta_i),$$

$$d_i = \frac{1}{4}(n_1+p)^2\xi_i-2(n_1+p)\xi_i^2+\frac{1}{2}(n_1+p)\xi_i+\frac{20}{3}\xi_i^3-3\xi_i^2,$$

$$d_{ij} = a_ia_j(\gamma_i-\gamma_j)^{-1}(\theta_i-\theta_j)^{-1}$$

$$c_{ij} = [(\theta_i\gamma_i)^{1/2} + (\theta_j\gamma_j)^{1/2}](\theta_i-\theta_j)^{-1}(\gamma_i-\gamma_j)^{-1} \quad , \quad 1 \le i < j \le q \ ,$$

and

$$c_{ij} = \theta_i^{1/2}\ \gamma_i^{-1/2}(\theta_i-\theta_j)^{-1} \quad , \quad i=1,\ldots,q \quad , \quad j=q+1,\ldots,p \ .$$

Letting

$$x_i = n_2^{1/2}(\theta_i-a_i(1+a_i)^{-1}) \quad , \quad i=1,\ldots,q \ ,$$

and

$$x_j = \theta_j\ n_2 \quad , \quad j=q+1,\ldots,p \ ,$$

we obtain the distribution of the normalized roots $x_1,\ldots,x_p$ as:

$$\pi^{\frac{1}{2}(p-q)^2}\ \Gamma_{p-q}^{-1}(\tfrac{1}{2}(p-q))\Gamma_{p-q}^{-1}(\tfrac{1}{2}(n_1-q))$$

$$\prod_{q<i<j\le p} (x_i-x_j) \prod_{j=q+1}^{p} x_j^{\frac{1}{2}(n_1-p-1)}\ e^{-\frac{1}{2}x_j}$$

$$(2\pi)^{-\frac{1}{2}q} \prod_{i=1}^{q} \{\sigma_i^{-1} \exp - \tfrac{1}{2}\ x_i^2/\sigma_i^2\}$$

$$\{1 + n_2^{-1/2}\ T_1 + n^{-1}(\tfrac{1}{2}\ T_1^2 + T_2 + T_3) + 0(n^{-3/2})\} \ ,$$

where

$$\sigma_i^2 = r(1+a_i)^{-4}\ a_i(2+a_i) \ ,$$

$$T_1 = \sum_{i=1}^{q} x_i^3[\tfrac{1}{24}\ (1+a_i)^3(2+a_i)^{-3} + \tfrac{1}{8}\ (1+a_i)^3(2+a_i)^{-1}a_i^{-2} - \tfrac{1}{6}\ (1+a_i)^3]$$

$$+ \sum_{i=1}^{q} x_i[\tfrac{1}{2}\ a_i^{-1}(2+a_i)^{-1}(1+a_i)(n_1-p) + \tfrac{1}{2}\ (1+a_i)$$

$$+ (1+a_i)(2+a_i)^{-2}a_i^{-2} - \tfrac{1}{2}\ (1+a_i)^3(2+a_i)^{-2}a_i^{-1}]$$

$$+ \tfrac{1}{2} \sum_{1\le i<j\le q} (x_i-x_j)(a_i-a_j)^{-1} + \tfrac{1}{2}\ (p-q) \sum_{i=1}^{q} x_i(1+a_i)a_i^{-1} \ ,$$

$$T_2 = \sum_{i=1}^{q} x_i^4[-\frac{1}{512}(1+a_i)^2\ a_i^3(2+a_i)^{-6}(10a_i^2 + 19a_i + 8)$$

$$+ \frac{1}{128}(1+a_i)^3\ a_i^2(2+a_i)^{-5}(a_i-2) - \frac{1}{64}(1+a_i)^{-4}(2+a_i)^{-4}$$

$$+ \frac{1}{256}(1+a_i)^2\ a_i^2(2+a_i)^{-3} - \frac{1}{64}(1+a_i)^3\ a_i^{-1}(2+a_i)^{-2}$$

$$+ \frac{1}{64}(1+a_i)^3(2+a_i)^{-1}\ a_i^{-1} - \frac{1}{4}(1+a_i)^4$$

$$- \frac{1}{512}(1+a_i)^2\ a_i^{-3}(4a_i^2 + 45a_i + 30)]$$

$$+ \sum_{i=1}^{q} x_i^2[\frac{3}{16} n_1(2+a_i)^{-4}\ a_i^2(1+a_i)^2 + \frac{1}{16} n_1(2+a_i)^{-4}\ a_i^2(1+a_i)$$

$$+ \frac{1}{2} n_1 a_i(1+a_i)^2(2+a_i)^{-4} - \frac{1}{8} n_1 a_i(1+a_i)^2(2+a_i)^{-3}$$

$$+ \frac{1}{4} n_1(1+a_i)^2(2+a_i)^{-3} - \frac{1}{16} n_1(1+a_i)(2+a_i)^{-1}$$

$$- \frac{1}{64}(n_1-p+2)(1+a_i)(a_i^{-1} + a_i^2(2+a_i)^{-3} + 2(1+a_i)(2+a_i)^{-2})$$

$$- \frac{1}{8}(n_1-p+2)(1+a_i)^2\ a_i^{-2} + \frac{1}{4}(p+1)(1+a_i)^2$$

$$+ \frac{1}{4}((1+a_i)a_i^{-1} - 2(1+a_i)^2(2+a_i)^{-2} - (1+a_i)a_i^{-1}(2+a_i)^{-1})^2$$

$$- (1+a_i)^3\ a_i^{-2}(2+a_i)^{-2} + (1+a_i)^4(2+a_i)^{-4}\ a_i^{-2}$$

$$- \frac{1}{2}(1+a_i)^2\ a_i^{-2}(2+a_i)^{-2} - \frac{1}{2}(1+a_i)^3\ a_i^{-2}(2+a_i)^{-3}$$

$$+ (1+a_i)^3\ a_i^{-2}(2+a_i)^{-2} + \frac{1}{2}(1+a_i)^2\ a_i^{-2}(2+a_i)^{-1}]$$

$$+ \frac{1}{4} n_1(n_1-p+2) \sum_{i=1}^{q}(2+a_i)^{-1} - \frac{1}{2} n_1^2 \sum_{i=1}^{q}(2+a_i)^{-2} + \frac{1}{2} n_1 \sum_{i=1}^{q} a_i(2+a_i)^{-2}\ ,$$

and

$$T_3 = -\frac{1}{4}\sum_{1\le i<j\le q}(x_i-x_j)^2(a_i-a_j)^{-2}$$

$$-\frac{1}{2}\left(\sum_{i=1}^{q}(a_i+1)a_i^{-1}\right)\left(\sum_{j=q+1}^{p}x_j\right)+\sum_{j=q+1}^{p}\left[(p+1)x_j-\frac{1}{4}x_j^2\right]$$

$$-\frac{1}{4}\sum_{i=1}^{q}x_i^2(1+a_i)^2\,a_i^{-2}(p-q)$$

$$+\sum_{i=1}^{q}\left[\left(\frac{1}{8}(n_1+p)^2+\frac{1}{4}(n_1+p)\right)(2+a_i)^{-1}-\frac{1}{2}(n_1+p)(2+a_i)^{-2}\right]$$

$$+\frac{1}{2}\sum_{1\le i<j\le q}(a_i+a_j+a_ia_j)(a_i-a_j)^{-2}(a_i+a_j+1)^{-1}(1+a_i)(1+a_j)$$

$$+\sum_{i=1}^{q}\left[(p-q)a_i^{-1}-\frac{1}{8}(n_1-p)(n_1+p-2q-2)a_i\right]$$

$$+\frac{1}{4}q(q-1)-\frac{1}{24}q(q-1)(2q-1)+\frac{1}{4}(p-q)n_1(n_1-p-q+3)\;.$$

4. ASYMPTOTIC DISTRIBUTION OF THE NORMALIZED CANONICAL CORRELATIONS.

4.1. Introduction. Let $\binom{X}{Y}$ be a $(p+\ell)\times n$ matrix, with $X_{p\times n}$, $Y_{\ell\times n}$, $p \le \ell$. Suppose the columns of $\binom{X}{Y}$ are independently distributed as $N_{p+\ell}(\underline{0},\Sigma)$ where Σ is partitioned as

$$\Sigma = \begin{pmatrix} \Sigma_{11} & \Sigma_{12} \\ \Sigma_{12} & \Sigma_{22} \end{pmatrix}.$$

To test for the independence of X and Y one tests $H: \Sigma_{12} = 0$. By invariance the tests should be based on the characteristic roots $r_1^2,\ldots,r_p^2$ of $(XX')^{-1}X'Y(YY')^{-1}Y'X$. The power of such a test would then only depend on the characteristic roots of $\rho_1^2 \ge \ldots \ge \rho_p^2$ of $\Sigma_{11}^{-1}\Sigma_{12}\Sigma_{22}^{-1}\Sigma_{12}'$. Four invariant test statistics are

(i) $T_1 = r_1^2$ (Roy's maximum root test)

(ii) $T_2 = \prod_{i=1}^{p}(1-r_i^2)$ (likelihood ratio test)

(iii) $T_3 = \sum_{i=1}^{p} r_i^2$ (Nanda-Pillai)

(iv) $T_4 = \sum_{i=1}^{p} r_i^2/(1-r_i^2)$ (Lawley-Hotelling) .

The distribution of $r_1^2,\dots,r_p^2$ is given from James (1964) by

$$\pi^{\frac{1}{2}p^2} \Gamma_p^{-1}(\tfrac{1}{2}p)\Gamma_p^{-1}(\tfrac{1}{2}\ell)\Gamma_p^{-1}(\tfrac{1}{2}(n-\ell))\Gamma_p(\tfrac{1}{2}n)$$

$$\alpha_p(D_r) \prod_{i=1}^{p} r_i^{(\ell-p-1)}(1-r_i^2)^{\frac{1}{2}(n-p-\ell-1)}$$

$${}_2F_1(\tfrac{1}{2}n , \tfrac{1}{2}n ; \tfrac{1}{2}\ell : D_r^2 , D_\rho^2) ,$$

where $D_f = \mathrm{diag}(f_1,\dots,f_p)$,

$$\Gamma_p(\tfrac{1}{2}b) = \pi^{\frac{1}{4}p(p-1)} \prod_{i=1}^{p} \Gamma(\tfrac{1}{2}(b-i+1)) ,$$

$$\alpha_p(D_f) = \prod_{1\le i<j\le p} (f_i-f_j) ,$$

and ${}_2F_1$ is as in James (1964). The asymptotic representation of the ${}_2F_1$ function is given in Srivastava and Carter (1979) up to $O(n^{-2})$ for $\rho_1^2 > \dots > \rho_q^2 > \rho_{q+1}^2 = \dots = \rho_p^2 = 0$. In the next section the asymptotic distribution of

$$x_i = n^{1/2}(r_i^2 - \rho_i^2) , i = 1,\dots,q$$

and

$$x_j = nr_j^2 , j = q+1,\dots,p$$

is derived up to $O(n^{-3/2})$.

From this result, the asymptotic distribution of the four statistics for independence can be obtained up to $O(n^{-3/2})$. We present the distribution of T_1 in Subsection 4.2, since this is the only one not available in the literature; others are given by Fujikoshi (1977) up to $O(n^{-1})$.

4.2. Asymptotic distribution of the transformed roots. Assuming that only q population canonical roots are non-zero, that is, $\rho_1^2 > \dots > \rho_q^2 > \rho_{q+1}^2 = \dots = \rho_p^2$, the asymptotic distribution of the sample roots $r_1^2 > \dots > r_p^2$ can be obtained using the representation of the ${}_2F_1$ function given in Theorem 5 of Carter and Srivastava (1979). That is, the density of $r_1^2,\dots,r_p^2$ is given by

$$f(r_1^2,\dots,r_p^2) = \Gamma_p^{-1}(\tfrac{1}{2}p)\Gamma_p^{-1}(\tfrac{1}{2}\ell)\Gamma_p^{-1}(\tfrac{1}{2}(n-\ell))\Gamma_p(\tfrac{1}{2}n)\Gamma_q(\tfrac{1}{2}\ell)\Gamma_q(\tfrac{1}{2}p)$$

$$[\alpha_q(D_r^2)/\alpha_q(D_\rho^2)]^{1/2} \prod_{i=1}^{q}\prod_{j=q+1}^{p} (r_i^2 - r_j^2)^{1/2}$$

$$\prod_{q\le i<j\le p} (r_i^2 - r_j^2) \prod_{i=q+1}^{p} (r_i^2)^{\frac{1}{2}(\ell-p-1)} (1-r_i^2)^{\frac{1}{2}(n-p-\ell-1)}$$

$$\prod_{i=1}^{q} [(r_i^2)^{\frac{1}{4}(\ell-p-2)} (\rho_i^2)^{-\frac{1}{4}(\ell+p-2q)} (1-r_i^2)^{\frac{1}{2}(n-p-\ell-1)} (1-\rho_i^2)^{\frac{1}{2}n}$$

$$(1-r_i\rho_i)^{-n+\frac{1}{2}(\ell+p-1)}]\{1+n^{-1}[\tfrac{1}{2}\sum_{1\le i<j\le q} ((r_i\rho_i)^{1/2} + (r_j\rho_j)^{1/2})$$

$$(1-r_i\rho_i)(1-r_j\rho_j)(r_i^2 - r_j^2)^{-1}(\rho_i^2 - \rho_j^2)^{-1}$$

$$+ \frac{1}{2}\sum_{i=1}^{q}\sum_{j=q+1}^{p} (r_i/\rho_i)(1-r_i\rho_i)(r_i^2-r_j^2)^{-1}$$

$$+ \frac{1}{8}\sum_{i=1}^{q} r_i\rho_i - \frac{1}{8}(\ell-p)(\ell-p-2)\sum_{i=1}^{q} (r_i\rho_i)^{-1}$$

$$+ \frac{1}{8}q(\ell-p)(\ell-p-2) + \frac{q}{3} + \frac{q}{8}(p+\ell-2)(p+\ell+2)] + O(n^{-2})\} ,$$

where

$$\alpha_q(D_a) = \prod_{1\le i<j\le q} (a_i - a_j) .$$

Making the transformation

$$r_i^2 = \rho_i^2 + x_i\, n^{-1/2} , \quad i = 1,\dots,q ,$$

and

$$r_j^2 = x_j\, n^{-1} , \quad j = q+1,\dots,p ,$$

and denoting the variance of x_i by $\sigma_i^2 = 4\rho_i^2(1-\rho_i^2)^2$, one obtains the density of the transformed roots $x_1,\ldots,x_p$ as $f(x_1,\ldots,x_p)$

$$\Gamma^{-1}_{p-q}(\tfrac{1}{2}(p-q))\Gamma^{-1}_{p-q}(\tfrac{1}{2}(\ell-q))\pi^{\frac{1}{2}(p-q)^2} \prod_{q\le i<j\le p} (x_i-x_j)$$

$$\prod_{i=q+1}^{p} [x_i^{\frac{1}{2}(\ell-p-1)} \exp - \tfrac{1}{2} x_i](2\pi)^{-\frac{1}{2}q} \prod_{i=1}^{q} [\sigma_1^{-1} \exp - \tfrac{1}{2} x_i^2\sigma_i^{-2}]$$

$$\{1 + n^{-1/2} P_1 + n^{-1}(\tfrac{1}{2} P_1^2 + P_2 + P_3) + 0(n^{-3/2})\},$$

where

$$P_1 = \sum_{i=1}^{q} [(\ell+p+3)\rho_i x_i(2\sigma_i)^{-1} + (1-3\rho_i^2)x_i^3(2\rho_i\sigma_i^3)^{-1}$$

$$+ (\ell-p-2)x_i(1-\rho_i^2)(2\rho_i\sigma_i)^{-1} + \tfrac{1}{2} (p-q)x_i(\rho_i^2)^{-1}]$$

$$+ \tfrac{1}{2} \sum_{1\le i<j\le q} (x_i-x_j)(\rho_i^2 - \rho_j^2)^{-1} ,$$

$$P_2 = \sum_{i=1}^{q} [(-29\rho_i^4 + 20\rho_i^2 - 5)(8\rho_i^2\sigma_i^4)^{-1}x_i^4$$

$$- (\ell-p-2)x_i^2(1-\rho_i^2)^2(2\sigma_i^2\rho_i^2)^{-1} + (\ell+p-1)(1+2\rho_i^2)x_i^2(4\sigma_i^2)^{-1}$$

$$- \tfrac{1}{4} (p-q)x_i^2(\rho_1^4)^{-1} + \tfrac{1}{2} x_i^2(1-\rho_i^2)^{-2}]$$

$$+ \sum_{i=q+1}^{p} [\tfrac{1}{2} (p+\ell+1)x_i - \tfrac{1}{4} x_i^2]$$

$$- \tfrac{1}{2} \sum_{i=1}^{q} \sum_{j=q+1}^{p} x_j(\rho_i^2)^{-1} - \tfrac{1}{4} \sum_{1\le i<j\le q} (x_i-x_j)^2(\rho_i^2 - \rho_j^2)^{-1} ,$$

and

$$P_3 = \tfrac{1}{2} \sum_{1\le i<j\le q} (\rho_i^2 + \rho_j^2)(1-\rho_i^2)(1-\rho_j^2)(\rho_i^2 - \rho_j^2)^{-2}$$

$$+ \sum_{i=1}^{q} [\tfrac{1}{2} (p-q)(1-\rho_i^2)(\rho_i^2)^{-1} + \tfrac{1}{8} \rho_i^2 - \tfrac{1}{8} (\ell-p)(\ell-p-2)(\rho_i^2)^{-1}]$$

$$- \tfrac{1}{8} q(\ell-p)(\ell-p-2) + \tfrac{q}{8} (\ell+p-2)(\ell+p+2) - \tfrac{1}{4} p\ell(\ell+p+1) .$$

4.3. Asymptotic distribution of T_1. The distribution of $T_1 = r_1^2$ can be obtained directly. That is, the density of $x_1 = n^{1/2}(T_1-\rho_1^2)$ is given by

$$(2\pi\sigma_1^2)^{-1/2}(\exp - \frac{1}{2} x_1^2/\sigma_1^2)\{1 + n^{-1/2}[(\ell+p+3)\rho_1 x_1(2\sigma_1)^{-1}$$

$$+ (1-3\rho_1^2)x_1^3(2\rho_1\sigma_1^3)^{-1} + (\ell-p-2)(1-\rho_1^2)x_1(2\rho_1\sigma_1)^{-1}$$

$$+ \frac{1}{2}(p-q)x_1(\rho_1^2)^{-1} + \frac{1}{2}x_1 \sum_{j=2}^{q} (\rho_1^2 - \rho_j^2)^{-1}] + O(n^{-1})\} ,$$

where $\sigma_1^2 = 4\rho_1^2(1-\rho_1^2)^2$. Equivalently, we have

$$P(n^{1/2}(r_1^2 - \rho_1^2)/\sigma_1 \leq \xi) = \Phi(\xi) - n^{-1/2}\psi(\xi) [\frac{(1-3\rho_1^2)}{2\rho_1} (\xi^2-1)$$

$$+ [\frac{\ell+p-2q+1}{2\rho} + (q-4)\rho + \rho_1(1-\rho_1^2) \sum_{j=2}^{q} (\rho_1^2 - \rho_j^2)^{-1}] + O(n^{-1}).$$

REFERENCES

[1] Anderson, T.W. (1951). Estimating linear restrictions on regression coefficients for multivariate normal distributions. Ann. Math. Statist. 22, 327-351.

[2] Carter, E.M. and Srivastava, M.S. (1979). Asymptotic Distribution of the Latent Roots of the Non-central Wishart Distribution and the Power of the L.R.T. for Non-additivity. To appear in Can. J. Statist. (1980).

[3] Corsten, L. and Van Eijnsbergen (1972). Multiplicative effects in two-way analysis of variance. Statistica Neerlandica 26, 61-68.

[4] Fujikoshi, Y. (1977). Asymptotic Expansions for the Distributions of Some Multivariate Tests. Multivariate Analysis IV, 55-71, North-Holland, New York.

[5] Hsu, L. (1948). A theorem on the asymptotic behaviour of a multiple integral. Duke Math. J. 15, 623-632.

[6] James, A.T. (1964). Distribution of matrix variates and latent roots derived from normal samples. Ann. Math. Statist. 35, 475-500.

[7] Johnson, D.E. and Graybill, F. (1972). An analysis of a two-way model with interaction and no replications. J. Amer. Statist. Assoc. 67, 862-868.

[8] Khatri, C.G. and Srivastava, M.S. (1978). Asymptotic expansions for distributions of characteristic roots of covariance matrices. South Afr. J. Statist. 12, 161-186.

[9] Kraft, C.J., Olkin, I. and van Eeden, C. (1972). Estimation and testing for differences in magnitude or displacement in the mean vectors of two multivariate normal populations. Ann. Math. Statist. 43, 455-467.

[10] Krishnaiah, P.R. and Schuurmann, F.J. (1974). On the evaluation of some distributions that arise in simultaneous tests for the equality of the latent roots of the covariance matrix.

[11] Srivastava, M.S. and Carter, E.M. (1979). Asymptotic expansions for hypergeometric functions. To appear in Multivariate Analysis V. North-Holland, New York.

[12] Srivastava, M.S. and Khatri, C.G. (1979). An Introduction to Multivariate Statistics. North-Holland, New York.

Multivariate Statistical Analysis
R.P. Gupta (ed.)

ON APPROXIMATING MULTIVARIATE DISTRIBUTIONS

W.Y. Tan
Department of Mathematical Sciences
Memphis State University
Memphis, Tennessee 38152
U.S.A.

This paper introduces two approximations to multivariate distributions: the generalized Gram-Charlier series expansion and the generalized Laguerre polynomial expansion. Some properties of these approximations are derived. A brief indication of the applications is also given.

1. INTRODUCTION

Multivariate probability distributions are usually very complicated even under normality assumption (James 1964). When the universe is not normal, exact multivariate distributions are usually extremely difficult to obtain if not impossible. Thus it is desirable to obtain some good and computable approximations to important multivariate distributions. The purpose of this paper is to present two such approximations: The generalized Gram-Charlier series expansion and the generalized Laguerre polynomial expansion.

In Section 2 we present the generalized Gram-Charlier series expansion. Some applications of the results are then given in Section 3. The generalized Laguerre polynomial expansion approximation is given in Section 4. The present paper deals only with theoretical aspects of the approximation, leaving numerical examples and Monte Carlo studies in the second paper.

2. THE GENERALIZED GRAM-CHARLIER SERIES EXPANSION

Suppose $f(\underset{\sim}{x})$ and $g(\underset{\sim}{x})$ are two p-dimensional absolutely continuous density functions having finite mixed cumulants $\lambda_1(\underset{\sim}{r}) = \lambda_1(r_1,r_2,\ldots,r_p)$ and $\lambda_2(\underset{\sim}{r}) = \lambda_2(r_1,r_2,\ldots,r_p)$ respectively. Let $C(\underset{\sim}{r})$ be obtained by letting $\Delta(\underset{\sim}{r}) = \lambda_1(\underset{\sim}{r}) - \lambda_2(\underset{\sim}{r})$ be the mixed cumulants in the expression $\mu'_{r_1 r_2 \ldots r_p}$ (the mixed moment around origin) in terms of the cumulants (see Cook (1951) p. 181, or Good (1962, p. 540)). The following theorem, given by Finey (1963), provides a definition of the generalized multivariate Gram-Charlier series expansion of $f(\underset{\sim}{x})$ in terms of $g(\underset{\sim}{x})$.

Theorem (2.1). Provided that the series on the right-hand side of (2.1) converges absolutely and that it is permissible to interchange differentiation and integration, we have:

$$f(\underset{\sim}{x}) = \exp\left\{\sum_{|\underset{\sim}{r}|=1}^{\infty} \frac{1}{\underset{\sim}{r}!}\Delta(\underset{\sim}{r})\left(-\frac{d}{d\underset{\sim}{x}}\right)^{\underset{\sim}{r}}\right\} g(\underset{\sim}{x})$$

$$= \left\{1+\sum_{|\underset{\sim}{r}|=1}^{\infty} \frac{1}{\underset{\sim}{r}!} C(\underset{\sim}{r})\left(-\frac{d}{d\underset{\sim}{x}}\right)^{\underset{\sim}{r}}\right\} g(\underset{\sim}{x}) , \tag{2.1}$$

where $|\underset{\sim}{r}| = r_1 + r_2 + \ldots + r_p$, $\underset{\sim}{r}! = \prod_{j=1}^{p}(r_j!)$ and $(-\frac{d}{d\underset{\sim}{x}})^{\underset{\sim}{r}} = \prod_{j=1}^{p}(-\frac{\partial}{\partial x_j})^{r_j}$

for any $\underset{\sim}{r}' = (r_1, r_2, \ldots, r_p)$, $r_j \geq 0$ being nonnegative integers.

Using (2.1), one has then an approximation $f_k(\underset{\sim}{x})$ of $f(\underset{\sim}{x})$ by $g(\underset{\sim}{x})$:

$$f_k(\underset{\sim}{x}) = \left\{1 + \sum_{|\underset{\sim}{r}|=1}^{k} \frac{1}{\underset{\sim}{r}!} C(\underset{\sim}{r})\left(-\frac{d}{d\underset{\sim}{x}}\right)^{\underset{\sim}{r}}\right\} g(\underset{\sim}{x}) , \tag{2.2}$$

For writing $C(\underset{\sim}{r})$ explicitly, it is convenient to rewrite $\lambda_1(\underset{\sim}{r})$, $\lambda_2(\underset{\sim}{r})$, $\Delta(\underset{\sim}{r})$ and $C(\underset{\sim}{r})$ as $\lambda_1(\underset{\sim}{r}) = \lambda_{(1)j_1j_2\cdots j_\ell}^{(i_1i_2\cdots i_\ell)}$,

$\lambda_2(\underset{\sim}{r}) = \lambda_{(2)j_1j_2\cdots j_\ell}^{(i_1i_2\cdots i_\ell)}$, $\Delta(\underset{\sim}{r}) = \Delta_{j_1j_2\cdots j_\ell}^{(i_1i_2\cdots i_\ell)}$ and $C(\underset{\sim}{r}) = C_{j_1j_2\cdots j_\ell}^{(i_1i_2\cdots i_\ell)}$,

$\ell=1,2,\ldots,p$, if in $\underset{\sim}{r}'$, the elements at the $i_1^{th}, i_2^{th}, \ldots, i_\ell^{th}$ positions are $j_1, j_2, \ldots, j_\ell$ respectively and are zeros at other positions. Using these notations, we have, for $k = 4$:

$$f_4(\underset{\sim}{x}) = [1 + \sum_{i=1}^{p} C_1^{(i)}(\frac{\partial}{\partial x_i}) + \frac{1}{2!}\{\sum_{i=1}^{p} C_2^{(i)}(-\frac{\partial}{\partial x_i})^2 + 2\sum_{i\neq j} C_{11}^{(ij)}(-\frac{\partial}{\partial x_i})(-\frac{\partial}{\partial x_j})\}$$

$$+ \frac{1}{3!}\{\sum_{i=1}^{p} C_3^{(i)}(-\frac{\partial}{\partial x_i})^3 + 3\sum_{i\neq j} C_{21}^{(ij)}(-\frac{\partial}{\partial x_i})^2(-\frac{\partial}{\partial x_j})$$

$$+ 6\sum_{i<j<s} C_{111}^{(ijs)}(-\frac{\partial}{\partial x_i})(-\frac{\partial}{\partial x_j})(-\frac{\partial}{\partial x_s})\} + \frac{1}{4!}\{\sum_{i=1}^{p} C_4^{(i)}(-\frac{\partial}{\partial x_i})^4$$

$$+ 4\sum_{i\neq j} C_{31}^{(i)}(-\frac{\partial}{\partial x_i})^3(-\frac{\partial}{\partial x_j}) + 6\sum_{i<j} C_{22}^{(ij)}(-\frac{\partial}{\partial x_i})^2(-\frac{\partial}{\partial x_j})^2$$

$$+ 12\sum_{i<j}\sum_{s\neq i\neq j} C_{112}^{(ijs)}(-\frac{\partial}{\partial x_i})(-\frac{\partial}{\partial x_j})(-\frac{\partial}{\partial x_s})^2$$

$$+ 24 \sum_{i<j<s<t} C_{1111}^{(ijst)}(-\frac{\partial}{\partial x_i})(-\frac{\partial}{\partial x_j})(-\frac{\partial}{\partial x_s})(-\frac{\partial}{\partial x_t})\}]g(\underset{\sim}{x}) \ , \tag{2.3}$$

where

$$C_1^{(i)} = \Delta_1^{(i)} \ ,$$

$$C_2^{(i)} = \Delta_2^{(i)} + (\Delta_1^{(i)})^2 \ ,$$

$$C_{11}^{(ij)} = \Delta_{11}^{(ij)} + \Delta_1^{(i)}\Delta_1^{(j)} \ ,$$

$$C_3^{(i)} = \Delta_3^{(i)} + 3\Delta_2^{(i)}\Delta_1^{(i)} + (\Delta_1^{(i)})^3 \ ,$$

$$C_{21}^{(ij)} = \Delta_{21}^{(ij)} + \Delta_2^{(i)}\Delta_1^{(j)} + 2\Delta_{11}^{(ij)}\,\Delta_1^{(i)} + (\Delta_1^{(i)})^2\,\Delta_1^{(j)} \ ,$$

$$C_{111}^{(iju)} = \Delta_{111}^{(iju)} + \Delta_{11}^{(ij)}\,\Delta_1^{(u)} + \Delta_{11}^{(iu)}\,\Delta_1^{(j)} + \Delta_{11}^{(ju)}\,\Delta_1^{(i)} + \Delta_1^{(i)}\,\Delta_1^{(j)}\,\Delta_1^{(u)}$$
$$(C_{111}^{(iju)} = 0 \ \text{ if } \ p = 2) \ ,$$

$$C_4^{(i)} = \Delta_4^{(i)} + 4\Delta_3^{(i)}\,\Delta_1^{(i)} + 3(\Delta_2^{(i)})^2 + 6\Delta_2^{(i)}(\Delta_1^{(i)})^2 + (\Delta_1^{(i)})^4 \ ,$$

$$C_{31}^{(ij)} = \Delta_{31}^{(ij)} + \Delta_3^{(i)}\Delta_1^{(j)} + 3\Delta_{21}^{(ij)}\Delta_1^{(i)} + 3\Delta_2^{(i)}\Delta_{11}^{(ij)} + 3\Delta_2^{(i)}\Delta_1^{(i)}\Delta_1^{(j)}$$
$$+ 3\Delta_{11}^{(ij)}(\Delta_1^{(i)})^2 + (\Delta_1^{(i)})^3\Delta_1^{(j)} \ ,$$

$$C_{22}^{(ij)} = \Delta_{22}^{(ij)} + 2\Delta_{21}^{(ij)}\Delta_1^{(j)} + 2\Delta_{12}^{(ij)}\Delta_1^{(i)} + \Delta_2^{(i)}(\Delta_1^{(j)})^2 + \Delta_2^{(j)}(\Delta_1^{(i)})^2$$
$$+ \Delta_2^{(i)}\Delta_2^{(j)} + 2(\Delta_{11}^{(ij)})^2 + 4\Delta_{11}^{(ij)}\Delta_1^{(i)}\Delta_1^{(j)} + (\Delta_1^{(i)}\Delta_1^{(j)})^2 \ ,$$

$$C_{211}^{(ijs)} = \Delta_{211}^{(ijs)} + \Delta_{21}^{(ij)}\Delta_1^{(s)} + \Delta_{21}^{(is)}\Delta_1^{(j)} + \Delta_{11}^{(js)}\Delta_2^{(i)} + \Delta_2^{(i)}\Delta_1^{(j)}\Delta_1^{(s)}$$
$$+ 2\Delta_1^{(i)}\Delta_{111}^{(ijs)} + 2\Delta_{11}^{(ij)}\Delta_{11}^{(is)} + 2\Delta_{11}^{(ij)}\Delta_1^{(i)}\Delta_1^{(s)} + 2\Delta_{11}^{(is)}\Delta_1^{(i)}\Delta_1^{(j)}$$
$$+ (\Delta_1^{(i)})^2\Delta_1^{(j)}\Delta_1^{(s)} + (\Delta_1^{(i)})^2\Delta_{11}^{(js)} \quad (C_{211}^{(ijs)} = 0 \ \text{ if } \ p = 2) \ ,$$

$$C_{1111}^{(ijst)} = \Delta_{1111}^{(ijst)} + \Delta_{111}^{(ijs)}\Delta_1^{(t)} + \Delta_{111}^{(ijt)}\Delta_1^{(s)} + \Delta_{111}^{(ist)}\Delta_1^{(j)}$$
$$+ \Delta_{111}^{(jst)}\Delta_1^{(i)} + \Delta_{11}^{(ij)}\Delta_{11}^{(st)} + \Delta_{11}^{(is)}\Delta_{11}^{(jt)}$$
$$+ \Delta_{11}^{(it)}\Delta_{11}^{(js)} + \Delta_{11}^{(ij)}\Delta_1^{(s)}\Delta_1^{(t)} + \Delta_{11}^{(is)}\Delta_1^{(j)}\Delta_1^{(t)} + \Delta_{11}^{(it)}\Delta_1^{(j)}\Delta_1^{(s)}$$

$$+ \Delta_{11}^{(js)}\Delta_1^{(i)}\Delta_1^{(t)} + \Delta_{11}^{(jt)}\Delta_1^{(i)}\Delta_1^{(s)} + \Delta_{11}^{(st)}\Delta_1^{(i)}\Delta_1^{(j)}$$

$$+ \Delta_1^{(i)}\Delta_1^{(j)}\Delta_1^{(s)}\Delta_1^{(t)} \qquad (C_{1111}^{(ijst)} = 0 \text{ if } p < 4) \ .$$

If $\lambda_1^{(i)} = \lambda_2^{(i)}$ (i.e., the expectations are equal) for $i = 1,2,\ldots,p$, then the above coefficients reduce to:

$$\left.\begin{aligned}
&C_{st}^{(ij)} = \Delta_{st}^{(ij)} \quad \text{for} \quad 1 \leq s+t \leq 3 \ , \\
&C_{40}^{(i)} = \Delta_4^{(i)} + 3(\Delta_2^{(i)})^2 \ , \quad C_{31}^{(ij)} = \Delta_{31}^{(ij)} + 3\Delta_2^{(i)}\Delta_{11}^{(ij)} \ , \\
&C_{22}^{(ij)} = \Delta_{22}^{(ij)} + \Delta_2^{(i)}\Delta_2^{(j)} + 2(\Delta_{11}^{(ij)})^2 \ , \\
&C_{211}^{(ijs)} = \Delta_{211}^{(ijs)} + \Delta_2^{(i)}\Delta_{11}^{(js)} + 2\Delta_{11}^{(ij)}\Delta_{11}^{(is)} \ , \\
&C_{1111}^{(ijst)} = \Delta_{1111}^{(ijst)} + \Delta_{11}^{(ij)}\Delta_{11}^{(st)} + \Delta_{11}^{(is)}\Delta_{11}^{(jt)} + \Delta_{11}^{(it)}\Delta_{11}^{(js)} \ .
\end{aligned}\right\} \qquad (2.4)$$

Note that when $p = 1$, (2.2) and (2.3) have been used successfully by Geary (1947) and by Bowman, Beauchamp and Shenton (1977) to approximate the distribution of t-statistic from non-normal universes.

<u>Theorem (2.2).</u> Suppose that (2.1) holds and that $g(\underset{\sim}{x})$ satisfies the condition

$$\{\prod_{j=1}^{p} (\frac{\partial}{\partial x_j})^{r_j}\}g(x) = 0 \quad \text{at the boundaries,} \qquad (2.5)$$

Then,

$$\int_{-\infty}^{\infty}\cdots\int\{\prod_{j=1}^{p} x_j^{r_j}\}f_k(\underset{\sim}{x})d\underset{\sim}{x} = E\{\prod_{j=1}^{p} x_j^{r_j}\} \quad \text{for all nonnegative integers}$$

r_j , $0 \leq \sum_{j=1}^{p} r_j \leq k$.

Proof.

By repeated integration by parts, the condition (2.5) leads immediately to the result that

$$I(s_j,r_j) = \int_{-\infty}^{-\infty}\cdots\int \prod_{j=1}^{p} \{x_j^{s_j}(-\frac{\partial}{\partial x_j})^{r_j}\}g(\underset{\sim}{x})d\underset{\sim}{x} = 0 \quad \text{if} \quad s_j < r_j \quad \text{for some}$$

$$i \leq j \leq p \qquad (2.6)$$

and

$$I(s_j,r_j) = \{\prod_{j=1}^{p}(s_j)_{r_j}\}\ V_{(s_1-r_1)(s_2-r_2)\dots(s_p-r_p)} \quad \text{if} \quad s_j \geq r_j \quad \text{for all}$$

$$j=1,2,\dots,p \ ,$$

From (2.1) and (2.6) it follows that, if r_j are nonnegative integers satisfying $0 \leq \sum_{j=1}^{p} r_j \leq k$, then

$$E\{\prod_{j=1}^{p} x_j^{r_j}\} = \int_{-\infty}^{\infty}\dots\int \{\prod_{j=1}^{p} x_j^{r_j}\}\ f_k(\underset{\sim}{x})d\underset{\sim}{x} \ .$$

QED.

Remark: By straightforward calculation, it can be shown that the results of Theroem (2.2) hold also for $k \leq 6$ even if (2.1) fails. It is conjectured that Theorem (2.2) would hold if $k > 6$ even if (2.1) fails; but general proof has not been available. The essence of Theorem (2.2) suggests, however, that if one approximates $f(\underset{\sim}{x})$ by $f_k(\underset{\sim}{x})$, and if condition (2.5) holds, then for $k \geq 6$, at least all mixed moments up to order 6 would be equal. This result holds even if (2.1) does not hold. If $g(\underset{\sim}{x})$ is a normal density or a Wishart density with d.f. $= f > 2k + 3$, it can readily be seen that condition (2.5) does hold.

3. SOME APPLICATIONS

(a) Approximation of multivariate distribution by normal density.

If $-\infty < x_i < \infty$, (2.1) holds and $g(\underset{\sim}{x})$ is p-dimensional multivariate normal, then (2.2) and (2.3) represent a multivariate extension of Edgeworth series expansion. This special case has been studied in detail by Chamber (1967). Using this method, Chamber (1967) obtained approximations to many multivariate distributions from normal universes.

(b) Approximation of non-central Wishart distribution by central Wishart distribution.

Let S be a $r \times r$ symmetric positive definite matrix distributed as non-central Wishart with degrees of freedom f, parameter matrix Σ and non-centrality $\theta\Sigma^{-1}$ (θ symmetric non-negative definite), $S \sim W_r(f;\Sigma,\theta\Sigma^{-1})$, and let $f(S)$ be the density of $S = (s_{ij})$. Put $x_1 = s_{11}$, $x_2 = s_{12},\dots,x_p = s_{(r-1)r}$, $p = \frac{1}{2}r(r+1)$, and let $g(\underset{\sim}{x}) = g(S)$ be the density of $W_r(f,\Sigma+\frac{1}{f}\theta)$, an r-dimensional central Wishart with degrees of freedom f and parameter matrix $\Sigma+\frac{1}{f}\theta$. Then (2.2) and (2.3) represent an approximation of non-central Wishart density by central Wishart density. This is the approximation considered by Tan

(1978). Note that the first term of this approximation turns out to be the approximation considered by Steyn and Roux (1972). Note also that, if f is large (say $f > 20$) and if $6 \geq k \geq 4$, then for the approximation (2.2), all mixed moments up to order k of the approximation equal to the mixed moments of order k of $S \sim W_r(f;\Sigma,\theta\Sigma^{-1})$. Hence, if f is sufficiently large and if $tr(\theta\theta')$ is not big, (2.2) would provide a close approximation to $f(S)$ if $k \geq 4$. The mixed cumulants of the central and non-central Wishart distributions and the partial derivatives of the central Wishart density have been given by Tan (1978). Using these results one has immediately an approximation as given in (2.2). For detail the readers are referred to Tan (1978).

(c) Approximating the distribution of sample covariance matrix from non-normal universes.

Suppose that $\underset{\sim}{y}_1,\ldots,\underset{\sim}{y}_n$ is a random sample from an r-dimensional universe Π (not necessarily normal) with covariance matrix Σ. Put $\bar{y} = \frac{1}{n}\sum_{j=1}^{n} \underset{\sim}{y}_j$ and $S = \sum_{j=1}^{n} (\underset{\sim}{y}_j-\bar{\underset{\sim}{y}})(\underset{\sim}{y}_j-\bar{\underset{\sim}{y}})'$ and let $p = \frac{1}{2}r(r+1)$, $x_1 = s_{11}, x_2 = s_{12},\ldots,x_p=s_{(r-1)r}$. Let $g(\underset{\sim}{x}) = g(\underset{\sim}{S})$ be the density of $W_r(n-1,\Sigma)$. Then, using (2.2) and (2.3), we have an approximation to the density $f(S)$ of S by the central Wishart distribution. This approximation can readily be implemented by the observation that the elements of $\frac{1}{n-1}S$ are second order k-statistics so that their mixed cumulants are readily available (Cook 1951, Kendall and Stuart 1958 Vol I); further, the mixed cumulants and partial derivatives of the central Wishart density are available in Tan (1978). Note that, if $k \geq 4$, then Theorem (2.2) provides that all mixed moments up to at least order four of the approximation equal to those of $f(S)$. Thus, if n is very large while k is taken to be ≥ 4 and if the mixed higher cumulants of Π are small, it is expected that (2.2) would provide a close approximation to $f(S)$.

4. APPROXIMATION OF MULTIVARIATE DISTRIBUTIONS BY GENERALIZED LAGUERRE POLYNOMIALS

It is well-known that classical Laguerre polynomial expansion provide close approximations to many probability distributions from non-normal universes (Tiku 1964, 1966, 1971, Tan and Wong 1977, 1978, 1980); further, many multivariate distributions such as non-central Wishart (Roux and Raath 1973), generalized Hotelling's T^2(Constantine 1966) and multivariate quadratic forms (James 1964) can be expanded as infinite series involving generalized Laguerre polynomials in matrix argument. It is therefore logical to consider approximating many multivariate distribution by generalized Laguerre polynomial expansions. In this paper, an attempt is made to approximate the distribution of sample covariance matrix from bivariate non-normal universes. This is motivated by the desire to approximate distribution of sample correlation coefficient and to evaluate the effect of departure from normality in multivariate analysis of variance procedures.

4.1 The Generalized Laguerre Polynomials.

Let W be a pxp symmetric matrix of real numbers and $(k) = (k_1, k_2, \ldots, k_p)$ be a partition of integer k into p parts satisfying $k_1 \geq k_2 \geq \ldots \geq k_p \geq 0$ and $\sum_{j=1}^{p} k_j = k$. Then, for $\gamma > 0$, a generalized Laguerre polynomial $L^{\gamma}_{(k)}(W)$ of degree (k) in the matrix argument W is defined as (Constantine 1966)

$$L^{\gamma}_{(k)}(W) = \sum_{v=0}^{k} (-1)^v \sum_{(v)} a_{(k),(v)} \frac{\Gamma_p(\gamma+\ell,(k))C_{(v)}(W)C_{(k)}(I_p)}{\Gamma_p(\gamma+\ell,(v))C_{(v)}(I_p)} \qquad (4.1)$$

where $\ell = (p+1)/2$, $\Gamma_p(\gamma+m,(k)) = \pi^{p(p-1)/4} \prod_{j=1}^{p} \Gamma(\gamma+m+k_j - (j-1)/2)$, and $C_{(k)}(W)$ is the zonal polynomial of degree (k) in W as defined in James (1964).

In (4.1), the coefficients $a_{(k),(v)}$ are derived in the expansion $C_{(k)}(I_p+W)/C_{(k)}(I_p) = \sum_{v=0}^{k} \sum_{(v)} a_{(k),(v)} C_{(v)}(W)/C_{(v)}(I_p)$ and have been tabulated by Constantine (1966) up to order $k = 4$. As shown in James (1964), the zonal polynomials $C_{(k)}(W)$ depends on W through elementary symmetric functions of W. When $p = 2$, the $C_{(k)}(W)$'s are then functions of $\mathrm{tr}W$ (trace of W) and $|W|$ (determinant of W). Using results given in James (1964) and Constantine (1966), one may then write $L^{\gamma}_{(k)}(W)$ explicitly at least up to order $k = 4$. Given below are the explicit formulas for $L^{\gamma}_{(k)}(W)$ in $p = 2$ up to order $k = 4$:

$$L^{\gamma}_{(00)}(W) = 1 ,$$

$$L^{\gamma}_{(10)}(W) = (2\gamma + 3)\{1 - \frac{\mathrm{tr}W}{2\gamma+3}\} ,$$

$$L^{\gamma}_{(20)}(W) = \frac{2}{3}(2\gamma+3)(2\gamma+5)\{1 - \frac{2\mathrm{tr}W}{2\gamma+3} + \frac{3(\mathrm{tr}W)^2-4|W|}{2(2\gamma+3)(2\gamma+5)}\}$$

$$L^{\gamma}_{(11)}(W) = \frac{2}{3}(\gamma+1)(2\gamma+3)\{1 - \frac{2\mathrm{tr}W}{2\gamma+3} + \frac{2|W|}{(\gamma+1)(2\gamma+3)}\}$$

$$L^{\gamma}_{(30)}(W) = \frac{2}{5}(2\gamma+3)(2\gamma+5)(2\gamma+7)\{1 - \frac{3\mathrm{tr}W}{2\gamma+3} + \frac{3[3(\mathrm{tr}W)^2-4|W|]}{2(2\gamma+3)(2\gamma+5)}$$

$$- \frac{5(\mathrm{tr}W)^3-12|W|\mathrm{tr}W}{2(2\gamma+3)(2\gamma+5)(2\gamma+7)}\} ,$$

$$L^{\gamma}_{(21)}(W) = \frac{6}{5}(\gamma+1)(2\gamma+3)(2\gamma+5)\{1 - \frac{3trW}{2\gamma+3} + \frac{2[3(trW)^2-4|W|]}{3(2\gamma+3)(2\gamma+5)}$$

$$+ \frac{10|W|}{3(\ +1)(2\ +3)} - \frac{2|W|trW}{(\ +1)(2\ +3)(2\ +5)}\}$$

$$L^{\gamma}_{(40)}(W) = \frac{8}{35}(2\gamma+3)(2\gamma+5)(2\gamma+7)(2\gamma+9)\{1 - \frac{4trW}{2\gamma+3} + \frac{3[3(trW)^2 - 4|W|]}{(2\gamma+3)(2\gamma+5)}$$

$$- \frac{2[5(trW)^3 - 12|W|trW]}{(2\gamma+3)(2\gamma+5)(2\gamma+7)} + \frac{35(trW)^4 - 120|W|(trW)^2 + 48|W|^2}{8(2\gamma+3)(2\gamma+5)(2\gamma+7)(2\gamma+9)}\}$$

$$L^{\gamma}_{(31)}(W) = \frac{8}{7}(\gamma+1)(2\gamma+3)(2\gamma+5)(2\gamma+7)\{1 - \frac{4trW}{2\gamma+3} + \frac{11[3(trW)^2 - 4|W|]}{6(2\gamma+3)(2\gamma+5)}$$

$$+ \frac{14|W|}{3(\gamma+1)(2\gamma+3)} - \frac{3[5(trW)^3 - 12|W|trW]}{5(2\gamma+3)(2\gamma+5)(2\gamma+7)}$$

$$- \frac{28|W|trW}{5(\gamma+1)(2\gamma+3)(2\gamma+5)} + \frac{3|W|(trW)^2 - 4|W|^2}{(\gamma+1)(2\gamma+3)(2\gamma+5)(2\gamma+7)}\}$$

and

$$L^{\gamma}_{(22)}(W) = \frac{4}{5}(\gamma+1)(\gamma+2)(2\gamma+3)(2\gamma+5)\{1 - \frac{4trW}{2\gamma+3} + \frac{4[3(trW)^2 - 4|W|]}{3(2\gamma+3)(2\gamma+5)}$$

$$+ \frac{20|W|}{3(\gamma+1)(2\gamma+3)} - \frac{8|W|trW}{(\gamma+1)(2\gamma+3)(2\gamma+5)}$$

$$+ \frac{4|W|^2}{(\gamma+1)(\gamma+2)(2\gamma+3)(2\gamma+5)}\} .$$

One may note that, if we write $P_r(w) = \frac{1}{\Gamma_2(\frac{2\gamma+3}{2})}|W|^{\gamma}\exp\{-trW\}$, where $\Gamma_p(n) = \Gamma_p(N,(0))$ with $(0) = (0,\ldots,0)$, then, as shown in Constantine (1966),

$$\iint_{W>0} L^{\gamma}_{(k)}(W)L^{\gamma}_{(v)}(W)P_r(W)dW = 0 \quad \text{if} \quad (k) \neq (v) \quad , \qquad (4.2)$$

where $W > 0$ denotes that W is symmetric positive definite and dW is the differential of W .

4.2 The Generalized Laguerre Polynomial Approximation to the Distribution of Two-Dimensional Sample Covariance Matrix.

Let $\underset{\sim}{x}_1, \underset{\sim}{x}_2, \ldots, \underset{\sim}{x}_n$ be a random sample from some bivariate universe where the (r,s)th cumulant is $\lambda_{rs}(\lambda_{00}=0)$, r, s = 0,1,2,..., with $\lambda_{20}=\sigma_1^2$, $\lambda_{02}=\sigma_2^2$ and $\lambda_{11}=\rho\sigma_1\sigma_2$. Put $\bar{\underset{\sim}{x}} = \frac{1}{n}\sum_{j=1}^{n}\underset{\sim}{x}_j$, $R = \sum_{j=1}^{n}(\underset{\sim}{x}_j-\bar{\underset{\sim}{x}})(\underset{\sim}{x}_j-\bar{\underset{\sim}{x}})'$ and $S = \frac{1}{2}\Lambda R\Lambda' = (s_{ij})$ where Λ is a 2x2 diagonal matrix with diagonal elements σ_1^{-1} and $\sigma_2^{-1}(\sigma_i>0)$, i.e., $\Lambda = \mathrm{Diag}(\sigma_1^{-1},\sigma_2^{-1})$. Further, we let $k_{rs} = \lambda_{rs}/(\sigma_1^r\sigma_2^s)$, $m = \frac{1}{2}(n-1) = \gamma + \frac{3}{2}$, and $G_{(k)}^m(S) = L_{(k)}^{\gamma}(S\Sigma^{-1})$, where $\Sigma = \begin{pmatrix}1 & \rho\\ \rho & 1\end{pmatrix}$. (Note that $\mathrm{tr}(S\Sigma^{-1}) = \frac{1}{(1-\rho^2)}(s_{11}+s_{22}-2\rho s_{12})$ and $|S\Sigma^{-1}| = |S|/|\Sigma|$).
If $|k_{rs}| < \infty$, one may then extend Roy and Tiku (1962) to approximate the sampling distribution of S by a finite series $\phi(S)$ involving $G_{(k)}^m(S)$ up to order 4:

$$\phi(S) = f_m(S)\{1 + \sum_{k=1}^{4}\sum_{(k)} C_{(k)}\, G_{(k)}^m(S)\} , \tag{4.3}$$

where $f_m(S) = \frac{1}{\Gamma_2(m)}|\Sigma|^{-m}|S|^{m-3/2}\exp\{-\mathrm{tr}S\Sigma^{-1}\}$, and $\sum_{(k)}$ sums over all partitions $(k) = (k_1,k_2)$ of k with $k_1 + k_2 = k$ and $k_1 \geq k_2 \geq 0$.

Making the transformation $W = \Sigma^{-1/2}S\Sigma^{-1/2}$, where $\Sigma^{-1/2}\Sigma^{-1/2} = \Sigma^{-1}$, and using (2.2) , it is obvious that $\int\int_{S>0} G_{(k)}^m(S)G_{(v)}^m(S)f_m(S)dS = 0$ if $(k) \neq (v)$. It follows that

$$C_{(k)} = E\{G_{(k)}^m(S)\}/d_{(k)}^2 , \tag{4.4}$$

where $d_{(k)}^2 = \int\int_{S>0}\{G_{(k)}^m(S)\}^2 f_m(S)dS$.

Making use of the results given in Cook (1951), the coefficients $C_{(k)}$ can readily be derived. After some tedious algebra, we obtain:

$$C_{(10)} = 0;$$

$$C_{(20)} = \frac{3(n-1)D_4}{4n(n+1)} ,$$

where $D_4 = \frac{1}{4(1-\rho^2)^2}[(k_{40}+k_{04}) - 4\rho(k_{31}+k_{13}) + 2(1+2\rho^2)k_{22}]$;

$$C_{(11)} = 0 \ ;$$

$$C_{(30)} = -\frac{1}{12n(n+1)(n+3)}\left[\frac{5(n-1)^2 D_6}{n} + 4(n-2)D_{61}\right] ,$$

where $D_6 = \frac{1}{8(1-\rho^2)^3}\{(k_{60}+k_{06}) - 6\rho(k_{51}+k_{15})$

$$+ 3(1+4\rho^2)(k_{42}+k_{24}) - 4\rho(3+2\rho^2)k_{33}\} ,$$

$$D_{61} = \frac{1}{8(1-\rho^2)^3}[5(k_{30}^2+k_{03}^2) - 2\rho(3+2\rho^2)k_{30}k_{03}$$

$$- 18\rho(3+2\rho^2)k_{21}k_{12} + 9(1+4\rho^2)(k_{21}^2+k_{12}^2) + 5(1+4\rho^2)$$

$$\times (k_{30}k_{12}+k_{03}k_{21}) - 30\rho(k_{30}k_{21}+k_{03}k_{12})];$$

$$C_{(21)} = -\frac{2D_{62}}{3n(n+1)} ,$$

where $D_{62} = \frac{1}{4(1-\rho^2)^2}[(k_{21}^2+k_{12}^2) - \rho k_{21}k_{12} - (k_{30}k_{12}+k_{03}k_{21})$

$$+ \rho k_{30}k_{03}];$$

$$C_{(40)} = \frac{1}{24n^2}\left\{\frac{35(n-1)^3 D_8}{8(n+1)(n+3)(n+5)} + \frac{30(n-2)(n-1)D_{81}}{(n+1)(n+3)(n+5)}\right.$$

$$\left. + \frac{(3n^3+23n^2-63n+45)D_{82}}{8(n-1)^3} - \frac{75(n^2-3n+3)\rho^2 k_{40}k_{04}}{4(1-\rho^2)^3(n-1)^3}\right\} ,$$

where $D_8 = \frac{1}{16(1-\rho^2)^4}[(k_{80}+k_{08}) - 8\rho(k_{71}+k_{17}) + 4(1+6\rho^2)(k_{62}+k_{26})$

$$- 8\rho(3+4\rho^2)(k_{53}+k_{35}) + 2(3+24\rho^2+8\rho^4)k_{44}] ,$$

$$D_{81} = \frac{1}{16(1-\rho^2)^4}\{k_{50}k_{30}+k_{05}k_{03}) - \rho[5(k_{41}k_{30}+k_{14}k_{03})$$

$$+ 3(k_{50}k_{21}+k_{05}k_{12})] + (1+6\rho^2)[15(k_{41}k_{21}+k_{14}k_{12})$$

$$+ 10(k_{32}l_{30}+k_{23}k_{03}) + 3(k_{50}k_{12}+k_{05}k_{21})]$$

$$- \rho(3+4\rho^2)[30(k_{32}k_{21}+k_{23}k_{12}) + 15(k_{41}k_{12}+k_{14}k_{21})$$

$$+ 10(k_{30}k_{23}+k_{03}k_{32}) + (k_{50}k_{03}+k_{05}k_{30})]$$

$$+ (3+24\rho^2+8\rho^4)[6(k_{32}k_{12}+k_{23}k_{21}) + (k_{41}k_{03}+k_{14}k_{30})]\} ,$$

and

$$D_{82} = \frac{1}{16(1-\rho^2)^4} \{35[(k_{40}^2+k_{04}^2) - 8\rho(k_{40}k_{31}+k_{04}k_{13})]$$
$$- 40\rho(3+4\rho^2)[6k_{22}(k_{31}+k_{13}) + (k_{40}k_{13}+k_{04}k_{31})]$$
$$+ 20(1+6\rho^2)[3k_{22}(k_{40}+k_{04}) + 4(k_{31}^2+k_{13}^2)]$$
$$+ 2(3+24\rho^2+8\rho^4)(k_{40}k_{04}+16k_{31}k_{13}+18k_{22}^2)\} ;$$

$$C_{(31)} = \frac{1}{12n^2(n+1)(n+3)} \{6(n-1)D_{83} - \frac{(n^3-4n^2+7n-6)D_{84}}{(n-2)}$$
$$- \frac{3\rho^2(n^2-3n+3)k_{40}k_{04})}{4(n-2)(1-\rho^2)^3}\} ,$$

where

$$D_{83} = \frac{1}{8(1-\rho^2)^3} \{2(k_{41}k_{21}+k_{14}k_{12}) - (k_{32}k_{30}+k_{23}k_{03}+k_{50}k_{12}+k_{05}k_{21})$$
$$+ (1+2\rho^2)(k_{23}k_{21}+k_{32}k_{12} - k_{41}k_{03} - k_{14}k_{30}) - 5\rho(k_{32}k_{21}+k_{23}k_{12})$$
$$+ 3\rho(k_{30}k_{23}+k_{03}k_{32}) + \rho(k_{41}k_{12}+k_{14}k_{21}+k_{50}k_{03}+k_{05}k_{30})\} ,$$

and

$$D_{84} = \frac{1}{8(1-\rho^2)^3} \{3[(k_{31}^2+k_{13}^2) - k_{22}(k_{40}+k_{04})]$$
$$+ 6\rho[(k_{40}k_{13}+k_{04}k_{31}) - k_{22}(k_{31}+k_{13})]$$
$$+ (1+2\rho^2)(3k_{22}^2-2k_{31}k_{13}+k_{40}k_{04})\} ;$$

and

$$C_{(22)} = \frac{5(n-2)(3k_{22}^2-4k_{31}k_{13}+k_{40}k_{04})}{48(1-\rho^2)^2n^2(n+1)} .$$

It follows that

$$\phi(S) = f_m(S)\{1 + \sum_{k=2}^{4} \sum_{(k)} C_{(k)}G_{(k)}^m(S)\} , \qquad (4.4)$$

As in the univariate case, it can be shown that, if the density of S can be expanded in Laguerre polynomials in matrix argument,

$$E(S_{11}^{\ell_1}S_{22}^{\ell_2}S_{12}^{\ell_3}) = \int\int_{S>0} S_{11}^{\ell_1}S_{22}^{\ell_2}S_{12}^{\ell_3}\,\phi(S)dS \quad \text{for}$$

all $\ell_1 + \ell_2 + \ell_3 \leq 4$ and nonnegative integer ℓ_i. In these cases, if one uses $\phi(S)$ to approximate the distribution of S, then mixed moments up to the fourth order using $\phi(S)$ equal to the respective exact mixed moments. Thus, if $\phi(S) \geq 0$, $\phi(S)$ would provide a very close approximation to the distribution of S. Further studies on this point are needed, however.

REFERENCES

[1] Bowman, K.O., Beauchamp, J.J. and Shenton, L.R. (1977). "The distribution of the t-statistic under non-normality." Inst. Statist. Review 45, 233-242.

[2] Chambers, J.M. (1967). "On methods of asymptotic approximation for multivariate distributions", Biometrika 54, 367-383.

[3] Constantine, A.G. (1966). "The distribution of Hotelling's generalized T_0^2. Ann. Math. Statist. 37, 215-25.

[4] Cook, M.B. (1951). "Bivariate k-statistics and cumulants of their joint distribution. Biometrika 38, 179-95.

[5] Finney, D.J. (1963). "Some properties of a distribution specified by its cumulants". Technometrics 5, 63-69.

[6] Geary, R.C. (1947). "Testing for normality". Biometrika 34, 209-242.

[7] Good, I.J. (1962). "The multivariate saddlepoint method and chisquared for the multinomial distribution", Ann. Math. Statist. 33, 535-548.

[8] James, A.T. (1964). "Distributions of matrix variates and latent roots derived from normal samples", Ann. Math. Statist. 35, 475-497.

[9] Kendall, M.G. and Studart, A. (1958). "The Advanced Theory of Statistics", Vol. I, Griffin.

[10] Roux, J.J.J. and Raath, E.L. (1973). "Generalized Laguerre series forms of Wishart distributions", S. Afri. Statist. J. 7, 23-34.

[11] Steyn H.S. and Roux, J.J.J. (1972). "Approximations for the non-central Wishart distribution", S. Afri. Statist. J. 6, 165-173.

[12] Tan, W.Y. (1978). "On the approximation of non-central Wishart distribution by Wishart distribution", I. General Theory. (To appear in Metron).

[13] Tan, W.Y. and Wong, S.P. (1977). "On the Roy-Tiku approximation to the distribution of sample variance from nonnormal universes", J. Amer. Statist. Association 72, 875-81.

[14] Tan, W.Y. and Wong, S.P. (1978). "On approximating the central and non-central multivariate gamma distributions. Communication in Statistics, B7, No. 3 (To appear).

[15] Tan, W.Y. and Wong, S. P. (1980). "On approximating the null and nonnull distributions of the F-ratio in unbalanced random effect models from non-normal universes". Technical Report N. 78-8, Memphis State University, Memphis, Tennessee.

[16] Tiku, M.L. (1964). "Approximating the general nonnormal variance-ratio sampling distributions. Biometrika 51, 83-95.

[17] Tiku, M.L. (1965). "Laguerre series forms of noncentral χ^2 and F distributions". Biometrika 52, 415-27.

[18] Tiku, M.L. (1971). "Power function of F-test under nonnormal situations". J. Amer. Statist. Association 66, 913-16.

Multivariate Statistical Analysis
R.P. Gupta (ed.)

ROBUST ESTIMATION OF THE VARIANCE-COVARIANCE MATRIX OF SYMMETRIC MULTIVARIATE DISTRIBUTIONS

M.L. Tiku

Department of Mathematical Sciences
McMaster University, Hamilton, Ontario
Canada

Tiku's (1967, 1978, 1979) modified maximum likelihood estimators of the location and scale parameters, based on symmetrically censored samples with about 10% observations censored on either side, are shown to provide robust-estimators of the variance-covariance matrices of bivariate symmetric distributions. Generalization to symmetric multivariate distributions is quite straightforward.

INTRODUCTION

Let $x_1, x_2, \ldots, x_n$ be a random sample from the normal $N(\mu,\sigma)$ distribution, and let

$$X_{r+1}, X_{r+2}, \ldots, X_{n-r} \tag{1}$$

be the Type II symmetrically censored sample obtained by arranging the above n random observations in ascending order of magnitude and censoring the r smallest and the r largest observations; the value of r will be specified later. Tiku's (1967) MML estimators (defined formally by Tiku and Stewart, 1977) of μ and σ are given by

$$\hat{\mu} = \{\sum_{i=r+1}^{n-r} X_i + r\beta(X_{r+1} + X_{n-r})\}/m \tag{2}$$

and

$$\hat{\sigma} = \{B + \surd(B^2 + 4AC)\}/2\surd\{A(A-1)\} , \tag{3}$$

where

$$m = n - 2r + 2r\beta,\ A = n - 2r,\ B = r\alpha(X_{n-r} - X_{r+1})$$

and

$$C = \sum_{i=r+1}^{n-r} X_i^2 + r\beta(X_{r+1}^2 + X_{n-r}^2) - m\hat{\mu}^2 ;$$

α and β are simple constants and are given by Tiku (1978, p.1220). For $n \geq 10$, however, α and β are obtained form the following (asymptotic) equations,

$$\beta = f(t)\{t - f(t)/q\}/q \quad \text{and} \quad \alpha = \{f(t)/q\} - \beta t,\ Q(t) = q \ ; \tag{4}$$

$Q(t) = 1 - F(t)$, $F(t) = \int_{-\infty}^{t} f(z)dz$ and $f(z) = (2\pi)^{-1/2}\exp(-\frac{1}{2}z^2)$.

For normal samples, the efficiencies and the distributions of μ_c and σ_c are investigated by Tiku (1978). Suffice it to say here that μ_c and σ_c provide an ideal pair of estimators of the mean μ and the standard deviation σ of the $N(\mu,\sigma)$ distribution, based on Type II censored samples; see also Tiku (1968, 1973) and Smith, Zeiss and Syler (1973). Note that for normal samples, $(\mu_c-\mu)\sqrt{m}/\sigma$ and $(A-1)\sigma_c^2/\sigma^2$ are independently distributed as normal $N(0,1)$ and chi-square with $A-1$ df, respectively, for large $A = n - 2r$; see Tiku (1978, Lemmas 1 and 2).

Tiku (1979) investigated the efficiencies of $\hat{\mu}$ and $\hat{\sigma}$ for estimating the location parameter μ and scale parameter σ of symmetric non-normal distributions and concluded that, (i) for long-tailed distributions with existent mean and standard deviation, $\hat{\mu}$ and $\hat{\sigma}$ based on the censored sample (1) with $r = [1/2 + 0.1n]$, i.e. about 10% censoring on either side, are jointly as efficient as some of the most prominent and celebrated 'robust' estimators (wave, Hampel and bisquare estimators, Gross 1976), and (ii) for 'disaster' situations (which can hopefully be easily spotted through plotting), that is, when the sample contains a large proportion (more than 20%)of outliers or the sample comes from a distribution with nonexistent standard deviation (Cauchy for example), $\hat{\mu}$ and $\hat{\sigma}$ based on the censored sample (1) with $r = [1/2 + 0.3n]$, i.e. about 30% censoring, are jointly more efficient than the celebrated 'robust' estimators mentioned above.

If one has no knowledge whether the situation is that of (i) or (ii) above, the adaptive estimators $\hat{\mu}_{adp}$ and $\hat{\sigma}_{adp}$ may be used and these are defined to be exactly the same as $\hat{\mu}$ and $\hat{\sigma}$ but with the constant fraction $q = r/n$ in (4) replaced by the random fraction $q^* = r^*/n$, where $r^* = 0$ if $m^*/n = 0$, $r^* = [1/2 + 0.1n]$ if $m^*/n \leq 0.1$, and $r^* = [1/2 + 0.3n]$ if $m^*/n > 0.1$; m^* is the number of values of $|z_i| = |x_i - \text{median}\ (x_i)|/1.48\ \text{median}\ |x_i - \text{median}\ (x_i)|$, $i = 1,2,\ldots,n$, which exceed 3.0. The adaptive esitmators $\hat{\mu}_{adp}$ and $\hat{\sigma}_{adp}$ are on the whole jointly as efficient as the celebrated 'robust' estimators (Gross, 1976) mentioned above; see Tiku (1979).

We now use the MML estimators $\hat{\mu}$ and $\hat{\sigma}$ to develop 'robust' estimators of the variance-covariance matrix of bivariate symmetric distributions; generalization to multivariate symmetric distributions is straightforward.

ROBUST ESTIMATORS OF VARIANCES AND COVARIANCES

Let (x_i,y_i) , $i = 1,2,\ldots,n$, be a sample of size n from a symmetric bivariate distribution with $E(x) = \mu_1$, $E(y) = \mu_2$, $V(x) = \sigma_1^2$, $V(y) = \sigma_2^2$ and the correlation coefficient $Cov(x,y)/\sigma_1\sigma_2 = \rho$(positive). Note that for bivariate distributions with nonexistent means and standard deviations, μ_1 and

μ_2 and σ_1 and σ_2 are exclusively the location and scale parameters. Define $z_i = x_i - y_i$, $i = 1,2,\ldots,n$, and note that for $\rho > 0$ the ordered sample $(Z_1, Z_2, \ldots, Z_n)$ will hopefully have fewer extremely large or extremely small observations than either of the two ordered samples $(X_1, X_2, \ldots, X_n)$ and $(Y_1, Y_2, \ldots, Y_n)$. Like the censored sample (1), let

$$X_{r+1},\ X_{r+2},\ldots,X_{n-r}, \tag{5}$$

$$Y_{r+1},\ Y_{r+2},\ldots,Y_{n-r}, \tag{6}$$

and

$$Z_{r+1},\ Z_{r+2},\ldots,Z_{n-r}, \tag{7}$$

be the three Type II symmetrically censored samples, obtained from the three random samples (x_i), (y_i) and (z_i) , $i = 1,2,\ldots,n$, respectively. The robust estimators $\hat{\mu}_1$ and $\hat{\mu}_2$ of μ_1 and μ_2 are given by the equation (2) and the robust estimators $\hat{\sigma}_1$, $\hat{\sigma}_2$ and $\hat{\sigma}_3$ of σ_1 , σ_2 and $\sigma_3 = \sqrt{V(z)}$ are given by the equation (3), with the observations (1) replaced by the observations (5), (6) and (7), respectively. The robust estimator of ρ is given by

$$\hat{\rho} = (\hat{\sigma}_1^2 + \hat{\sigma}_2^2 - \hat{\sigma}_3^2)/2\hat{\sigma}_1\hat{\sigma}_2 \quad . \tag{8}$$

It is possible that a value of $\hat{\rho}$ greater than 1.0 (less than 0.0) may occur but this is replaced by 1.0(0.0). Note that for $r = 0$, the above estimators reduce to the ordinary estimators based on the complete random samples (x_i) , (y_i) and (z_i) , $i = 1,2,\ldots,n$.

Since symmetric bivariate distributions have symmetric marginal distributions, $\hat{\mu}_1$ and $\hat{\mu}_2$ are clearly unbiased.

If the joint distribution of x and y is bivariate normal then the marginal distributions of x and y are normal and the distribution of $z = x - y$ is also normal and, therefore, the distributions of $(\hat{\mu}_1 - \mu_1)\sqrt{m}/\sigma_1$ and $(\hat{\mu}_2 - \mu_2)\sqrt{m}/\sigma_2$ are normal $N(0,1)$ and the distributions of $(A - 1)\hat{\sigma}_1^2/\sigma_1^2$, $(A - 1)\hat{\sigma}_2^2/\sigma_2^2$ and $(A - 1)\hat{\sigma}_3^2/\sigma_3^2$ are chi-square with $A - 1$ df, for large A; see Tiku (1978, Lemmas 1 and 2). Consequently, the distribution of $\hat{\rho}$ for the bivariate normal will be the same as the distribution of the ordinary correlation coefficient based on a random sample of size $A = n - 2r$, for large A .

To investigate the bias and the efficiencies of the above estimators, we generated a number of bivariate distributions(with correlation coefficient ρ) through the equations

$$x = \sqrt{(\tfrac{1}{2})}[u_1 + \sqrt{\{(1-\rho)/(1+\rho)\}}\ u_2]\sqrt{(1+\rho)}$$

and

$$y = \sqrt{(\tfrac{1}{2})}[u_1 - \sqrt{\{(1-\rho)/(1+\rho)\}}\ u_2]\sqrt{(1+\rho)} \quad , \tag{9}$$

where u_1 and u_2 are iid, with mean (location parameter) 0 and standard deviation (scale parameter) 1, and have one of the following distributions, (1) Normal, (2) Logistic, (3) Student's t_4, (4) (n - 1)N(0,1) & 1N(0,3), (5) .90N(0,1) + .20N(0,3), and (6) Student's t_2, and simulated (from 5000 Monte Carlo runs) the means of $\hat{\sigma}_1$, $\hat{\sigma}_2$ and $\hat{\rho}$, and mean square errors (MSE) of $\hat{\mu}_1$, $\hat{\mu}_2$, $\hat{\sigma}_1$, $\hat{\sigma}_2$ and $\hat{\rho}$. These values are given in Table I. Since $\hat{\mu}_1$ and $\hat{\mu}_2$ are unbiased, their MSE are equal to their variances. Note that for the bivariate distributions generated through the equations (9), the expected values $E(\hat{\sigma}_1)$ and $E(\hat{\sigma}_2)$ are essentially equal and equal to $E(\hat{\sigma})$, given in Table I, and similarly for the MSE of $\hat{\mu}_1$ and $\hat{\mu}_2$ and the MSE of $\hat{\sigma}_1$ and $\hat{\sigma}_2$. It is clear from these values that the stimators $\hat{\mu}_1$, $\hat{\mu}_2$, $\hat{\sigma}_1$, $\hat{\sigma}_2$ and $\hat{\rho}$ are on the whole considerably more efficient than the ordinary estimators based on complete samples (obtained by putting $r = 0$ in (2), (3) and (8)). We also simulated the values for $n = 10$ and $r = [1/2 + 0.1n] = 1$, and found no unexpected features.

For 'disaster' situations, the efficiencies of the estimators $\hat{\mu}_1$, $\hat{\mu}_2$, $\hat{\sigma}_1$ and $\hat{\sigma}_2$, but unfortunately not of $\hat{\rho}$, can be enhanced by censoring a greater proportion of observations. For example for the distribution (6) with $\rho = .50$, the simulated values (based on 5000 runs) are, $E(\hat{\sigma}) = 1.31$, $E(\hat{\rho}) = .61$, $V(\hat{\mu}) = .123$, MSE $(\hat{\sigma}) = .303$ and MSE $(\hat{\rho}) = .105$; $n = 20$ and $r = [1/2 + 0.3n]=6$. The sum of MSE is considerably reduced, however.

We also used the adaptive estimators $\hat{\mu}_{adp}$ and $\hat{\sigma}_{adp}$ and the robust estimators (wave, Hampel and bisquare, Gross 1976) in equation (8) to produce a robust estimator of ρ, but this resulted into prohabitively large bias and MSE. This approach was, therefore, abandoned.

Note that if the correlation coefficient ρ between x and y, is negative, one may calculate the estimate of the correlation coefficient between x and $-y$ which is positive from the equation (8) and change its sign to obtain an estimate of a negative ρ.

ACKNOWLEDGEMENT

Thanks are due to the NSERC for a research grant to support this research.

TABLE I

Simulated values of the Means and MSE; n = 20, q = r/n, $\sigma_1 = \sigma_2 = 1$.

		(1)		(2)		(3)		(4)		(5)		(6)	
		q=0	q=.1	q=0	q=.1	q=0	q=.1	q=0	q=.1	q=0	q=.1	q=0	q=.1
						ρ=0.8							
Mean	$\hat{\sigma}$	.99	.98	.98	.93	.97	.84	.97	.87	.96	.84	2.39	1.51
	$\hat{\rho}$	.79	.79	.78	.79	.77	.80	.78	.79	.76	.80	.72	.81
MSE	$\hat{\mu}$	.050	.051	.050	.048	.050	.039	.050	.040	.050	.040	.552	.150
	$\hat{\sigma}$	.026	.037	.035	.043	.085	.063	.053	.046	.067	.060	6.70	.460
	$\hat{\rho}$	.008	.013	.013	.016	.025	.016	.017	.014	.028	.015	.076	.023
SUM MSE*		.084	.101	.098	.107	.160	.118	.120	.100	.145	.115	7.30	.633
						ρ=0.5							
Mean	$\hat{\sigma}$	.99	.98	.98	.94	.97	.86	.97	.87	.97	.85	2.43	1.58
	$\hat{\rho}$	.49	.49	.48	.50	.48	.54	.49	.49	.47	.52	.48	.58
MSE	$\hat{\mu}$	.050	.051	.050	.049	.050	.041	.050	.041	.050	.041	.522	.163
	$\hat{\sigma}$	.026	.026	.036	.041	.084	.057	.053	.046	.058	.058	6.59	.544
	$\hat{\rho}$	.032	.048	.043	.051	.062	.051	.051	.047	.066	.049	.104	.067
SUM MSE*		.108	.135	.126	.141	.196	.149	.154	.134	.174	.148	7.22	.774
						ρ=0.2							
Mean	$\hat{\sigma}$	.99	.98	.99	.94	.96	.86	.97	.87	.97	.86	2.50	1.62
	$\hat{\rho}$	.22	.24	.23	.27	.25	.31	.25	.24	.25	.30	.30	.39
MSE	$\hat{\mu}$	.050	.051	.050	.049	.050	.043	.050	.041	.050	.042	.522	.171
	$\hat{\sigma}$	.027	.036	.032	.041	.063	.056	.053	.046	.053	.057	5.85	.604
	$\hat{\rho}$	.034	.048	.042	.058	.061	.069	.052	.049	.062	.063	.107	.108
SUM MSE*		.111	.135	.124	.148	.174	.168	.155	.136	.165	.162	6.49	.883

*The reciprocal of these values is the trace-efficiency of the estimators $\hat{\mu}$, $\hat{\sigma}$ and $\hat{\rho}$.

REFERENCES

[1] Gross, A.M. (1976). Confidence interval robustness with long-tailed symmetric distributions. J. Amer. Statist. Assoc. 71, 409-16.

[2] Smith, W.B., Zeis, C.D. and Syler, G.W. (1973). Three parameter lognormal estimation from censored data. J. Indian Statist. Assoc. 11, 15-31.

[3] Tiku, M.L. (1967). Estimating the mean and standard deviation from censored normal samples. Biometrika 54, 155-65.

[4] Tiku, M.L. (1968). Estimating the parameters of log-normal distribution from censored samples. J. Amer. Statist. Assoc. 63, 134-40.

[5] Tiku, M.L. (1973). Testing group effects from Type II censored normal samples in experimental design. Biometrics 29, 25-33.

[6] Tiku, M.L. (1978). Linear regression model with censored observations. Commun. Statist. A7 (13), 1219-32.

[7] Tiku, M.L. (1979). Robustness of MML estimators based on censored samples and robust test statistics (submitted for publication).

[8] Tiku, M.L. (1979). Robust regression via MML estimators (submitted for publication).

[9] Tiku, M.L. and Stewart D. (1977). Estimating and testing group effects from Type I censored normal samples in experimental design. Commun. Statist. A6 (15), 1485-1501.

Multivariate Statistical Analysis
R.P. Gupta (ed.)

THE ANALYSIS OF NONORTHOGONAL MANOVA DESIGNS EMPLOYING A RESTRICTED FULL RANK MULTIVARIATE LINEAR MODEL

Neil H. Timm
School of Education
University of Pittsburgh
Pittsburgh, Pennsylvania

In this paper, a restricted full rank multivariate linear model is developed that may be used to analyze data from a fixed effect balanced or unbalanced experimental design with or without covariates and repeated measurements. To illustrate the advantages of the model over the classical less than full rank model, several designs are selected for analysis.

1. INTRODUCTION

Fisher's early development of the analysis of variance was first conceptualized by his colleagues using an equal number of observations per cell, cell means. and a full rank linear model, Urquhart, Weeks and Henderson (1973). Since the development of the analysis of variance technique, statisticians and researchers have been debating, discussing and writing about hypothesis testing when nonorthogonal experimental designs arise in practice. Some of the earliest papers are those written by Brant (1933); Snedecor (1934) and Yates (1934). Even with the introduction of the texts by Bancroft (1968), Searle (1971) and Graybill (1976), which devote several chapters to the analysis of nonorthogonal designs, papers on the topic continue to appear in the literature. Some of the more recent papers include Eccleston and Russell (1977), Francis (1973), Herr and Gaebelein (1978), Hocking and Speed (1975), Overall, Spiegel and Cohen (1975), and Timm and Carlson (1975), to name a few.

The attention given to the analysis of designs that have unequal and disproportionate cell frequencies can be attributed to the classical less than full rank model which requires an understanding of the concepts of estimable parametric functions, testable hypotheses, generalized inverse matrices, reparameterization, and side conditions. Hocking and Speed (1975) and Timm and Carlson (1975) propose that much of the confusion associated with testing hypotheses, when confronted with a nonorthogonal design, may be removed if the analysis is presented employing a full rank model.

The problems associated with hypothesis testing in nonorthogonal univariate designs extend in a natural manner to the analysis of nonorthogonal multivariate designs, Bock (1975) and Timm (1975). In these texts and even in the more recent text by Srivastava and Khatri (1979), little attention is given to MANOVA employing a full rank model. To help clarify and associate the classical approach of the analysis of multivariate nonorthogonal designs with a full rank model that is not obtained by imposing side conditions or by using the technique of reparameterization, a restricted full rank multivariate linear model is developed in this paper. The model may be used to analyze any balanced or unbalanced fixed effect univariate or multivariate experimental design with or without covariates and repeated measurements data.

Since most previous articles on the analysis of nonorthogonal designs have dealt with two-factor designs that have observations in every cell, we shall discuss the analysis of these designs when empty cells occur. In addition, the analysis of a three-factor design is presented. Finally, the MANOVA design with repeated measurements is considered.

2. THE RESTRICTED FULL RANK MULTIVARIATE LINEAR MODEL (RFRMLM)

The RFRMLM examined in this paper is written

$$\tilde{\Omega} : E(Y) = WU$$
$$V(Y) = I_N \otimes \Sigma \tag{2.1}$$

subject to the restrictions

$$R'UA = \Theta$$

where

$Y(N\times p)$ is a matrix of observations,

$U(q\times p)$ is a nonrandom matrix of parameters, usually populations means,

$W(N\times q)$ is a full rank design matrix of rank $q \le N$ which for standard MANOVA designs has only one unit element in each row, as many units in each column as there are observations in each population, and zeros elsewhere,

$R'(r\times q)$ is a known matrix of real numbers of rank $r \le q$,

$\Theta(r\times p)$ is a specified matrix of known constants, usually zero, and

$A(p\times v)$ is a known matrix of real numbers of rank $v \le p$.

Removing the matrix A from the expression $R'UA = \Theta$, we have the partially restricted full rank model (PRFRMLM) discussed by Timm and Carlson (1973).

Eliminating the restrictions $R'UA = \Theta$ from the model, we have the unrestricted full rank linear model (UFRMLM):

$$\Omega : E(Y) = WU$$
$$V(Y) = I_N \otimes \Sigma \tag{2.2}$$

To test hypotheses under $\tilde{\Omega}$, we shall assume that each row vector of the matrix Y has a nonsingular multivariate normal distribution with mean vector implicite in the expression for $E(Y)$ and nonsingular $p \times p$ variance-covariance matrix Σ.

3. ESTIMATING U AND Σ

Minimizing the least squares criterion,

$$Tr[(Y-WU)'(Y-WU)]$$

under $\tilde{\Omega}$, where Tr denotes the trace operator, the restricted least squares estimator of U, which is the best linear unbiased estimator (BLUE) of U is obtained. Letting

$$F = Tr[(Y-WU)'(Y-WU)] + 2\ Tr[(\Delta'(R'UA-\Theta)]$$

where Δ' is a matrix of Lagrange multipliers, and differentiating F with respect to U and Δ', the restricted model normal equations are directly obtained.

$$(W'W)\hat{U}_{\tilde{\Omega}} + R\Delta A' = W'Y$$
$$R'\hat{U}_{\tilde{\Omega}}A = \Theta \tag{3.1}$$

Letting $\hat{U}_{\tilde{\Omega}} = (W'W)^{-1}W'Y$, the unrestricted estimator of U under (2.2), the first equation in (3.1) yields

$$\hat{U}_{\tilde{\Omega}} = \hat{U}_{\Omega} - (W'W)^{-1}R\Delta A'$$

Premultiplying $\hat{U}_{\tilde{\Omega}}$ by R' and postmultiplying by A, we have that

$$\Theta = R'\hat{U}_{\Omega}A - R'(W'W)^{-1}R\Delta A'A$$

or

$$R'(W'W)^{-1}R\Delta(A'A) = R'\hat{U}_{\Omega}A - \Theta$$

so that

$$\Delta = (R'(W'W)^{-1}R)^{-1}(R'\hat{U}_\Omega A-\Theta)(A'A)^{-1}$$

Hence, from (3.1) the restricted least squares estimator of U under $\hat{\Omega}$ is given by the expression

$$\hat{U}_{\tilde{\Omega}} = \hat{U}_\Omega - (W'W)^{-1}R(R'(W'W)^{-1}R)^{-1}(R'\hat{U}_\Omega A-\Theta)(A'A)^{-1}A' \quad . \tag{3.2}$$

Letting $D = W'W$, a diagonal matrix for MANOVA designs and $E' = R'D^{-1}$, the restricted estimator of U is written as

$$\hat{U}_{\tilde{\Omega}} = \hat{U}_\Omega - E(E'DE)^{-1}R'\hat{U}_\Omega A(A'A)^{-1}A' + E(E'DE)^{-1}\Theta(A'A)^{-1}A' \tag{3.3}$$

To obtain an unbiased estimator of Σ under $\tilde{\Omega}$, the estimated restricted error sum of squares and products matrix (SSP)

$$\begin{aligned}\hat{Q}_{\tilde{\Omega}} &= (Y-W\hat{U}_{\tilde{\Omega}})/(Y-W\hat{U}_{\tilde{\Omega}}) \\ &= (Y-W\hat{U}_\Omega)'(Y-W\hat{U}_\Omega) + (\hat{U}_\Omega-\hat{U}_{\tilde{\Omega}})'(W'W)(\hat{U}_\Omega-\hat{U}_{\tilde{\Omega}}) \\ &= \hat{Q}_\Omega + A(A'A)^{-1}(R'\hat{U}_\Omega A-\Theta)'(R'(W'W)^{-1}R)^{-1}(R'\hat{U}_\Omega A-\Theta)(A'A)^{-1}A' \end{aligned} \tag{3.4}$$

where $\hat{Q}_\Omega = Y'(I-W(W'W)^{-1}W')Y$ is the unrestricted SSP matrix, is divided by $N-q+r$ where $p \leq N-q+r$. That is, the matrix

$$S_{\tilde{\Omega}} = \hat{Q}_{\tilde{\Omega}}/(N-q+r) \tag{3.5}$$

is an unbiased estimator of Σ under $\tilde{\Omega}$.

4. TESTING LINEAR HYPOTHESES EMPLOYING THE RFRMLM

To test the hypothesis

$$\tilde{\omega} : H_0{:}C'UA = \Gamma \tag{4.1}$$

under $\tilde{\Omega}$ where $C'(\nu_h \times q)$ is a known matrix of real numbers of rank $\nu_h \leq q$, $A(p\times t)$ is a known real matrix of rank $t \leq p$, and $\Gamma(\nu_h \times t)$ is a matrix of known constants, the partitioned matrices

$$Q' = \begin{pmatrix} R' \\ C' \end{pmatrix} \quad \text{and} \quad \Psi = \begin{pmatrix} \Theta \\ \Gamma \end{pmatrix}$$

are formed and the expression $Tr[(Y-WU)'(Y-WU)]$ is minimized subject to the restrictions $Q'UA = \Psi$ under $\tilde{\omega}$.

Provided no row of C' is equal to any row of R' or dependent on the rows of R' and/or other rows of C' and that the rank of the matrix T is v , that the expression for the hypothesis is not inconsistent with the restrictions on the model, the independently distributed hypothesis and error sum of squares and products matrices for testing H_0: $C'UA = \Gamma$ are given by the following formulae.

$$S_h = (C'\hat{U}_{\tilde{\Omega}}A-\Gamma)'(C'(FD^{-1}F')C)^{-1}(C'\hat{U}_{\tilde{\Omega}}A-\Gamma)$$

$$S_e = A'\hat{Q}_{\tilde{\Omega}}A \tag{4.2}$$

When the null hypothesis is true. S_h and S_e are distributed as central Wishart matrices:

$$S_h \sim W_t(\nu_h, A'\Sigma A)$$

$$S_e \sim W_t(N-q+r, A'\Sigma A)$$

Letting $\lambda_1,\lambda_2,\ldots,\lambda_s$ with $s = \min(t,\nu_h)$ denote the roots of the determinantal equation

$$|S_h - \lambda S_e| = 0$$

hypotheses of the form H_0: $C'UA = \Gamma$ may be tested using the standard well known multivariate test criteria, c.f. Roy, Gnanadesikan and Srivastava (1971) or Srivastava and Khatri (1979).

5. MANOVA OF A THREE-FACTOR DESIGN WITH UNEQUAL, DISPROPORTIONATE CELL FREQUENCIES AND INTERACTION

Under either the classical multivariate linear model or the RFRMLM, the only hypotheses that can be tested are those that involve estimable functions. However, since the basic population parameters of the RFRMLM are usually population means, estimability is never a problem for the applied researcher or data analyst. To test hypotheses using the RFRMLM formulation, one merely has to express the hypothesis of interest as linearly independent contrasts of cell means. When other more traditional formulations are used, "the user is seldom aware of the hypothesis being tested, this important fact being frequently left to the whim of the particular computing method employed", Hocking and Speed (1975, p. 712). The problem is particularly troublesome when the reduction in sum of squares procedure, the R ( ) notation employed by Searle (1971, p. 246) is used.

Letting $\underline{y}_1', \underline{y}_2', \ldots, \underline{y}_N'$ represent the N p-variate row vectors in the data matrix Y and $\underline{\mu}_1', \underline{\mu}_2', \ldots, \underline{\mu}_q'$, the q row p-vectors in the parameter matrix U, the UFRMLM for a three-way classification design with interaction, disproportionate cell frequencies N_{ijk}, and data in all cells of the design is given by

$$E(\underline{y}_{ijkn}) = \underline{\mu}_{ijk}\ ,\ i = 1,2,\ldots,I\ ;\ j = 1,2,\ldots,J\ ;$$
$$k = 1,2,\ldots,K\ ;\ n = 1,2,\ldots,N_{ijk} > 1 \qquad (5.1)$$

where the p-variate observation vector is represented by

$$\underline{y}_{ijkn}' = (y_{ijkn}^{(1)}, y_{ijkn}^{(2)}, \ldots, y_{ijkn}^{(p)})$$

and the p-vector $\underline{\mu}_{ijk}$ is given by

$$\underline{\mu}_{ijk}' = (\mu_{ijk1}, \mu_{ijk2}, \ldots, \mu_{ijkp})$$

which is the ijk^{th} row in the parameter matrix U. For the classical multivariate linear model, $\underline{\mu}_{ijk}$ corresponds to $\underline{\mu} + \underline{\alpha}_i + \underline{\beta}_j + \underline{\gamma}_k + (\underline{\alpha\beta})_{ij} + (\underline{\alpha\gamma})_{ik} + (\underline{\beta\gamma})_{jk}(\underline{\alpha\beta\gamma})_{ijk}$.

Using (5.1), we will refer to the three factors of the design as A, B and C and we will use factor A to construct some main effect hypotheses. To do this, we define four population averages using the I levels of factor A:

$$\tilde{\bar{\underline{\mu}}}_{i\cdot\cdot} = \frac{1}{N_{i++}} \sum_j \sum_k N_{ijk}\, \underline{\mu}_{ijk} \qquad (5.2)$$

$$\bar{\underline{\mu}}_{i\cdot\cdot} = \frac{1}{J} \sum_j \frac{1}{N_{ij+}} \sum_k N_{ijk}\, \underline{\mu}_{ijk} \qquad (5.3)$$

$$\tilde{\underline{\mu}}_{i\cdot\cdot} = \frac{1}{K} \sum_k \frac{1}{N_{i+k}} \sum_j N_{ijk}\, \underline{\mu}_{ijk} \qquad (5.4)$$

$$\underline{\mu}_{i\cdot\cdot} = \frac{1}{JK} \sum_j \sum_k \underline{\mu}_{ijk} \qquad (5.5)$$

where the plus notation represents the sum of the cell frequencies over the + subscript.

In the construction of $\tilde{\bar{\underline{\mu}}}_{i\cdot\cdot}$, we are ignoring the fact that there are JK populations at the ith level of A so that the hypothesis of no difference is referred to as the hypothesis of A ignoring B and C, written as

$$H_{A\ ig \cdot BC} : \text{all } \tilde{\bar{\underline{\mu}}}_{i\cdot\cdot} \text{ are equal .} \tag{5.6}$$

The means $\bar{\underline{\mu}}_{i\cdot\cdot}$ are determined by first finding the mean of all N_{ij+} cases at each level of factor B within the ith level of factor A , and then taking the simple average of the resulting means so that we are ignoring the fact that there are K levels of C , but take into account that there are J levels for factor B . Thus, the hypotheses is labeled the hypothesis of A ignoring C given B and we write it:

$$H_{A\ ig \cdot C|B} : \text{all } \bar{\underline{\mu}}_{i\cdot\cdot} \text{ are equal .} \tag{5.7}$$

By analogy, testing the hypothesis involving the means $\bar{\underline{\mu}}_{i\cdot\cdot}$ leads to the hypothesis of A ignoring B given C :

$$H_{A\ ig \cdot B|C} : \text{all } \tilde{\underline{\mu}}_{i\cdot\cdot} \text{ are equal .} \tag{5.8}$$

Finally, since in the construction of $\underline{\mu}_{i\cdot\cdot}$ we are merely taking the simple average of JK means at the ith level, we refer to the hypothesis as the test of A given B and C :

$$H_{A|BC} : \text{all } \underline{\mu}_{i\cdot\cdot} \text{ are equal} \tag{5.9}$$

Similar formulations hold for main effect hypotheses for the factors B and C .

To test for two-factor interactions in the three-factor design, means are again constructed by either including or ignoring factors entering into the hypothesis. For example, consider calculating the mean of all N_{ij+} cases at the ith level of A and the jth level of B :

$$\underline{\mu}_{ij\cdot} = \frac{1}{N_{ij+}} \sum_k N_{ijk} \underline{\mu}_{ijk} \tag{5.10}$$

and testing

$$H_{AB\ ig \cdot C} : \text{all } \underline{\mu}_{ij\cdot} - \underline{\mu}_{i'j\cdot} - \underline{\mu}_{ij'\cdot} + \underline{\mu}_{i'j'\cdot} = \underline{0} \ . \tag{5.11}$$

Alternatively, we may form

$$\underline{\mu}_{ij\cdot} = \frac{1}{K} \sum_k \underline{\mu}_{ijk} \tag{5.12}$$

and test

$$H_{AB|C} : \underline{\mu}_{ij\cdot} - \underline{\mu}_{i'j\cdot} - \underline{\mu}_{ij'\cdot} + \underline{\mu}_{i'j'\cdot} = \underline{0}$$

The same procedure may be used for the other two-factor interactions. Finally, the three-factor interaction is written:

$$H_{ABC} : \underline{\mu}_{ijk} - \underline{\mu}_{i'jk} - \underline{\mu}_{ij'k} + \underline{\mu}_{i'j'k} - \underline{\mu}_{i'jk'} + \underline{\mu}_{ij'k'} - \underline{\mu}_{i'j'k'} = \underline{0} .$$

For a given set of data, selection of the appropriate hypothesis to test will depend on whether the sample sizes are proportional to the population sample sizes. When the sample sizes have no meaning relative to the population values, the experimenter would select a hypothesis that does not depend on the cell frequencies.

As discussed by Kutner (1974), there are problems associated with the interpretation of various hypotheses that may be the reduction notation employed by Searle (1971). For the hypotheses discussed so far, there is a natural correspondence between certain reductions in sums of squares and the full rank hypothesis, provided the reductions are constructed by imposing equal weight side conditions on the classical model parameters. The correspondence is shown in Table 5.1.

Table 5.1

Full Rank Tests and Classical Corresponding Reduction

Full Rank	R( )
$H_{A\ ig\cdot BC}$	$R(\underline{\alpha}\|\underline{\mu})$
$H_{A\ ig\cdot C\|B}$	$R(\underline{\alpha}\|\underline{\mu},\underline{\beta},\underline{\alpha\beta})$
$H_{A\ ig\cdot B\|C}$	$R(\underline{\alpha}\|\underline{\mu},\underline{\gamma},\underline{\alpha\gamma})$
$H_{A\|BC}$	$R(\underline{\alpha}\|\underline{\mu},\underline{\beta},\underline{\gamma},\underline{\alpha\beta},\underline{\alpha\gamma},\underline{\beta\gamma},\underline{\alpha\beta\gamma})$
$H_{AB\ ig\cdot C}$	$R(\underline{\alpha\beta}\|\underline{\mu},\underline{\alpha},\underline{\beta})$
$H_{AB\|C}$	$R(\underline{\alpha\beta}\|\underline{\mu},\underline{\alpha},\underline{\beta},\underline{\gamma},\underline{\alpha\gamma},\underline{\beta\gamma},\underline{\alpha\beta\gamma})$
H_{ABC}	$R(\underline{\alpha\beta\gamma}\|\text{all other effects})$

6. MANOVA OF A TWO-FACTOR DESIGN WITH UNEQUAL, DISPROPORTIONATE CELL FREQUENCIES, INTERACTION, NO INTERACTION, AND EMPTY CELLS

To illustrate the analysis of a two-factor design with interaction, unequal cell frequencies and empty cells, the data pattern in Table 6.1 is utilized.

Table 6.1

Data Pattern for a Nonorthogonal Two-Factor Classification Design with Empty Cells

		Factor B			
		B_1	B_2	B_3	B_4
	A_1	$\underline{\mu}_{11}$ $N_{11}=3$	Empty	$\underline{\mu}_{13}$ $N_{13}=1$	$\underline{\mu}_{14}$ $N_{14}=2$
Factor A	A_2	$\underline{\mu}_{21}$ $N_{21}=2$	$\underline{\mu}_{22}$ $N_{22}=2$	Empty	Empty
	A_3	Empty	$\underline{\mu}_{32}$ $N_{32}=2$	$\underline{\mu}_{33}$ $N_{33}=2$	$\underline{\mu}_{34}$ $N_{34}=4$

The UFRMLM for a two-factor design with interaction, an unequal number of observations per cell, and empty cells is

$$E(\underline{y}_{ijk}) = \underline{\mu}_{ij} \ , \ i = 1,2,\ldots,I \ ; \ j = 1,2,\ldots,J \ ;$$
$$h = 1,2,\ldots,K_{ij} \geq 0 \ . \qquad (6.1)$$

For the data pattern in Table 6.1, several hypotheses employing the full rank model are investigated.

The 8×p parameter matrix U has the form

$$\underset{8\times p}{U} = \begin{pmatrix} \mu'_{11} \\ \mu'_{13} \\ \mu'_{14} \\ \mu'_{21} \\ \mu'_{22} \\ \mu'_{32} \\ \mu'_{33} \\ \mu'_{34} \end{pmatrix} = \begin{pmatrix} \mu_{111} & \mu_{112} & \cdots & \mu_{11p} \\ \mu_{131} & \mu_{132} & \cdots & \mu_{13p} \\ \cdot & & \cdots & \cdot \\ \cdot & & \cdots & \cdot \\ \cdot & & \cdots & \cdot \\ \cdot & & \cdots & \cdot \\ \cdot & & \cdots & \cdot \\ \mu_{341} & \mu_{342} & \cdots & \mu_{34p} \end{pmatrix}$$

since the cells (1,2), (2,3), (2,4) and (3,1) are empty. For a two-factor design with unequal cell frequencies and empty cells, we can apply either simple unequal weights or complex unequal weights to the means to test main effect type hypotheses. With empty cells in a design, equal weights proportional to the number of levels of a factor are not appropriate. Main effect hypotheses which might be of interest are

$$H_A^* : \text{all } \bar{\underline{\mu}}_{i\cdot} \text{ are equal}$$

$$H_B^* : \text{all } \bar{\underline{\mu}}_{\cdot j} \text{ are equal} \tag{6.2}$$

which for the data in Table 6.1 become

$$H_A^* : \quad \frac{3\mu_{11}+\mu_{13}+2\mu_{14}}{6} = \frac{2\mu_{21}+2\mu_{22}}{4} = \frac{2\mu_{32}+2\mu_{33}+4\mu_{34}}{8}$$

$$H_B^* : \quad \frac{3\mu_{11}+2\mu_{21}}{5} = \frac{2\mu_{22}+2\mu_{32}}{4} = \frac{\mu_{13}+2\mu_{33}}{3} = \frac{2\mu_{14}+4\mu_{34}}{6} \quad .$$

Hypothesis test matrices C' for these hypotheses are

$$C'_{A*} = \begin{pmatrix} 3/6 & 1/6 & 2/6 & -2/4 & -2/4 & 0 & 0 & 0 \\ 3/6 & 1/6 & 2/6 & 0 & 0 & -2/8 & -2/8 & -4/8 \end{pmatrix}$$

$$C'_{B*} = \begin{pmatrix} 3/6 & 0 & 0 & 2/5 & -2/4 & -2/4 & 0 & 0 \\ 3/5 & -1/3 & 0 & 2/5 & 0 & 0 & -2/3 & 0 \\ 3/5 & 0 & -2/6 & 2/5 & 0 & 0 & 0 & -4/6 \end{pmatrix}$$

and the matrix $A = I$. Another set of main effect hypotheses for the data are

$$H_{A_1} : \quad \frac{\underline{\mu}_{11}+\underline{\mu}_{13}+\underline{\mu}_{14}}{3} + \frac{\underline{\mu}_{21}+\underline{\mu}_{22}}{2} = \frac{\underline{\mu}_{32}+\underline{\mu}_{33}+\underline{\mu}_{34}}{3}$$

$$H_{B_1} : \quad \frac{\underline{\mu}_{11}+\underline{\mu}_{21}}{2} = \frac{\underline{\mu}_{22}+\underline{\mu}_{32}}{2} = \frac{\underline{\mu}_{13}+\underline{\mu}_{33}}{2} = \frac{\underline{\mu}_{14}+\underline{\mu}_{34}}{2}$$

involving simple averages, but not equal weights since some cells in the design are empty. These hypotheses are represented as

$$H_{A_1}: \text{ all } \underset{\sim}{\mu}_{i\bullet} \text{ are equal}$$

$$H_{B_1}: \text{ all } \underset{\sim}{\mu}_{\bullet j} \text{ are equal} \qquad (6.3)$$

where the means $\underset{\sim}{\mu}_{i\bullet}$ and $\underset{\sim}{\mu}_{\bullet j}$ are unweighted averages over all cells with data. The hypothesis test matrices for evaluating H_{A_1} and H_{B_1} are:

$$C'_{A_1}: \begin{pmatrix} 1/3 & 1/3 & 1/3 & 0 & 0 & -1/3 & -1/3 & -1/3 \\ 0 & 0 & 0 & 1/2 & 1/2 & -1/3 & -1/3 & -1/3 \end{pmatrix}$$

$$C'_{B_1}: \begin{pmatrix} 1/2 & 0 & -1/2 & 1/2 & 0 & 0 & 0 & -1/2 \\ 0 & 0 & -1/2 & 0 & 1/2 & 1/2 & 0 & -1/2 \\ 0 & 1/2 & -1/2 & 0 & 0 & 0 & 1/2 & -1/2 \end{pmatrix}$$

To test for significant interactions in a two-factor design with empty cells, f-I-J+1 linearly independent individual sums or differences of the parametric functions $\underset{\sim}{\mu}_{ij} - \underset{\sim}{\mu}_{i'j} - \underset{\sim}{\mu}_{ij'} + \underset{\sim}{\mu}_{i'j'}$ are constructed to eliminate parameters $\underset{\sim}{\mu}_{ij}$ associated with the empty cells and are equated to zero (where f denotes the number of cells with data). For the data in Table 6.1, f-I-J+1 = 8-4-3+1 = 2 so that only two degrees of freedom are associated with testing the interaction hypothesis instead of 6, which would be the case if all cells in the design contained data. The hypothesis text matrix to test the interaction hypothesis.

$$\begin{aligned} H_{AB}: \underset{\sim}{\mu}_{13} - \underset{\sim}{\mu}_{33} - \underset{\sim}{\mu}_{14} + \underset{\sim}{\mu}_{34} &= \underset{\sim}{0} \\ \underset{\sim}{\mu}_{11} - \underset{\sim}{\mu}_{21} + \underset{\sim}{\mu}_{22} - \underset{\sim}{\mu}_{34} - \underset{\sim}{\mu}_{13} + \underset{\sim}{\mu}_{33} &= \underset{\sim}{0} \end{aligned} \qquad (6.4)$$

is

$$C'_{AB} = \begin{pmatrix} 0 & 1 & -1 & 0 & 0 & 0 & -1 & 1 \\ 1 & -1 & 0 & -1 & 1 & -1 & 1 & 0 \end{pmatrix} .$$

Under the less than full rank classical model, the hypotheses H_{A*}, H_{B*}, H_{A_1} and H_{B_1} are equivalent to testing

$$H_{A*}: \text{ all } \underset{\sim}{\alpha}_i + \sum_j K_{ij}(\underset{\sim}{\beta}_j + \underset{\sim}{\gamma}_{ij})/K_{i+} \text{ are equal}$$

$$H_{B*}: \text{ all } \underset{\sim}{\beta}_j + \sum_i K_{ij}(\underset{\sim}{\alpha}_i + \underset{\sim}{\gamma}_{ij})/K_{+j} \text{ are equal}$$

$$H_{A_1}: \text{ all } \underline{\alpha}_i + \sum_j (\underline{\beta}_j + \underline{\gamma}_{ij})/I_i \text{ are equal}$$

$$H_{B_1}: \text{ all } \underline{\beta}_j + \sum_i (\underline{\alpha}_i + \underline{\gamma}_{ij})/J_j \text{ are equal} \tag{6.5}$$

where I_i and J_j denote the number of nonempty cells in the ith row and the jth column of the design respectively. For a design with data in every cell, H_{A_1} and H_{B_1} are identical to testing H_A and H_B ; this is not the case for designs with empty cells.

Assuming $\underline{\gamma}_{ij} = \underline{0}$ for all i and j , so that we have a two-factor classification design with unequal, disproportionate cell frequencies, no interaction, and empty cells, we might test the following main effect hypotheses:

$$H_{A*}: \text{ all } \bar{\underline{\mu}}_{i\cdot}\text{'s are equal}$$

$$H_{B*}: \text{ all } \bar{\underline{\mu}}_{\cdot j}\text{'s are equal}$$

$$H_A : \delta_{ij}\,\underline{\mu}_{ij} = \delta_{i'j}\,\underline{\mu}_{i'j} \text{ for all } i, i' \text{ and } j'$$

$$H_B : \delta_{ij}\,\underline{\mu}_{ij} = \delta_{ij'}\,\underline{\mu}_{ij'} \text{ for all } j, j' \text{ and } i \tag{6.6}$$

where $\delta_{ij} = 1$ if $K_{ij} \neq 0$ and $\delta_{ij} = 0$ if $K_{ij} = 0$. The equivalent hypotheses for the classical model are:

$$H_A : \text{ all } \underline{\alpha}_i \text{ are equal}$$

$$H_B : \text{ all } \underline{\beta}_j \text{ are equal}$$

$$H_{A*}: \text{ all } \underline{\alpha}_i + \sum_j K_{ij}\,\underline{\beta}_j/K_{i+} \text{ are equal}$$

$$H_{B*}: \text{ all } \underline{\beta}_j + \sum_j K_{ij}\,\underline{\alpha}_i/K_{+j} \text{ are equal .}$$

Assuming no interaction for the data in Table 6.1, we must associate with the model given in (6.1) a set of restrictions. The matrix R' for the PRMFRLM is obtained from testing for no interaction in the UMFRLM. That is,

$$R' = \begin{pmatrix} 0 & 1 & -1 & 0 & 0 & 0 & -1 & 1 \\ 1 & -1 & 0 & -1 & 1 & -1 & 1 & 0 \end{pmatrix}$$

the same as C'_{AB} in (6.4) and $T = I$. To test H_A and H_B in (6.6) under the restrictions $R'U = 0$, the hypothesis test matrices $\tilde{C}'_A$ and $\tilde{C}'_B$ are constructed:

$$\tilde{C}'_A = \begin{pmatrix} 1 & 0 & 0 & -1 & 0 & 0 & 0 & 0 \\ 0 & 0 & 0 & 0 & 1 & -1 & 0 & 0 \\ 0 & 1 & 0 & 0 & 0 & 0 & -1 & 0 \\ 0 & 0 & 1 & 0 & 0 & 0 & 0 & -1 \end{pmatrix}$$

$$\tilde{C}'_B = \begin{pmatrix} 1 & 0 & -1 & 0 & 0 & 0 & 0 & 0 \\ 1 & -1 & 0 & 0 & 0 & 0 & 0 & 0 \\ 0 & 0 & 0 & 1 & -1 & 0 & 0 & 0 \\ 0 & 0 & 0 & 0 & 0 & 1 & -1 & 0 \\ 0 & 0 & 0 & 0 & 0 & 1 & 0 & -1 \end{pmatrix}$$

for the data in Table 6.1.

Although the matrices $\tilde{C}'_A$ and $\tilde{C}'_B$ are of full row rank, the matrices

$$Q'_1 = \begin{pmatrix} R' \\ \tilde{C}'_A \end{pmatrix} \quad \text{and} \quad Q'_2 = \begin{pmatrix} R' \\ \tilde{C}'_B \end{pmatrix}$$

are not of full row rank. Reducing Q'_1 and Q'_2 to full row rank leaving R' fixed, the matrices C'_A and C'_B used to test H_A and H_B are:

$$C'_A = \begin{pmatrix} 1 & 0 & 0 & -1 & 0 & 0 & 0 & 0 \\ 0 & 0 & 0 & 0 & 1 & -1 & 0 & 0 \end{pmatrix}$$

$$C'_B = \begin{pmatrix} 1 & 0 & -1 & 0 & 0 & 0 & 0 & 0 \\ 1 & -1 & 0 & 0 & 0 & 0 & 0 & 0 \\ 0 & 0 & 0 & 1 & -1 & 0 & 0 & 0 \end{pmatrix}$$

these matrices are inserted into formula (4.2) for S_h .

7. MANCOVA WITH REPEATED MEASUREMENTS

In the previous sections we reviewed some applications of the UFRMLM and the PRFRMLM for some common MANOVA designs. The full rank model may also be used to analyze data obtained using a multivariate analysis of covariance (MANCOVA) design.

Assume for the moment that we have a two group MANCOVA design where for a set of p dependent variates an associated vector of covariates is obtained. In this situation, we let $X = WZ$ where X is a matrix of full rank and Z is a matrix of covariates. In particular, assume that the number of covariates is one. Then, the PRFRMLM may be represented:

$$E(\underline{y}_1\ \underline{y}_2\ \underline{y}_3) = \begin{bmatrix} \underline{1}_{N_1} & \underline{0}_{N_1} & Z_1 & 0 \\ & & & \\ \underline{0}_{N_2} & \underline{1}_{N_2} & 0 & Z_2 \end{bmatrix} \begin{bmatrix} \mu_{11} & \mu_{12} & \mu_{13} \\ \mu_{21} & \mu_{22} & \mu_{23} \\ \beta_{11} & \beta_{12} & \beta_{13} \\ \beta_{21} & \beta_{22} & \beta_{23} \end{bmatrix}$$

subject to the restrictions that the regression lines are parallel,

$$\begin{pmatrix} \beta_{11} \\ \beta_{12} \\ \beta_{13} \end{pmatrix} = \begin{pmatrix} \beta_{21} \\ \beta_{22} \\ \beta_{23} \end{pmatrix} .$$

This two group model with one covariate has the general form of the PRFRMLM, that is

$$E(Y) = XB$$
$$R'B = \Theta$$

where for the two group design the matrix of restrictions R' takes the form, $R' = (0\ 0\ 1\ {-1})$. The restrictions are used to ensure parallelism of slopes. Letting $X - WU$ and $B' = (U'\Delta')$, the general representation takes the form

$$E(Y) = WU + Z\Delta$$
$$(R_1'\ R_2')\binom{U}{\Delta} = \Theta$$

where R_1' is the matrix of restrictions on W and R_2' is the matrix of restrictions on Δ .

As discussed by Timm (1980, in press), assume that the vector of dependent variates are obtained repeatedly over p conditions or are repeated measurements and that for each repeated observation there is associated a unique covariate. A design of this type for two groups may take the form shown in Table 7.1 where Y is the criterion measure and Z is the covariate.

For the data in Table 7.1, we may want to impose the restrictions

$$\beta_{11} - \beta_{13} - \beta_{12} + \beta_{23} = 0 \qquad \beta_{12} - \beta_{13} - \beta_{22} + \beta_{23} = 0$$

$$\beta_{12} - \beta_{23} - \beta_{13} + \beta_{33} = 0 \qquad \beta_{22} - \beta_{23} - \beta_{32} + \beta_{33} = 0$$

and test the hypothesis that there is no difference in intercepts across conditions:

$$H_o: \begin{pmatrix}\mu_{11}\\ \mu_{21}\end{pmatrix} = \begin{pmatrix}\mu_{12}\\ \mu_{22}\end{pmatrix} = \begin{pmatrix}\mu_{13}\\ \mu_{23}\end{pmatrix} .$$

For this design we must employ the RFRMLM:

$$E(Y) = WU + Z\Delta$$

$$(R_1' \; R_2')\binom{U}{\Delta}A = \Theta \quad .$$

Setting $R_1' = 0$, the matrices R_2' and A take the form:

$$R_2' = (1 \quad -1) \quad , \quad A = \begin{pmatrix}1 & 0\\ -1 & 1\\ 0 & -1\end{pmatrix} .$$

Table 7.1

MANCOVA WITH REPEATED MEASUREMENTS CONDITIONS

		C_1		C_2		C_3	
G_1	S_1	y_{111}	z_{111}	y_{112}	z_{112}	y_{113}	z_{113}
	S_2	y_{121}	z_{121}	y_{122}	z_{122}	y_{123}	z_{123}
	⋮	⋮	⋮	⋮	⋮	⋮	⋮
	S_{N_1}	$y_{1N_1 1}$	$z_{1N_1 1}$	$y_{1N_1 2}$	$z_{1N_1 2}$	$y_{1N_1 3}$	$z_{1N_1 3}$
G_2	S_1'	y_{211}	z_{211}	y_{212}	z_{212}	y_{213}	z_{213}
	S_2'	y_{221}	z_{221}	y_{222}	z_{222}	y_{223}	z_{223}
	⋮	⋮	⋮	⋮	⋮	⋮	⋮
	S_{N_2}'	$y_{2N_2 1}$	$z_{2N_2 1}$	$y_{2N_2 2}$	$z_{2N_2 2}$	$y_{2N_2 3}$	$z_{2N_2 3}$

8. SUMMARY

The coverage given the general full rank multivariate linear model given in this paper extend the results developed by Hocking and Speed (1975) and Timm and Carlson (1975) to the analysis of multivariate MANOVA and MANCOVA designs. With the development and implementation of the model in practice, we hope that the approach will permit applied researchers to construct hypotheses from their data and that it will provide instructors of experimental design with an alternative approach to the analysis of fixed effect multivariate experiments. We feel that the approach presented may facilitate our understanding of nonorthogonal designs since the complications of side conditions, reparameterization and generalized inverses have been eliminated.

REFERENCES

[1] Bancroft, T.A. (1968). Topics in Intermediate Methods, Vol. I., Ames, Iowa. Iowa University Press.

[2] Bock, R.D. (1975). Multivariate statistical methods in behavioral research, New York, McGraw-Hill.

[3] Brant, A.E. (1933). The Analysis of Variance in a '2xs' Table with Disproportionate Frequencies. J. Amer. Stat. Assoc. 28, 164-173.

[4] Eccleston, J.A. and K.G. Russell (1977). Adjusted orthogonality in non-orthogonal designs, Biometrika 64, No. 2, 339-346.

[5] Francis, I. (1973). A Comparison of Several Analysis of Variance Programs. J. Amer. Stat. Assoc. 70, 706-712.

[6] Graybill, F.A. (1976). Theory and Application of the linear model, Massachusetts, Duxbury Press.

[7] Herr, D.G. and J. Gaebelein (1976). Nonorthogonal two-way analysis of variance. Psychological Bull. 85, 207-216.

[8] Hocking, R.R. and F.M. Speed (1975). A full rank analysis of some linear model problems. J. Amer. Stat. Assoc. 70, 706-712.

[9] Kutner, M.H. (1974). Hypothesis Testing in Linear Models (Eisenhart Model I). The American Statistician 28, 90-100.

[10] Overall, J.E., D.K. Spiegel and J. Cohen (1975). Equivalence of Orthogonal and Nonorthogonal Analysis of Variance. Psychological Bull. 82, 182-186.

[11] Roy, S.N., R. Gnanadesikan and J.N. Srivastava (1971). Analysis and design of certain quantitative multiresponse experiments. New York, Pergman Press.

[12] Searle, S.R. (1971). Linear Models. New York, John Wiley.

[13] Snedecor, G.W. (1934). The Method of Expected Numbers for Tables of Multiple Classification with Disproportionate Subclass Numbers. J. Amer. Stat. Assoc. 29, 389-393.

[14] Srivastava, M.S. and C.G. Khatri (1979). An Introduction to Multivariate Statistics. New York, North Holland.

[15] Timm, N.H. (1975). Multivariate Analysis with Applications in Education and Psychology. Monterey, California, Brooks Cole.

[16] Timm, N.H. (1980). Multivariate Analysis of Variance of Repeated Measurements. Handbook of Statistics 1, 1-47, P.R. Krishnaiah, ed., New York, North-Holland.

[17] Timm, N.H. and J.E. Carlson (1973). Multivariate Analysis of Nonorthogonal Experimental Designs using a Multivariate Full Rank Model. Paper presented at the Annual Meeting of the American Statistical Association, New York, December

[18] Timm, N.H. and J.E. Carlson (1975). An Analysis of Variance Through Full Rank Models. Multivariate Behavioral Research Monographs, No. 75-1.

[19] Urquhart, N.S., D.L. Weeks, C.R. Henderson (1973). Estimation Associated with Linear Models: A Revisitation. Communications in Statistics 1, 303-330.

[20] Yates, F. (1934). The Analysis of Multiple Classifications with Unequal Numbers in the Different Classes. J. Amer. Stat. Assoc. 29, 51-66.

Multivariate Statistical Analysis
R.P. Gupta (ed.)

ON THE LIMITING BEHAVIOUR OF SAMPLE CORRELATION COEFFICIENTS*

R.J. Tomkins
Department of Mathematics and Statistics
University of Regina
Regina, Saskatchewan, Canada S4S 0A2

This paper presents several attempts to define a generalized correlation coefficient, one which would be defined for every bivariate population and would equal the familiar correlation coefficient when variances are finite. Two of the approaches involve the limiting behaviour of sequences of sample correlation coefficients.

INTRODUCTION

Let X and Y be random variables defined on some probability space; let σ_X^2 and σ_Y^2 be their respective variances, if they exist.

If $E(X^2) < \infty$ and $E(Y^2) < \infty$ then the correlation coefficient $\rho = \rho(X,Y)$ of X and Y is well-defined:

$$(1)\quad \rho(X,Y) \equiv \frac{E(XY)-E(X)E(Y)}{\sigma_X \sigma_Y} .$$

Many properties of ρ are well-known; in particular, a perfect linear relationship between X and Y exists if and only if $\rho^2 = 1$.

However, the existence of a perfect linear relationship and such concepts as independence and uncorrelatedness are not inherently contingent on the finitude of second moments. With this fact in mind, this paper will address the following question: Is it possible to define a generalized correlation coefficient ρ^* which would be well-defined for every pair of random variables, would satisfy $\rho^* = \rho$ whenever ρ is defined, and would possess the "usual" properties of ρ ?

Here are two simple examples in which a value for such a ρ^* is plausible.

Example 1. Suppose $E(X^2) = \infty$ and $Y = a_0 + a_1 X$ for some numbers a_0 and a_1. Would it not be reasonable to define $\rho^* = 1$ if $a_1 > 0$ and $\rho^* = -1$ if $a_1 < 0$?

Example 2. If X and Y are independent or uncorrelated (i.e. $E(XY) = E(X)E(Y)$), why not let $\rho^* = 0$?

Section 2 will present two approaches to the quest for a generalized correlation coefficient, both based on the limiting behaviour of sample correl-

*This work was supported by a grant from the Natural Sciences and Engineering Research Council of Canada.

ation coefficients. Section 3 will provide three illustrative examples which shed some light on the interpretation of ρ, and indicate some applications of the methods in Section 2. An outline of a new approach to regression and correlation will appear in Section 4.

TWO APPROACHES USING SAMPLE CORRELATION COEFFICIENTS

Consider a sequence $(X_1,Y_1),(X_2,Y_2),\ldots$ of independent random vectors representing paired observations on X and Y. Let $r_n = r_n(X,Y)$ denote the sample correlation coefficient of X and Y based on the first n pairs in the sequence.

Approach 1. It is well-known that $r_n \to \rho$ almost surely (a.s.) as $n \to \infty$ whenever ρ is defined. Moreover, since $|r_n| \leq 1$, it is plausible that, in general, the sequence $\{r_n\}$ may converge in some sense (in distribution, in probability or a.s.). If this were indeed so, one could define ρ^* to be the resulting limit.

Notice that the desired values of ρ^* in examples 1 and 2 would obtain if this approach were valid, and $\rho^* = \rho$ when ρ is defined. However, establishing the validity of this method looks to be difficult. In the case when both X and Y have some finite moments, it may be possible to use a strong law of large numbers (such as that of Marcinkiewicz and Zygmund (see, for example, p. 122 of [1])) to prove the existence of ρ^*, but this would not apply to all pairs (X,Y).

Approach 2. For any number $a > 0$, define the truncated random variables

$$(2)\quad X_a = \begin{cases} X \text{ if } |X| \leq a \\ 0 \text{ if } |X| > a \end{cases} \quad\text{and}\quad Y_a = \begin{cases} Y \text{ if } |Y| \leq a \\ 0 \text{ if } |Y| > a \end{cases}.$$

Then $\rho(a) \equiv \rho(X_a,Y_a)$ is well-defined; moreover, $r_n(X_a,Y_a) \to \rho(a)$ a.s. as $n \to \infty$. Approach 1 looked at $\lim_{n\to\infty} r_n = \lim_{n\to\infty}\lim_{a\to\infty} r_n(X_a,Y_a)$. In this second approach, consider $\lim_{a\to\infty}\lim_{n\to\infty} r_n(X_a,Y_a) = \lim_{a\to\infty} \rho(a)$ a.s. If $\lim_{a\to\infty} \rho(a)$ were always extant, one could let ρ^* be that limiting value. However, the following example shows that $\lim_{a\to\infty} \rho(a)$ need not always exist.

Example 3. Suppose $P[X=\pm 2^k] = 2^{-k-2}$ for each $k \geq 0$. Then define

$$Y = \begin{cases} X \text{ if } \log_2 |X| \text{ is even} \\ -X \text{ otherwise.} \end{cases}$$

where "$\log_2 x$" represents the logarithm of x to the base 2.

For $a > 1$, define $n = n_a = [\log_2 a]$, where $[x]$ stands for the integer part of x, and let $m = m_a = [n/2]$. If n is even, then

$$E(X_a Y_a) = (4^m-1)/3 = (2^n-1)/3 ,$$

whereas, if n is odd,

$$E(X_a Y_a) = -(1+2(4^m))/3 = -(1+2^n)/3 .$$

Since $E(X_a) = E(Y_a) = 0$ by symmetry and $E(X_a^2) = E(Y_a^2) = 2^n - 1$, it is evident that $\rho(a) = 1/3$ if n_a is even and $\rho(a) < -1/3$ if n_a is odd. Indeed,

$$\liminf_{a\to\infty} \rho(a) = -1/3 \quad \text{and} \quad \limsup_{a\to\infty} \rho(a) = 1/3 .$$

SOME INTERESTING EXAMPLES

Throughout this section, let X be a random variable with density function

$$f(x) = \begin{cases} |x|^{-3} & \text{if } |x| > 1 \\ 0 & \text{if } |x| \le 1 \end{cases} .$$

Then X is symmetric, $P[X>x] = (2x^2)^{-1}$ for $x > 1$, and $E(X) = 0$, but $E(X^2) = \infty$. Fix any number $q > 1$.

Example 4. Define $Y = X_q$ (cf. (2)). Then $\rho(a) = (E(X_q^2)/E(X_a^2))^{1/2}$ for $a \ge q$, so $\rho(a) \to 0$ as $a \to \infty$.

Moreover, $E(XY) = E(X_q^2) < \infty$, so use of Kolmogorov's strong law of large numbers (p. 122 of 1], for instance) shows that $r_n \to 0$ a.s. So $\rho^* = 0$ using approach 1 or 2.

In some senses, this value of ρ^* defies one of the "usual" properties of a correlation coefficient. A value of ρ "near" zero is commonly interpreted as meaning that no prediction of Y based on a linear function of X could be considered to be very reliable. But, in this example, the linear function $Y = X$ has probability $1 - q^{-2}$ of predicting Y perfectly given X ... and yet $\rho^*= 0$. In fact, this example need not hinge on the use of an X with infinite variance; replacing X by X_a for an a so large that $|\rho(X_a,Y)| < .001$, note that $P[X_a \ne Y] = q^{-2} - a^{-2}$.

Example 5. Let $Y = X$ if $|X| \le q$ and let $Y = -X$ when $|X| > q$. Then $\rho(a) = -1 + 2E(X_q^2)/E(X_a^2) \to -1$ as $a \to \infty$, and $r_n \to -1$ a.s. as $n \to \infty$. Hence, approaches 1 and 2 both yield $\rho^* = -1$.

Example 6. Let $Y = X - X_q$. In this case, $\rho(a) = (1+E(X_q^2)/E(Y_a^2))^{-1/2} \to 1$ as $a \to \infty$ and $r_n \to 1$ a.s., so $\rho^* = 1$ by approaches 1 and 2.

Remark. In examples 4 and 5, the prediction equation $Y = X$ is very accurate in the sense that the probability of an error is only q^{-2}, and yet approaches 1 and 2 yield values of ρ^*(0 and -1 respectively) other than the value $\rho^* = +1$ that might be expected. This fact, taken together with example 6, leads one to surmise that the functional relationship between X and Y in the tails of the distribution of X (e.g., for $|X| > q$) is the determining factor in the value of ρ^* using approaches 1 and 2.

A NEW APPROACH TO LINEAR REGRESSION AND CORRELATION.

At the foundation of the method to be presented in this section is the fact (see, for example, Theorem 2 on p. 215 of [1]) that, for <u>every</u> pair X, Y of random variables, there exists a regular conditional distribution of Y given (the sigma-field generated by) X. Armed with this information, certain steps may be followed to find a regression line for Y on X.

For any pair X, Y with finite variances, the regression lines $Y' = \beta_0 + \beta_1 X$ and $X' = \beta_0' + \beta_1' Y$ can be found. In such a case, $\rho^2 = \beta_1 \beta_1'$. A definition of ρ^* motivated by this fact will be provided.

To illustrate the ideas behind this third approach, it will be assumed that Y has a well-defined (conditional) distribution function (d.f.) given $X = x$, for all x in S, the support of the d.f. of X.

<u>Approach 3</u>. Define the d.f. $F_0(y) = 1$ if $y \geq 0$, $F_0(y) = 0$ if $y < 0$. If $Y \equiv a_0 + a_1 X$, then F_0 is the d.f. of $Y - a_0 - a_1 X$. This suggests, more generally, an attempt to compare some d.f. of $Y - a_0 - a_1 X$ with F_0.

Here is an outline of the proposed method.

1. For any numbers y, c_0 and c_1, define

$$G(y) \equiv \sup_{x \in S} \left| P(Y \leq c_0 + c_1 x + y \mid X = x) - F_0(y) \right| \text{, and then}$$

$$\Delta(c_0, c_1) \equiv \sup_{-\infty < y < \infty} G(y)$$

2. If possible, find β_0^* and β_1^* such that $\Delta(\beta_0^*, \beta_1^*) \leq \Delta(c_0, c_1)$ for every pair (c_0, c_1). The line $Y' = \beta_0^* + \beta_1^* X$ will be called the <u>generalized regression line of Y on X</u>.

3. Reverse the roles of X and Y to find the generalized regression line $X' = \alpha_0^* + \alpha_1^* Y$ of X on Y.

4. Define the <u>generalized correlation coefficient</u> of X and Y to be the number ρ^* having the same sign as β_1^* and obeying $(\rho^*)^2 = \alpha_1^* \beta_1^*$.

<u>Remarks</u>.1. If $Y \equiv a_0 + a_1 X$ (cf. example 1) then $\Delta(c_0, c_1) = 0$ if and only if $c_0 = a_0$ and $c_1 = a_1$. If $a_1 \neq 0$ then $\alpha_1^* = a_1^{-1}$, so $\rho^* = 1$ or -1 accordingly as $a_1 > 0$ or $a_1 < 0$.

2. If X and Y are independent (cf. example 2), it can be shown that $\beta_1^* = 0$ and $\beta_0^* = \text{med}(X)$, where $\text{med}(X)$ represents any median of X. Clearly $\rho^* = 0$ in this case, as desired.

3. While this method can be applied to any pair X, Y, the values of β_0^*, β_1^* and ρ^* are dependent on the specific bivariate distribution under consideration; it is not clear that there will be any simple formulae which will yield these values without having to carry out the steps given above. The preceding remark shows that β_0^* need not be unique. Still open are several questions: Can a pair (β_0^*, β_1^*) always be found which <u>minimizes</u> Δ? If so, is β_1^* unique?

And is $\alpha_1^* \beta_1^* \geq 0$ in every case?

Nevertheless, it is interesting to note that the proposed method leads to the "usual" results when X and Y are normally distributed.

Theorem 1. Let X and Y have the bivariate normal distribution with respective means μ_X, μ_Y, variances σ_X^2, σ_Y^2 and correlation coefficient ρ (as in (1)). Then $\beta_1^* = \rho\sigma_Y/\sigma_X \equiv \beta_1$, $\beta_0^* = \mu_Y - \beta_1^*\mu_X \equiv \beta_0$, and $\rho^* = \rho$.

Proof. In view of remark 1 above, assume without loss of generality that $|\rho| < 1$. Then it is well-known that the conditional distribution of Y given $X = x$ is normal with mean $\beta_0 + \beta_1 x$ and variance $\sigma^2 \equiv (1-\rho^2)\sigma_Y{}^2$. Therefore, letting Z be a standard normal variate,

$$P(Y \leq c_0 + c_1 x + y \ X=x) = P[\sigma Z \leq (c_0-\beta_0)+(c_1-\beta_1)x+y].$$

If $c_1 \neq \beta_1$ then $G(y) = 1$ for every y, whereas $G(y) < 1$ for all y if $c_1 = \beta_1$, so $\beta_1^* = \beta_1$. Moreover, when $c_1 = \beta_1$,

$$\sup_{y<0} G(y) = \sup_{y<0} P[\sigma Z \leq (c_0-\beta_0)+y] = P[\sigma Z < c_0 - \beta_0] \text{ and}$$

$$\sup_{y\geq 0} G(y) = \sup_{y \geq 0} P[\sigma Z > (c_0-\beta_0)+y] = P[\sigma Z > c_0-\beta_0] .$$

Consequently $\Delta(c_0,\beta_1^*) > 1/2$ unless $c_0 = \beta_0$, in which case $\Delta(c_0,\beta_1^*) = 1/2$. Hence $\beta_0^* = \beta_0$.

Reversing the roles of X and Y, one finds $\alpha_1^* = \rho\sigma_X/\sigma_Y$, so $(\rho^*)^2 = \alpha_1^*\beta_1^* = \rho^2$. Since ρ and β_1 have the same sign, $\rho^* = \rho$. Q.E.D.

REFERENCES

[1] Y.S. Chow and T. Teicher, "Probability Theory: Independence, Interchangeability, Martingales", Springer-Verlag, New York, 1978.

Multivariate Statistical Analysis
R.P. Gupta (ed.)

A CHARACTERIZATION OF MULTIVECTOR MULTINOMIAL AND NEGATIVE MULTINOMIAL DISTRIBUTIONS

Derrick S. Tracy
Department of Mathematics
University of Windsor
Windsor, Ontario

Deva C. Doss
Department of Mathematics
University of Alabama at Huntsville
Huntsville, Alabama

A multivector distribution $P_{\theta_1,\ldots,\theta_k}$ of a multivector $(x_1,\ldots,x_k)$, where x_i is a vector with p_i components, is said to be linear exponential in $\theta_1,\ldots,\theta_k$ if

$$dP_{\theta_1,\ldots,\theta_k}(x_1,\ldots,x_k) = \exp\left(\sum_{i=1}^{k} \theta_i' x_i\right)d\pi(x_1,\ldots,x_k)/f(\theta_1,\ldots,\theta_k).$$

In this paper it is shown that a multivector linear exponential family is multinomial if and only if its vector marginals are multinomial. This is also true with negative multinomial provided its usual definition is adequately extended.

1. INTRODUCTION

The multinomial distribution may be viewed upon as a binomial whose random variable is replaced by a random vector. This becomes evident if we consider their probability generating functions (p.g.f.). This is also true of the negative multinomial and negative binomial distributions. This observation helps us generalize the known properties of binomial and negative binomial distributions to multinomial and negative multinomial distributions. Multivariate versions of binomial and negative binomial are known [5], [6], [8]. In this paper we study their vector counterparts.

If a set of random variables is partitioned into k subsets, denoted as vectors $x_1,\ldots,x_k$ with $p_1,\ldots,p_k$ components, and the multivariate distribution is looked upon as a joint distribution of these vectors, then it is called a multivector distribution (of multivector $(x_1,\ldots,x_k)$). The marginal distributions of vectors $x_1,\ldots,x_k$ are termed vector marginals. The characterizations on multivector distributions presented in this paper are done through their vector marginals.

It has been shown [2], [3] that a multivariate linear exponential family of distributions is binomial (negative binomial) if and only if its univariate marginal families are binomial (negative binomial). In this paper we establish similar characterization for multivector multinomial distributions [6]. This type of characterization has generalized the well-known definition of multivariate negative binomial distribution, e.g. [6], to a larger family of distributions with negative binomial marginals [2]. Here, the linear exponential characterization enables us to extend the definition of a multivector negative multinomial distribution to a wider class which admits independent vector marginals.

A family $P_e = \{P_\omega : \omega \in \Omega_\nu\}$ of probability distributions of a real vector x is said to be linear exponential in a vector ω if

$$dP_\omega(x) = \exp(\omega' x)d\nu(x)/f(\omega) \tag{1.1}$$

where Ω_ν is assumed to be the nonvoid interior of the natural parameter space. The natural parameter space consists of all vectors ω such that

$$f(\omega) = \int e^{\omega x}\, d\nu(x) \tag{1.2}$$

is positive and finite. The function $f(\omega)$ is called generating function (g.f.) of the linear exponential family. Since two g.f.'s differ only by a positive constant multiple, we can ignore or adjust the constant multiple of a g.f.

It is well-known [1] for a linear exponential family that the moment generating function (m.g.f.) is

$$M(t;\omega) = f(\omega+t)/f(\omega) \quad . \tag{1.3}$$

From (1.3) we can obtain the mean vector

$$\mu(\omega) = \left(\frac{\partial \log f(\omega)}{\partial \omega_1}, \ldots, \frac{\partial \log f(\omega)}{\partial \omega_k}\right) \tag{1.4}$$

and the dispersion matrix

$$\Sigma(\omega) = \left(\frac{\partial \mu_\alpha(\omega)}{\partial \omega_\beta}\right) \quad . \tag{1.5}$$

We can easily extend P_e to a family of distributions of several vectors $x_1,\ldots,x_k$, in which case

$$dP_{\omega_1,\ldots,\omega_k}(x_1,\ldots,x_k) = \exp\left(\sum_{i=1}^{k} \omega_i' x_i\right)d\nu(x_1,\ldots,x_k)/f(\omega_1,\ldots,\omega_k). \tag{1.6}$$

We can represent (1.6) more compactly in the form (1.1) by considering ω and x as multivectors with vector components, i.e. $\omega = (\omega_1,\ldots,\omega_k)$ and $x = (x_1,\ldots,x_k)$. We call this family multivector linear exponential family. It follows immediately from (1.6) that the vector marginal of any component x_i is linear exponential in ω_i .

In Section 3 we establish the linear exponential family characterization for multivector multinomial distribution, while in Section 4 we do the same for multivector negative multinomial distribution after extending its definition. In order to establish these characterizations we need a construction procedure for multivector linear exponential family from the given vector marginals. Since such a procedure is similar to the one presented in [4] for multivariate linear exponential family with given univariate marginals, we simply present its salient features in Section 2.

2. CONSTRUCTION OF MULTIVECTOR DISTRIBUTIONS

We assume throughout the paper that the probability distribution of a vector is nonsingular. Then we have the following lemmas.

LEMMA 1. Let $\mathcal{P}_e = \{P_\omega : \omega \in \Omega_\nu\}$ be a multivariate linear exponential family of distributions in a vector ω. Then $P_\omega = P_{\omega^*}$ if and only if $\omega = \omega^*$ for all $\omega, \omega^* \in \Omega_\nu$.

LEMMA 2. If two multivariate linear exponential families $\{P_\omega : \omega \in \Omega_\nu\}$ and $\{P'_\theta : \theta \in \Theta_\eta\}$ are identical then for all $\omega \in \Omega_\nu$ and $\theta \in \Theta_\eta$, $\omega - \theta = \Phi$ for some constant vector Φ.

The proofs follow arguments similar to those for the univariate case presented in [4].

Suppose we are given k families $\{P_{\omega_i} : \omega_i \in \Omega_{\nu_i}\}$, $i = 1,\ldots,k$, of distributions of random vectors x_i, each being linear exponential in ω_i. Let their mean vectors and dispersion matrices be $\mu_i(\omega_i)$ and $\Sigma_i(\omega_i)$ respectively. Let $\{P'_{\theta_1,\ldots,\theta_k} : (\theta_1,\ldots,\theta_k) \in \Theta_\eta\}$ be any k-vector family of distributions linear exponential in $\theta_1,\ldots,\theta_k$, so that the families of vector marginals of x_i, $i = 1,\ldots,k$, coincide with $\{P_{\omega_i} : \omega_i \in \Omega_{\nu_i}\}$, $i = 1,\ldots,k$. Then it is easily seen that, $i = 1,\ldots,k$,

$$m_i(\theta_1,\ldots,\theta_k) = \mu_i(\omega_i) \tag{2.1}$$

$$D_{ii}(\theta_1,\ldots,\theta_k) = \Sigma_i(\omega_i) \tag{2.2}$$

where $m_i(\theta_1,\ldots,\theta_k)$ and $D_{ii}(\theta_1,\ldots,\theta_k)$ are the mean vector and dispersion matrix of x_i obtained from $P'_{\theta_1,\ldots,\theta_k}$.

If we write $x_i = (x_{i_1},\ldots,x_{i_{p_i}})$ and similarly ω_i and μ_i, then the vector derivative $\frac{\partial \mu_i}{\partial \omega_i}$ (see [7]), using (1.5), is

$$\left(\frac{\partial \mu_{i\alpha}}{\partial \omega_{i\beta}}\right) = \Sigma_i(\omega_i) \tag{2.3}$$

which is positive definite for nonsingular distributions. Therefore, the components of vector ω_i can be expressed as functions of $(\mu_{i1},\ldots,\mu_{ip_i})$,and in turn by (2.1), as functions of $(\theta_1,\ldots,\theta_k)$.

Then, by Lemma 2, it follows that

$$\omega_i = \theta_i + \Phi_i\ (\theta_1,\ldots,\theta_{i-1},\ldots,\theta_k)\ ,\ i = 1,\ldots,k \tag{2.4}$$

where vectors Φ_i are analytic.

Since

$$\operatorname{cov}(x_{i\alpha},x_{j\beta}) = \frac{\partial \mu_{i\alpha}(\theta_i+\Phi_i)}{\partial \theta_{j\beta}} = \frac{\partial \mu_{i\beta}(\theta_j+\Phi_j)}{\partial \theta_{i\alpha}}\ , \tag{2.5}$$

$$i,j = 1,\ldots,k\ ,\quad \alpha = 1,\ldots,p_i,\quad \beta = 1,\ldots,p_j$$

then it is the $(\alpha,\beta)^{th}$ element of the matrix $\frac{\partial \mu_i}{\partial \theta_j}$ and $(\beta,\alpha)^{th}$ element of the matrix $\frac{\partial \mu_j}{\partial \theta_i}$. Using the chain rule on vector derivatives, we obtain

$$\frac{\partial \mu_i}{\partial \theta_j} = \frac{\partial \mu_i(\theta_i+\Phi_i)}{\partial \theta_i} \cdot \frac{\partial(\theta_i+\Phi_i)}{\partial \theta_j}$$

$$= \Sigma_i(\theta_i+\Phi_i)\ \frac{\partial \Phi_i}{\partial \theta_j}\ . \tag{2.6}$$

Similarly

$$\frac{\partial \mu_j}{\partial \theta_i} = \Sigma_j(\theta_j+\Phi_j)\ \frac{\partial \Phi_j}{\partial \theta_i}\ . \tag{2.7}$$

Hence, we arrive at

$$\Sigma_i(\theta_i+\Phi_i)\ \frac{\partial \Phi_i}{\partial \theta_j} = (\Sigma_j(\theta_j+\Phi_j)\ \frac{\partial \Phi_j}{\partial \theta_i})'$$

$$= (\frac{\partial \Phi_j}{\partial \theta_i})'\ \Sigma_j(\theta_j+\Phi_j) \tag{2.8}$$

By integrating $\sum_{i=1}^{k}\ \sum_{\alpha=1}^{p_i} \mu_{i\alpha}(\theta_i+\Phi_i)d\theta_i$, which is an exact differential form (in virtue of (2.5)), as is done in [4], we arrive at

$$f(\theta_1,\ldots,\theta_k) = \prod_{i=1}^{k} \frac{f_i(\theta_i+\Phi_i(\theta_1,\ldots,\theta_{i-1},\ \theta_{i+1}^0,\ldots,\theta_k^0)}{f_i(\theta_i^0+\Phi_i(\theta_1,\ldots,\theta_{i-1},\ \theta_{i+1}^0,\ldots,\theta_k^0)} \tag{2.9}$$

where $(\theta_1^0,\ldots,\theta_k^0) \in \Theta_\eta$.

In order to construct a multivector linear exponential family from the given marginals with $f_i(\omega_i)$, we first solve the set of partial differential equations (2.8) for Φ's defined in (2.4). Then the g.f. of the joint distribution is given by (2.9), from which we obtain, using (1.3) , the m.g.f.

$$M(t_1,\ldots,t_k;\ \theta_1,\ldots,\theta_k) = \frac{f(\theta_1+t_1,\ldots,\theta_k+t_k)}{f(\theta_1,\ldots,\theta_k)} . \tag{2.10}$$

3. MULTIVECTOR MULTINOMIAL DISTRIBUTION

Let $x_i = (x_{i1},\ldots,x_{ip_i})$, $i = 1,\ldots, k$, have a multinomial distribution with the probability function

$$\frac{n!}{x_{i0}!x_{i1}!\ldots x_{ip_i}!}\ \pi_{i0}^{x_{i0}}\ \pi_{i1}^{x_{i1}}\ \ldots\ \pi_{ip_i}^{x_{ip_i}} \tag{3.1}$$

where $x_{i0} = n - \sum_{\alpha=1}^{p_i} x_{i\alpha}$, $\pi_{i0} = 1 - \sum_{\alpha=1}^{p_i} \pi_{i\alpha}$, $0<\pi_{i0},\pi_{i1},\ldots,\pi_{ip_i} < 1$,

and n is the same for all x_i . If we write $e^{\omega_{i\alpha}} = \frac{\pi_{i\alpha}}{\pi_{i0}}$, $\alpha = 1,\ldots,p_i$, in (3.1), then we can easily see that x_i is linear exponential in ω_i with the g.f.

$$f_i(\omega_i) = (1 + \sum_{\alpha=1}^{p_i} e^{\omega_i})^n \tag{3.2}$$

and its parameter space $\Omega_i = R^{p_i}$. Then the g.f. of the multivector family of $(x_1,\ldots,x_k)$, obtained from (2.9) with $(\theta_1^0,\ldots,\theta_k^0) = (0,\ldots,0)$, is

$$f(\theta_1,\ldots,\theta_k) = \prod_{i=1}^{k}\left[\frac{1+\sum_{\alpha=1}^{p_i} e^{\theta_{i\alpha}+\Phi_{i\alpha}}}{1+\sum_{\alpha=1}^{p_i} e^{\Phi_{i\alpha}}}\right]^n$$

$$= \prod_{i=1}^{k}\left[1+\frac{\sum_{\alpha=1}^{p_i} e^{\Phi_{i\alpha}}(e^{\theta_{i\alpha}}-1)}{1+\sum_{\alpha=1}^{p_i} e^{\Phi_{i\alpha}}}\right]^n \tag{3.3}$$

where $\Phi_{i\alpha}$'s are evaluated at $(\theta_1,\ldots,\theta_{i-1},0,\ldots,0)$. We can write (3.3) as

$$[g(\eta_1,\ldots,\eta_k)]^n = \left[\prod_{i=1}^{k}(1+\sum_{\alpha=1}^{p_i}\eta_{i\alpha}\,\psi_{i\alpha})\right]^n \tag{3.4}$$

where $\eta_{i\alpha} = e^{\theta_{i\alpha}} - 1$, i.e., $\theta_{i\alpha} = \log(\eta_{i\alpha}+1)$ and

$$\psi_{i\alpha} = \frac{e^{\Phi_{i\alpha}}}{1+\sum_{\alpha=1}^{p_i} e^{\Phi_{i\alpha}}} = \psi_{i\alpha}(\eta_1,\ldots,\eta_{i-1}) ,$$

$$i = 1,\ldots,k , \quad \alpha = 1,\ldots,p_i .$$

Observe that the last factor of $g(\eta_1,\ldots,\eta_k)$ is $(1+\sum_{\alpha=1}^{p_k}\eta_{k\alpha}\,\psi_{k\alpha}(\eta_1,\ldots,\eta_{k-1}))$ and that no other factors involve η_k . Then $g(\eta_1,\ldots,\eta_k)$ can be written as

$$h_0(\eta_1,\ldots,\eta_k) + \sum_{\alpha=1}^{p_k}\eta_{k\alpha}\,h_{k\alpha}(\eta_1,\ldots,\eta_{k-1}) . \tag{3.6}$$

Since the subscripts of x_i are arbitrary in constructing a multivector family, (3.6) must be true if k were to be replaced by any other subscript $1,2,\ldots,k-1$.

Then $g(\eta_1,\ldots,\eta_k)$ is a polynomial

$$c_0 + \sum_{i,\alpha} c_{i\alpha} \eta_{i\alpha} + \sum_{\substack{i,j,\alpha,\beta \\ i \neq j}} c_{ij\alpha\beta} \eta_{i\alpha} \eta_{j\beta} + \dots + \sum_{\alpha_1,\dots,\alpha_k} c_{\alpha_1\dots\alpha_k} \eta_{1\alpha_1} \dots \eta_{k\alpha_k} \tag{3.7}$$

of degree not exceeding k, linear in components of η_i for a given i. Then

$$f(\theta_1,\dots,\theta_k) = [d_0 + \sum_{i,\alpha} d_{i\alpha} e^{\theta_{i\alpha}} + \sum_{\substack{i,j,\alpha,\beta \\ i \neq j}} d_{ij\alpha\beta} e^{\theta_{i\alpha}} e^{\theta_{j\beta}} + \dots + \sum_{\alpha_1,\dots,\alpha_k} d_{\alpha_1\dots\alpha_k} e^{\theta_{1\alpha_1}} \dots e^{\theta_{k\alpha_k}}]^n \tag{3.8}$$

from which we obtain the m.g.f., using (2.11), as

$$[d_0 + \sum_{i,\alpha} d_{i\alpha} e^{\theta_{i\alpha}} e^{t_{i\alpha}} + \sum_{i,j,\alpha,\beta} d_{ij\alpha\beta} e^{\theta_{i\alpha}+\theta_{j\beta}} e^{t_{i\alpha}} e^{t_{j\beta}} + \dots + \sum_{\alpha_1,\dots,\alpha_k} d_{\alpha_1 \dots \alpha_k} e^{\theta_{1\alpha_1}+\dots+\theta_{k\alpha_k}} e^{t_{1\alpha_1}} \dots e^{t_{k\alpha k}}]^n / f(\theta_1,\dots,\theta_k) . \tag{3.9}$$

The p.g.f. is of the form

$$[a_0 + \sum_{1,\alpha} a_{i\alpha} s_{i\alpha} + \sum_{i,j,\alpha,\beta} a_{ij\alpha\beta} s_{i\alpha} s_{j\beta} + \dots + \sum_{\alpha_1,\dots,\alpha_k} a_{\alpha_1\dots\alpha_k} s_{1\alpha_1} \dots s_{k\alpha_k}]^n \tag{3.10}$$

where the coefficients are functions of $(\theta_1,\dots,\theta_k)$. By Lemma 3.1 of [3], the coefficients are nonnegative and add up to 1. Then (3.10) is the p.g.f. of a multivector multinomial distribution (see [6], p.86).

Theorem 1. A k-vector linear exponential distribution is multivector multinomial if and only if its vector marginals are multinomial.

4. MULTIVECTOR NEGATIVE MULTINOMIAL DISTRIBUTION

A random vector x is said to have a negative multinomial distribution if its p.g.f. is of the form

$$P(s_1,\dots,s_p) = (a_0 + a_1 s_1 + \dots + a_p s_p)^{-m} \tag{4.1}$$

where $a_1,\ldots a_p < 0$, and $a_0, m > 0$. In order for a multivector distribution to be negative multinomial, its vector marginals should be negative multinomial. However, this does not lead to a unique definition unless more restrictions are placed. Since the negative multinomial distribution is linear exponential, it is natural to require that multivector negative multinomial distribution be also linear exponential. Then we present

Definition. A multivector distribution of vectors $x_1,\ldots,x_k$ is said to be negative multinomial if its p.g.f. is of the form

$$\Big(a_0 + \sum_{i,\alpha} a_{i\alpha} s_{i\alpha} + \sum_{i,j,\alpha,\beta} a_{ij\alpha\beta} s_{i\alpha} s_{j\beta} + \ldots + \sum_{\alpha_1,\ldots\alpha_k} a_{\alpha_1\ldots\alpha_k} s_{1\alpha_1}\ldots s_{k\alpha_k}\Big)^{-m} \tag{4.2}$$

where $m > 0$.

It can be immediately seen that the marginal distribution of x_i is negative multinomial. Our definition does admit the case of some or all independent vector marginals, while the one defined in [8] does not admit independent vector marginals. This is due to the fact that the coefficients in (4.2) are restricted to $a_0 > 0$, all other a's < 0 , in the literature.

If each vector x_i is scalar, then the multivector distribution reduces to a multivariate negative binomial distribution defined in [2], which is a generalization of the usual definition found in literature (e.g. see [6], p.82).

Since the probability function of a negative multinomial distribution is similar to that of multinomial given by (3.1), the construction procedure for a multivector linear exponential distribution from negative multinomial vector marginals follows very closely the procedure given in Section 3 for multivector multinomial. Therefore we present without proof

Theorem 2. A k-vector linear exponential distribution is multivector negative multinomial if and only if its vector marginals are negative multinomial.

REFERENCES

[1] Bildikar, S. and Patil, G.P. (1968). Multivariate exponential-type distributions. Ann. Math. Statist. 39 1316-1326.

[2] Doss, D.C. Definition and characterization of multivariate negative binomial distribution. J. Multivariate Anal., to appear.

[3] Doss, D.C. and Graham, R.C. (1975). A characterization of multivariate binomial distribution by univariate marginals. Calcutta Stat. Assoc. Bull. 24 93-99.

[4] Doss, D.C. and Graham, R.C. (1975). Construction of multivariate linear exponential distributions from univariate marginals. Sankhyā A 37 257-268.

[5] Johnson, N.L. and Kotz, S. (1969). Distributions in Statistics: Discrete Distributions. Boston: Houghton Miffin Company.

[6] Patil, G.P. and Joshi, S.W. (1968). A Dictionary and Bibliography of Discrete Distributions. Edinburgh: Oliver and Boyd Ltd.

[7] Tracy, D.S. and Dwyer, P.S. (1968). Multivariate maxima and minima with matrix derivatives. J. Amer. Statist. Assoc. 64 1576-1594.

[8] Wishart, J. (1949). Cumulants of multivariate multinomial distributions. Biometrika 36 47-58.